Case Studies in Cell Biology

This volume is part of the
Problem Sets in Biological and Biomedical Sciences series
Series Editor: P. Michael Conn

A complete list of books in this series appears at the end of the volume

Case Studies in Cell Biology

Merri Lynn Casem, BA, PhD
Department of Biological Science
California State University

ELSEVIER

AMSTERDAM • BOSTON • HEIDELBERG • LONDON
NEW YORK • OXFORD • PARIS • SAN DIEGO
SAN FRANCISCO • SINGAPORE • SYDNEY • TOKYO
Academic Press is an Imprint of Elsevier

Academic Press is an imprint of Elsevier
125, London Wall, EC2Y 5AS, UK
525 B Street, Suite 1800, San Diego, CA 92101-4495, USA
50 Hampshire Street, 5th Floor, Cambridge, MA 02139, USA
The Boulevard, Langford Lane, Kidlington, Oxford OX5 1GB, UK

British Library Cataloguing-in-Publication Data
A catalogue record for this book is available from the British Library

Library of Congress Cataloging-in-Publication Data
A catalog record for this book is available from the Library of Congress

ISBN: 978-0-12-801394-6

For information on all Academic Press publications
visit our website at http://store.elsevier.com/

Typeset by Thomson Digital

Publisher: Sara Tenney
Acquisition Editor: Jill Leonard
Editorial Project Manager: Fenton Coulthurst
Production Project Manager: Lucía Pérez
Designer: Mark Rogers

Dedication

I dedicate this work to the memory of my husband, Edward P. Casem Sr, and to the two fine young gentlemen I am proud to call my sons. I would also like to express my gratitude to the professors that inspired my academic career; Marvin J. Rosenberg, C. Eugene Jones, Charles Lambert, and Leah T. Haimo.

Contents

BIOGRAPHY..xvii
PREFACE xix

CHAPTER 1	**Introduction to scientific method from a cellular perspective**...1	
Subchapter 1.1	Scientific method in action 1	
	Introduction 1	
	Background..2	
	Methods 3	
	Results..4	
	Reference 10	
CHAPTER 2	**Cellular biodiversity** **11**	
Subchapter 2.1	Cellular biodiversity on the high seas 11	
	Introduction 11	
	Background..12	
	Methods 13	
	Results..13	
Subchapter 2.2	**The mystery of the missing mitochondria**..............**17**	
	Introduction 17	
	Background..18	
	Methods 19	
	Results..20	
	References 22	
CHAPTER 3	**Proteins**..**23**	
Subchapter 3.1	Bite of the brown recluse spider: an introduction to protein gel electrophoresis....................................23	
	Introduction 23	

Background...25
Methods 25
Results..26

Subchapter 3.2   Where to start? a case of too many AUGs.................31
Introduction 31
Background...31
Methods 32
Results..34

Subchapter 3.3   Cotranslational translocation: gatekeepers
of the RER 39
Introduction 39
Background...40
Methods 40
Results..41

Subchapter 3.4   From sequence to function 47
Introduction 47
Background...48
Methods 48
Results..49

Subchapter 3.5   Visualizing protein conformation....................54
Introduction 54
Background...54
Methods 55
Results..56

Subchapter 3.6   Second chance chaperones: how misfolded
proteins get refolded..62
Introduction 62
Background...63
Methods 64
Results..65
References 71

CHAPTER 4   The nucleus..73

Subchapter 4.1   Nuclear pore complex assembly is a house
of cards...73
Introduction 73
Background...74
Methods 75
Results..76

Subchapter 4.2   Importin β: the important importin 80
                 Introduction 80
                 Background.. 80
                 Methods 81
                 Results.. 82

Subchapter 4.3   A tale of tRNA transport...................................... 88
                 Introduction 88
                 Background.. 89
                 Methods 89
                 Results.. 92

Subchapter 4.4   When it comes to the nucleus – size matters 96
                 Introduction 96
                 Background.. 97
                 Methods 98
                 Results.. 99
                 References 103

CHAPTER 5   **Membranes and membrane transport**................... **105**

Subchapter 5.1   Flip this lipid.. 105
                 Introduction 105
                 Background.. 106
                 Methods 107
                 Results.. 109

Subchapter 5.2   Navigating the bilayer: lipid rafts and caveolae..... 113
                 Introduction 113
                 Background.. 113
                 Methods 114
                 Results.. 115

Subchapter 5.3   Shifting gears: calcium transport in flagella........... 120
                 Introduction 120
                 Background.. 120
                 Methods 121
                 Results.. 122
                 References 125

CHAPTER 6   **Cytoskeleton and intracellular motility**............... **127**

Subchapter 6.1   Plakins: keeping the cytoskeleton safe 127
                 Introduction 127
                 Background.. 128

Methods 128
Results... 129

Subchapter 6.2   The *moving* story of a microtubule motor protein.... 134
Introduction 134
Background... 134
Methods 135
Results... 137

Subchapter 6.3   The WASP and the barbed end 142
Introduction 142
Background... 143
Methods 143
Results... 144

Subchapter 6.4   Cilia grow where vesicles go 150
Introduction 150
Background... 151
Methods 151
Results... 153
References 156

CHAPTER 7   **Organelles** **157**

Subchapter 7.1   Putting the "retic" in the endoplasmic reticulum.... 157
Introduction 157
Background... 158
Methods 159
Results... 160

Subchapter 7.2   How the Golgi stacks up....................................... 164
Introduction 164
Background... 164
Methods 166
Results... 167

Subchapter 7.3   Case of the coated vesicle 172
Introduction 172
Background... 173
Methods 174
Results... 176

Subchapter 7.4   How to build a peroxisome.................................... 180
Introduction 180
Background... 181

Methods 182
Results.. 183

Subchapter 7.5   Putting the squeeze on mitochondria..................... 187
Introduction 187
Background.. 188
Methods 189
Results.. 190
References 192

CHAPTER 8   **Exocytosis** **193**

Subchapter 8.1   Coat proteins and vesicle transport....................... 193
Introduction 193
Background.. 194
Methods 194
Results.. 195

Subchapter 8.2   Endomembrane transport in the absence of a cell 202
Introduction 202
Background.. 203
Methods 204
Results.. 205

Subchapter 8.3   Extra large export: a case for cisternal
maturation.. 209
Introduction 209
Background.. 210
Methods 211
Results.. 212
References 215

CHAPTER 9   **Endocytosis** **217**

Subchapter 9.1   Following the fate of a phagosome......................... 217
Introduction 217
Background.. 218
Methods 219
Results.. 220

Subchapter 9.2   Catching a receptor by the tail.............................. 224
Introduction 224
Background.. 224
Methods 225
Results.. 226

Subchapter 9.3 Can clathrin bend a membrane? 229
    Introduction 229
    Background.. 230
    Methods 231
    Results.. 234

Subchapter 9.4 Modeling membrane fission 235
    Introduction 235
    Background.. 235
    Methods 236
    Results.. 237
    References 240

CHAPTER 10 Cell walls and cell adhesion 241

Subchapter 10.1 Biofilms and Antibiotic Resistance 241
    Introduction 241
    Background.. 242
    Methods 242
    Results.. 243

Subchapter 10.2 DIY ECM: cortactin and the secretion
    of fibronectin.. 246
    Introduction 246
    Background.. 247
    Methods 247
    Results.. 248

Subchapter 10.3 Bundling the brush border 253
    Introduction 253
    Background.. 253
    Methods 254
    Results.. 256
    References 261

CHAPTER 11 Cell metabolism... 263

Subchapter 11.1 When glucose is low, something must go............... 263
    Introduction 263
    Background.. 264
    Methods 264
    Results.. 266

Subchapter 11.2 Do plants really need two photosystems?............... 270
    Introduction 270
    Background.. 270

Methods 271

Results.. 272

Subchapter 11.3 FREX: opening a window into cellular
metabolism... 275

Introduction 275

Background.. 275

Methods 276

Results.. 277

References 281

CHAPTER 12   **Cell signaling**.. **283**

Subchapter 12.1 How cells know when it's time to go....................... 283

Introduction 283

Background.. 284

Methods 285

Results.. 286

Subchapter 12.2 Can you "Ad" hear me now? signaling
and intraflagellar transport..................................... 290

Introduction 290

Background.. 291

Methods 292

Results.. 293

References 297

CHAPTER 13   **Cell cycle** **299**

Subchapter 13.1 Now you see it, now you don't: the discovery
of cyclin 299

Introduction 299

Background.. 300

Methods 301

Results.. 302

Subchapter 13.2 Sorting out cyclins.. 305

Introduction 305

Background.. 306

Methods 307

Results.. 308

Subchapter 13.2 Of centriole separation and cyclins........................ 312

Introduction 312

Background.. 313

Methods 314
Results... 315

Subchapter 13.4 The path to S phase is paved with
phosphorylation... 319
Introduction 319
Background... 320
Methods 321
Results... 322
References 326

CHAPTER 14   Cell division 327

Subchapter 14.1 Push and pull: how motor proteins help
build a spindle 327
Introduction 327
Background... 328
Methods 328
Results... 329

Subchapter 14.2 Ready, set, anaphase! 333
Introduction 333
Background... 333
Methods 334
Results... 335

Subchapter 14.3 Building cell walls: cytokinesis in a plant cell 338
Introduction 338
Background... 339
Methods 339
Results... 340
References 343

CHAPTER 15   Cell systems 345

Subchapter 15.1 Do bumblebees have B cells? a case
of insect immunity.. 345
Introduction 345
Background... 346
Methods 347
Results... 349

Subchapter 15.2 What happens when the endosymbionts
"bug out"? 353
Introduction 353

Background... 353

Methods 354

Results... 356

Subchapter 15.3 Parvovirus: hijacking endocytosis 362

Introduction 362

Background... 363

Methods 363

Results... 365

References 371

SUBJECT INDEX... 373

Biography

Merri Lynn Casem earned her doctoral degree in Cellular and Molecular Biology from the University of California, Riverside, working in the laboratory of Dr Leah Haimo, where she investigated the role of the microtubule motor protein cytoplasmic dynein, in synaptic vesicle transport. She spent 5 years as a visiting assistant professor at the Keck Science Department of the Claremont Colleges before joining the faculty of the Department of Biological Science at California State University, Fullerton. There, she has been actively involved in curriculum development and biology education research in addition to her work on cellular and molecular aspects of the biology of spider silk and early spider embryogenesis. Dr Casem has also been a member of the Education Committee of the American Society for Cell Biology, and has recently been appointed as the Director of Nonmajors Biology, at CSU Fullerton.

Preface

The ability to read and critically evaluate primary literature is a fundamental skill in the sciences. The best way to develop this skill is through practice. I began writing case studies, as a way to introduce primary literature to my freshmen Cell Biology classes. Students have reported that this early exposure to the elements of primary literature had a positive impact on their subsequent upper division coursework.

Each case study in this collection is built around one or more core concepts in Cell Biology. By presenting concepts outside the context of a traditional textbook, I hope to help students appreciate the dynamic nature of science and relevance of those concepts to a broader understanding of our world. Questions are provided throughout each case study to engage students, challenge them to think critically, and make connections between concepts. Most importantly, each case study allows students to gain practice in the evaluation and interpretation of graphs, figures, and tables excerpted from the original article.

I use case studies for in-class, small group activities in my lecture sections. Group work in a large (200+) class is greatly facilitated by additional support personnel; either graduate teaching assistants, or as is the case for my course, advanced undergraduate supplemental instruction leaders. Additionally, the use of "clickers" or similar technology can provide immediate assessment of student understanding. I encourage you to be creative in the adaptation and implementation of these case studies in your courses.

Best regards

Merri Lynn Casem

CHAPTER 1

Introduction to Scientific Method from a Cellular Perspective

Scientific Method in Action [1]

INTRODUCTION

The **scientific method** is the process of asking and answering questions about the world through the collection of data from carefully designed and controlled experiments. Communication of the results of scientific research, through peer-reviewed **primary literature**, allows those results to be challenged and reexamined by others. **Replication** of scientific results validates the answers to our questions. At its best, science reveals truths, independent of human bias or prejudice.

Science starts with observations that lead to questions. A **research question** sets the context for a scientific study. How that study proceeds is determined by a **hypothesis**. A hypothesis is a "best guess" of the answer to the research question. Hypotheses draw on our existing knowledge, but are not limited by it. Most importantly, a hypothesis can be wrong! We can learn as much, and possibly more, from wrong hypotheses as we can from right ones. Hypothesis testing is the fuel that drives scientific discovery.

With a hypothesis in hand, it is now possible to make some **predictions**. Predictions are important as they set the stage for the experiments that will be conducted. **Experimental design** is another critical component of the scientific method. A good experiment tests a prediction by manipulating one variable, while keeping all other variables constant. Experiments must produce **data** that can be documented, measured, analyzed, and presented in a form that allows others to critique and form their own conclusions. Experiments must

1

also use an appropriately large **sample size** and be **replicated** multiple times within a study to ensure that a result is not a product of chance.

The use of **statistical tests** helps to support conclusions drawn from a study by determining whether the data can be considered significant. In standard practice, a result is considered to be significant if the probability, or *P* **value**, generated by a statistical test is 0.5 or less. Rather than directly test an experimental hypothesis, statistics are used to test the **null hypothesis**. The null hypothesis states that there is no significant relationship between sets of data. A statistical test resulting in a *P* value greater than 0.5 means that there is a greater than 5% chance that the null hypothesis is true. *P* values less than or equal to 0.5 are an indication of statistical significance and allow the researcher to reject the null hypothesis. Either outcome should lead to the "next question," and the process continues.

- Discuss how science differs from faith.
- Scientific communication can take many forms. What distinguishes the presentation of scientific discovery in popular media from peer-reviewed primary literature?
- Explain how the concept of sample size and replication relates to statistics.

BACKGROUND

Transport of macromolecules between the nucleus and the cytoplasm is a critical feature of eukaryotic cells. Proteins that function within the nucleus, such as **nucleoplasmin**, are targeted to the nucleus by a short amino acid sequence known as the nuclear localization sequence. Nuclear proteins interact with cytoplasmic proteins that bind to the nuclear localization sequence and then facilitate the movement of the protein to a **nuclear pore complex**. The proteins of the nuclear pore complex form a channel that spans the double membrane of the nuclear envelope, allowing the protein to pass from the cytoplasm and into the nucleoplasm. One of the unique features of the nuclear pore complex is its ability to change size to accommodate the transport of large or bulky cargo.

Investigations into cellular processes often take advantage of established **tissue culture cell lines**. Tissue culture cells can be grown in large numbers under highly controlled conditions. The use of tissue culture cells also helps limit variables since cells grown from the original inoculation are genetically identical. The source of most of the cell lines used in research is the American Type Culture Collection (ATCC). Mouse 3T3 cells, an embryonic mouse fibroblast cell line, were used in the following experiments.

Researchers have observed that transport from the cytoplasm to the nucleus is different in 3T3 cells that are **proliferating** than in cells that are **confluent**. This observation leads to the research question: how does a change in cellular growth affect transport from the cytoplasm to the nucleus? The focus of this case study will be on the process involved in answering the question more than the answer itself. Look for the hallmarks of a good scientific study.

- List some proteins that would need to move from the cytoplasm into the nucleus.
- List some macromolecules that would need to be transported from the nucleus into the cytoplasm.
- Conduct a search on the ATCC website for the mouse 3T3 cell line. What information is provided about these cells?
- Define the terms proliferating and confluent in the context of tissue culture cells.

METHODS

Cell cultures

Mouse fibroblast 3T3 cell cultures were obtained from the ATCC. The cells were grown in culture media supplemented with 4.5 g/L glucose and 10% calf serum. Cell cultures were maintained in flasks at 37°C in 5% CO_2. Experimental cell cultures were grown on coverslips in 35 mm Petri dishes. Cells were either injected 24 h after culturing (1-day proliferating cells) or after 10, 14–17, or 21 days (confluent cells). Serum-starved cells were prepared by changing the culture media after the cells had attached to coverslips containing 0.5 or 0.1% serum. Cultures were maintained at these serum levels for 4 or 7–8 days, respectively.

Microinjection

Small (20–50 Å), intermediate (20–120 Å) and large (80–240 Å) diameter gold particles were prepared in lab. The size range of the particles varied slightly between preparations. The gold particles were coated with either nucleoplasmin isolated from the frog *Xenopus* or bovine serum albumin (BSA). Microinjection was performed using a micromanipulator while viewing the sample using an inverted microscope.

Electron microscopy

Cells were fixed using 2% glutaraldehyde in buffer 30 min after microinjection. Cells were postfixed using osmium tetroxide and dehydrated using a graded ethanol series prior to embedding in resin. Resin blocks were cut using a microtome, sections were placed onto formvar-coated grids, and the grids were examined using a transmission electron microscope. Sections were not stained with the usual uranyl acetate or lead poststains to facilitate detection of the gold particles.

Calculation of nuclear uptake and size distribution

Transport into the nucleus, or nuclear uptake, is expressed as the ratio of the number of gold particles present inside the nucleus to the number in the cytoplasm. These values were determined by counting gold particles in equal and adjacent areas of nucleoplasm and cytoplasm in electron micrographs.

The size distribution of the gold particles was determined by measuring all gold particles present in randomly selected, but unequal, areas of nucleoplasm and cytoplasm in electron micrographs. Particles were categorized by size range. The number of particles in each size category was divided by the total number of particles counted in either the cytoplasm or nucleoplasm to arrive at a value for the percent of total particles.

Statistical tests were applied to the various experiments by comparing the numbers and sizes of gold particles in the experimental groups with the data for the proliferating cells.

- Calculate the molarity of glucose in the tissue culture media.
- Serum provides essential nutrients to support cell growth. Propose an explanation for why the 0.5% serum-starved cells were cultured for only 4 days while the 1% serum-starved cells were cultured for 7–8 days.
- Discuss the significance of frog nucleoplasmin supporting nuclear transport in a mouse cell?
- Conduct a search for any of the equipment or techniques you are not familiar with.
- Calculate the N/C ratio for an imaginary cell in which the number of nuclear gold particles is greater than the cytoplasmic gold particles and another for a cell in which the numbers are reversed. How does the N/C ratio change with increased nuclear transport?
- Justify the use of proliferating cultures as the control condition for these experiments.

RESULTS

The research question for this study asked how changes in 3T3 cell growth from proliferating to confluent states influences transport from the cytoplasm to the nucleus. The working hypothesis that was proposed is that the functional size of the nuclear pore changes with the growth of the cells.

- List a prediction that would be consistent with this hypothesis.
- What prediction is being tested in Figure 1.1.1 and Table 1.1.1?
- Summarize the conclusion that is supported by the data in Table 1.1.1.
- How well supported is that conclusion? Justify your answer using the data in Table 1.1.1.
- Would the data presented in Table 1.1.1 have been as convincing if you had not seen the micrographs shown in Figure 1.1.1? Explain your answer.
- Explain the problem that is being addressed by the experiment in Figure 1.1.2.

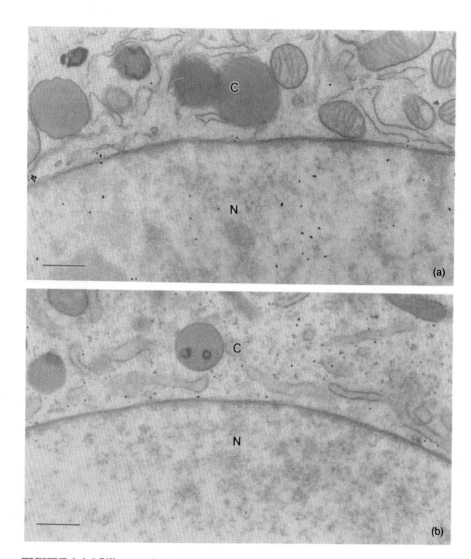

FIGURE 1.1.1 Differences in the intracellular distribution of large, nucleoplasmin-coated gold particles.
a. 80–240 Å nucleoplasmin-coated gold particles are accumulated in nucleoplasm (N) of a proliferating cell. b. Large nucleoplasmin-coated gold particles are present in the cytoplasm (C), but not the nucleoplasm of the 21-day confluent cell. Scale bar = 0.5 μm.

- No statistical test was applied to the data in Figure 1.1.2. Discuss whether statistics would have contributed to your analysis of this experiment.
- Is the hypothesis that the functional size of the nuclear pore changes with the growth of cells accepted or rejected based on the data presented? Justify your answer.

Table 1.1.1 Nuclear Uptake of Large, Nucleoplasmin-Coated Gold Particles in Proliferating and Confluent Cells

Experiment	Cells (*n*)	Total Particles Counted	N/C Ratio ± SE	Significance*
Proliferating	39	2904	2.49 ± 0.19	
10-day confluent	10	694	0.45 ± 0.06	s P << 0.001
14–17-day confluent	39	3116	0.08 ± 0.01	s P << 0.001
21-day confluent	15	677	0.03 ± 0.01	s P << 0.001

s, significantly different.
*The results of each experimental group were compared with the data obtained for proliferating cells.

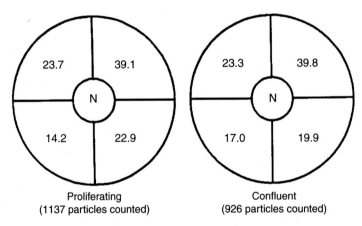

Proliferating
(1137 particles counted)

Confluent
(926 particles counted)

FIGURE 1.1.2 Variation in nuclear uptake is not a consequence of differences in cytoplasmic viscosity.
The cytoplasmic distribution of large nucleoplasmin-coated gold particles following microinjection was compared between proliferating and 21-day confluent cells. Particles were counted in randomly assigned quadrants of fourteen proliferating cells and thirteen 21-day confluent cells. The percent of total particles counted in each quadrant is reported. The quadrant with the highest count is presumably the site of microinjection.

- What prediction would you make about the size distribution of the 80–240 Å gold particles in the cell that would be consistent with the working hypothesis?
- Explain what the value "percentage of total particles" means.
- Construct a graph of your prediction using Figure 1.1.3 as a model.

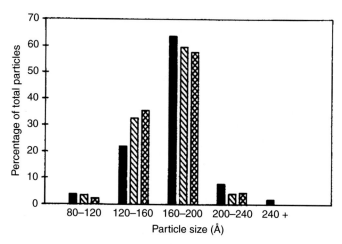

FIGURE 1.1.3 Size distribution of large, nucleoplasmin-coated gold particles in proliferating and confluent cells.

The size of the gold particles located within the cytoplasm (filled bars) or nucleoplasm of proliferating (hatched bars) or confluent (crossed bars) cell was measured. A single value for cytoplasmic particles is reported since the same gold fraction was used for proliferating and confluent cells. Totals of 1956 and 792 particles were counted in proliferating and confluent cells, respectively. A total of 2712 particles was counted in the cytoplasm.

- Summarize the conclusion that is supported by the data in Figure 1.1.3.
- Is the hypothesis that the functional size of the nuclear pore changes with the growth of cells accepted or rejected based on these data? Justify your answer.

The observation of reduced nuclear transport of large nucleoplasmin-coated gold particles in confluent cells compared with proliferating cells is still valid even if the original working hypothesis was rejected. A second hypothesis was developed. This hypothesis states that the number of nuclear pores available for transport decreases in confluent cells. This hypothesis was tested by microinjecting small (50–80 Å) gold particles coated with either nucleoplasmin protein or a nonnuclear protein, BSA.

- Describe the logic behind the use of small (50–80 Å) gold particles.
- Propose a different type of experiment that would test whether the number of nuclear pores differed in proliferating and confluent cells.
- Predict what the relative N/C ratios would be for proliferating and confluent cells if there were fewer nuclear pores in confluent cells.
- The possibility exists that small gold particles might enter the nucleus through passive diffusion. How did the researchers control for this variable?

Table 1.1.2 Nuclear Uptake of Small Nucleoplasmin-Coated and BSA-Coated Gold Particles in Proliferating and Confluent Cells

Experiment	Coating Agent	Cells (n)	Total Particles Counted	N/C Ratio ± SE	Significance*
Proliferating	NP	28	4795	2.33 ± 0.17	
14–17-day confluent	NP	25	4365	2.05 ± 0.19	ns P = 0.284
Proliferating	BSA	13	2516	0.016 ± 0.003	s P << 0.001

NP, nucleoplasmin; ns, not significant; s, significantly different.
*The results of each experimental group were compared with the data obtained for proliferating cells.

■ Is the hypothesis that the number of nuclear pores available for transport is decreased in confluent cells accepted or rejected based on the data in Table 1.1.2? Justify your answer.

The data generated by the experimental tests of the two previous hypotheses lead the researchers to yet another hypothesis. This new hypothesis proposed that the ability of the nuclear pore complex to accommodate large cargo is reduced as cells become confluent. To test this hypothesis, nucleoplasmin-coated gold particles of an intermediate size range (40–140 Å) were microinjected into proliferating 14-day and 19-day confluent cells.

■ How does the nuclear transport behavior of confluent cells compare between 14 days and 19 days?
■ How does the nuclear transport behavior of proliferating cells differ from that of confluent cells?
■ Describe the relationship between percentage of total particles found in the nucleus of proliferating cells and those found in the cytoplasm.
■ Critique Figure 1.1.4. Is the presentation of the data clear and compelling? What additional information, if any, should be included in the figure or figure legend?
■ Is the hypothesis that the ability of the nuclear pore complex to accommodate large cargo is reduced as cells become confluent accepted or rejected based on the data in Figure 1.1.4? Justify your answer.
■ What additional experiment(s) would you propose to further test this hypothesis?

Having established that nuclear transport in confluent 3T3 cells is different from transport in proliferating 3T3 cells, the researchers arrived at a new research question; what is it about the confluent cells that triggers the change in transport? To answer this question it is necessary to consider what happens when a culture becomes confluent. One of the hallmarks of a confluent cell culture is the cessation

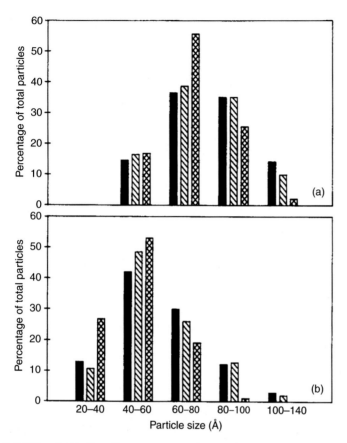

FIGURE 1.1.4 Size distribution of intermediate, nucleoplasmin-coated gold particles in proliferating and confluent cells.

The size of gold particles located within the cytoplasm (filled bars) or nucleoplasm of proliferating (hatched bars) or confluent (crossed bars) cells was measured. A single value for cytoplasmic particles is reported. a. Size distribution comparing proliferating and 14-day confluent cells (634 and 862 particles, respectively) were counted, as were 555 cytoplasmic particles. b. Size distribution comparing proliferating and 19-day confluent cells (645 and 773 particles, respectively) were counted, as were 819 cytoplasmic particles.

of mitotic cell division due to contact inhibition. Cell division can be inhibited in otherwise actively proliferating cells by reducing the amount of serum present in the tissue culture medium, a condition known as serum starvation.

- Use your textbook or online resources to learn more about contact inhibition.
- Refer back to the "Methods" section to determine the amount of serum normally included in tissue culture media.

Table 1.1.3 N/C Ratios of Proliferating and Serum-Starved Cells

Experiment	Gold Particle Size (Å)	Cells (n)	Total Particles Counted	N/C Ratio ± SE	Significance*
Proliferating	20–50	23	3911	1.98 ± 0.16	
4 days, 0.5% CS	20–50	8	1742	1.74 ± 0.14	ns $P = 0.389$
7 days, 1.0% CS	20–50	13	3726	2.19 ± 0.23	ns $P = 0.438$
Proliferating	80–240	19	1027	1.90 ± 0.20	
4 days, 0.5% CS	80–240	29	1274	0.20 ± 0.04	s $P \ll 0.001$
7–8 days, 1% CS	80–240	21	933	0.31 ± 0.05	s $P \ll 0.001$

Experimental cells were switched to a 0.5% serum media for 4 days or a 1% serum media for 7 days. Experimental and proliferating (1-day) cells were microinjected with small (20–50 Å) or large (80–240 Å) nucleoplasmin-coated gold particles. s, significantly different; ns, not significant.
*The data from each experimental group were compared with the data obtained for proliferating cells.

■ Propose your own hypothesis about why nuclear transport changes in confluent cells.

■ Outline the design of the experiment used to generate the data in Table 1.1.3.

■ Propose a hypothesis that would be consistent with the experiments in Table 1.1.3.

■ Which of the experiment(s) listed in the Table 1.1.3 could be considered the control experiment(s)?

■ Critique the data provided in Table 1.1.3. What are the strengths and weaknesses of the experiments?

■ Discuss the conclusions supported by these data in the context of the other experiments presented in this case study.

■ Formulate a new research question about nuclear transport.

■ Working from your research question, develop a hypothesis and a prediction.

Reference

[1] Feldherr CM, Akin D. Signal-mediated nuclear transport in proliferating and growth-arrested BALB/c3T3 cells. J Cell Biol 1991;115:933–9.

CHAPTER 2

Cellular Biodiversity

Cellular Biodiversity on the High Seas [1]

INTRODUCTION

Cellular diversity is the concept that many different types of cells have evolved specializations that allow them to function in unique ways and live in a wide variety of habitats. The cells that make up your body are one example of cellular diversity. The cells of your brain (neurons) are different, both in form and function, from the cells that make up your muscles (myoblasts). As multicellular organisms, it is easy for us to forget that a vast majority of biological organisms are *unicellular* (single celled). One place that unicellular diversity exists is within the aquatic phytoplankton that fills the oceans of the globe. Phytoplankton can be prokaryotic, such as *cyanobacteria* and *picocyanobacteria*, or eukaryotic, such as *algae* and *picoeukaryotes*.

Phytoplanktons are **photoautotrophs**. These cells use light energy to drive the process of photosynthesis to generate the ATP energy they need to chemically link together CO_2 molecules to form glucose. Photosynthetic cells capture light energy using specialized molecules called **pigments**. Different pigment molecules have slightly different chemical structures that allow them to absorb a specific wavelength of light. The wavelengths of light not absorbed by a pigment are reflected. It is these reflected wavelengths that we perceive as the color of an organism. The following is a list of pigments typically found in photosynthetic cells and the wavelengths of light that they can absorb:

- **Chlorophyll *a*** – peak absorbance at ~430 and 680 nm
- **Phycoerythrin** – peak absorbance at ~560–570 nm
- **Phycocyanin** – peak absorbance at ~620–630 nm

11

The combination of pigments found within an organism determines which wavelengths of light the organism can use to harvest energy for photosynthesis. The ability of a cell to efficiently absorb light and therefore gain more ATP energy and make more glucose represents an important trait that would be subject to the pressures of natural selection. This case study examines the competition for light by photosynthetic bacteria.

- What are the key differences between eukaryotic and prokaryotic cells?
- Describe how neurons and muscle cells illustrate the concept of cellular diversity.
- Research the absorbance spectra for each of the pigments listed previously.
- Explain how to use the absorbance spectra to predict the color you would observe when looking at each of the pigments.
- Connect the ability of a cell to absorb sunlight with its ability to reproduce.

BACKGROUND

A research expedition collected water samples from a depth of 10 m below the surface of the Baltic Sea. The scientists were able to isolate and culture several types of phytoplankton from the seawater samples. Two of these, labeled BS4 and BS5, were similar to cells in the genus *Synechococcus*, a type of **picocyanobacteria**. Comparison of the genomes of the two cultures revealed that they were very similar with less than 1% difference in their gene sequences. Despite this similarity, the researchers discovered an interesting difference between the two samples.

Competition for sunlight between unicellular, photosynthetic organisms was also examined by culturing picocyanobacteria in the presence of *Tolypothrix*, a filamentous, marine cyanobacterium. Photosynthesis in *Tolypothrix* uses the pigments phycocyanin and phycoerythrin. *Tolypothrix* has the intriguing ability to alter the ratio of these two pigments within its cytoplasm while keeping the total amount of pigment the same, a property known as *chromatic adaptation*.

- Scientists are notorious for using abbreviations, which make sense to other scientists, but make it difficult for everyone else. What would you guess "BS" stands for in this study?
- Scientists also use complicated terminology. Break down the term "picocyanobacteria" into its constituent parts. Define each part of the word and then use that information to formulate a definition for the whole word.
- What can you conclude based on the statement that the genetic sequences found within the DNA of these two cell types are similar?
- Suggest a scenario in which chromatic adaptation would provide an advantage to *Tolypothrix*.

METHODS

Cell culture and population density

BS4, BS5, and *Tolypothrix* cell cultures were grown in culture tubes containing a liquid medium that mimicked the salinity and mineral content of the Baltic Sea. Full spectrum white light, simulating sunlight, was used to stimulate photosynthesis in the cell cultures. For some experiments, the color of light the cultures received was modified using filters that absorbed some wavelengths and let others through. BS4 and BS5 cells were cultured, either alone (**monoculture**) or together (**competition**), under different wavelengths of light to test how the two cell types might interact in the wild. BS4 and BS5 cells were also cocultured with *Tolypothrix* cells under white light to examine how competition would affect growth of the cyanobacteria.

Cells that were able to absorb light grew and reproduced resulting in an increase in the number of cells in the culture tube. The number of cells in a given volume or area is defined as the **population density**. The population density of the cultures was measured using a *flow cytometer*. A flow cytometer is a machine that uses a laser beam and light detectors to not only count, but also characterize cells in solution.

Characterization of photosynthetic pigments

The optical properties of pigments within these photosynthetic cells were assayed using a spectrophotometer. A sample of the cell culture is exposed to a range of wavelengths of light that represent the visible spectrum. Sensors within the spectrophotometer measure the amount of light that is absorbed by the sample.

- Predict the effect of the use of either full spectrum (white) light or a single color of light on the growth of these cell cultures.
- How would the experimental design be affected if the researchers used a medium that did not mimic seawater?

RESULTS

- Describe what each of the graphs in this case study are illustrating based only on the axis labels.
- Use the data in Figure 2.1.1 to predict the color of the BS4 cell culture. What about the color of BS5? Support your prediction with information from the graphs.
- Use the list of peak pigment absorbance provided earlier to predict the photosynthetic pigment(s) in BS4. Do the same for BS5.
- The researchers determined that there was less than 1% difference in the gene sequences of BS4 and BS5. Describe the relationship between gene sequences and the pigments found inside these cells.

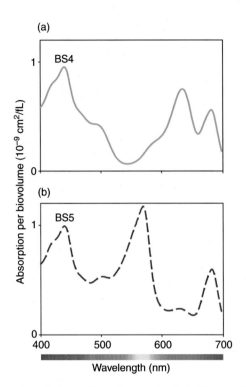

FIGURE 2.1.1 Optical characteristics of the picocyanobacteria BS4 and BS5.
Light absorption spectra of a culture of BS4 (a) and BS5 (b).

- Which color of light (wavelength) would you use if you wanted the cells to grow rapidly?
- Where inside the picocyanobacterial cells would you expect to find pigment molecules?
- Explain how a picocyanobacterium resembles, and how it differs from a picoeukaryote.
- Examine the data from the monoculture experiments shown in graphs (a) and (b) in Figure 2.1.2. What conclusions can you draw from these data?
- Use graphs (a) and (b) from Figure 2.1.2 to estimate the rate of increase in population density for each of the individual monocultures.
- Use the information in the legend to Figure 2.1.2 to describe, in your own words, the experimental design used to collect the data shown in graphs (c) and (d).
- Refer back to your choice for the best color (wavelength) of light to generate rapid cell growth. Is your choice supported by the data shown in Figure 2.1.2? Explain.
- Explain why a mismatch between the wavelength of light and pigment molecules would affect the ability of a cell to grow and reproduce.

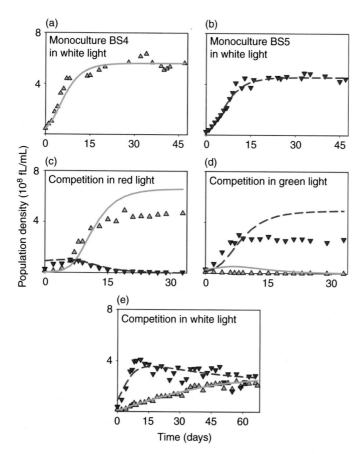

FIGURE 2.1.2 Monoculture and competition experiments with BS4 and BS5.
Time course of growth of monocultures of BS4 (a) and BS5 (b) in white light. Competition between BS4 and BS5 in red light (c), green light (d), and white light (e). Symbols represent the observed population densities of BS4 (green triangles) and BS5 (red triangles). Lines represent predicted population densities: green line for BS4, red dashed line for BS5. Population densities are expressed in biovolumes (in femtoliters) per milliliter of water.

- Study the graph in Figure 2.1.2e. How does this graph resemble or differ from the other graphs in Figure 2.1.2?
- Use the data provided in Figure 2.1.2e to answer whether BS4 and BS5 are competing for light in their natural environment.
- Predict the color of the *Tolypothrix* cells in the experiment shown in Figure 2.1.3b.
- How does the data in Figure 2.1.3b relate to the graph in Figure 2.1.3a?
- What color is the *Tolypothrix* when it is grown in the presence of BS5?
- Develop a hypothesis regarding competition and the chromatic adaptive behavior of *Tolypothrix*.

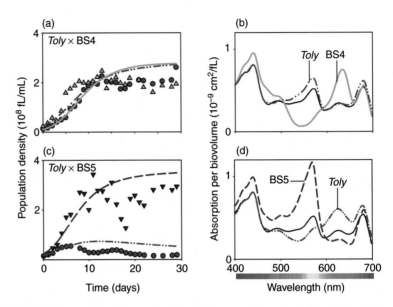

FIGURE 2.1.3 Competition between the filamentous cyanobacterium *Tolypothrix* and the picocyanobacteria BS4 and BS5.

a. Competition between *Tolypothrix* (circles) and BS4 (triangles) under white light. b. Absorption spectra of BS4 (green solid line) and *Tolypothrix* at day 0 (black solid line) and day 30 (brown dash-dotted line) of the competition experiment. c. Competition between *Tolypothrix* (circles) and BS5 (triangles) under white light. d. Absorption spectra of BS5 (red-dashed line) and *Tolypothrix* at day 0 (black solid line) and day 30 (brown dash-dotted line) of the competition experiment. Lines in (a) and (c) represent predicted changes in population density. Population densities are expressed in biovolumnes (in femtoliters) per milliliter of water.

- Develop an argument (for or against) the following statement, "natural selection favors the divergence of pigment composition increasing the biodiversity of phototrophic microorganisms."
- Describe or illustrate the connection between the concepts of *photosynthesis, cellular reproduction, cellular diversity,* and *natural selection* in the context of this study.

The Mystery of the Missing Mitochondria [2]

INTRODUCTION

One of the defining characteristics of a eukaryotic cell is the presence of a membrane-bound nucleus and other intracellular membrane-bound organelles. Chief among these are mitochondria, the organelles associated with high levels of ATP production. According to the **endosymbiotic theory**, mitochondria are the remnants of an alpha-proteobacterial endosymbiont. The ability of this prokaryotic endosymbiont to use O_2 in the production of ATP energy, provided an important selective advantage for the host protoeukaryote. However, not all eukaryotic cells live in the presence of O_2. **Anaerobic** organisms can only survive in environments with low or no oxygen. Would such anaerobic eukaryotes have mitochondria?

One method used to answer this question is **differential centrifugation**. Differential centrifugation uses centrifugal force to separate the contents of a sample based on their relative size. Samples are placed in a *centrifuge tube*, the tube is then placed into a *rotor*, and the rotor is placed inside a machine called a *centrifuge* that can spin the rotor and sample at various speeds, expressed in units of gravity (*g*) causing larger material in the sample to sediment or **pellet** to the bottom of the tube, but leaving the remaining smaller material in the overlying solution or **supernatant**. Researchers use the variables of speed and time to control the type of material that sediments. Short low-speed centrifugations will sediment large particles, while smaller particles require higher speeds and/or longer centrifugation times. The contents of a eukaryotic cell (nucleus, endoplasmic reticulum, mitochondria, ribosome, proteins, etc.) differ in size. Lysing or breaking open the cell, will release these subcellular components. Differential centrifugation then allows the isolation of specific subcellular contents from the cellular lysate.

Another important method is **immunoblotting**, also known as **Western blotting**. The first step in this technique is the isolation and separation of proteins from a sample using the SDS-polyacrylamide gel electrophoresis (SDS-PAGE) method. SDS-PAGE separates proteins within the matrix of a gel based on their relative sizes, with smaller proteins migrating more quickly toward the bottom of the gel than larger proteins. The proteins are then transferred from the gel onto the surface of a special type of "paper" known as a *nitrocellulose* *membrane.* The properties of nitrocellulose allow proteins to stick to its surface, making them available to react with various reagents including antibodies. You are probably familiar with the idea that antibodies in your immune system can recognize and bind to germs in your body. The same principle applies to immunoblotting, but now the antibodies are used to recognize and bind to specific proteins on the surface of the nitrocellulose membrane. In order to detect the binding of *primary antibodies* to a protein band, researchers must incubate the nitrocellulose blot with a *secondary antibody* that will recognize the presence of the primary antibody. Secondary antibodies are also chemically modified so they will trigger a color-forming reaction, producing a black band any place that the primary antibody (and hence the specific protein) is located on the blot.

- Research/review the evidence that supports the endosymbiotic origin of mitochondria.
- How would you explain the existence of a eukaryote that lacked mitochondria?
- Research/review the properties associated with mitochondria. Consider both mitochondrial structure and function.
- Predict the centrifugation conditions required to sediment nuclei versus ribosomes, from a cell lysate.
- Research/review the role of protein structure in the function of an antibody.

BACKGROUND

Protozoa are free-living unicellular eukaryotes. Like all other eukaryotic cells, protozoa have a nucleus and other membrane-bound organelles. Some protozoa, however, appear to be missing some of the organelles you would expect to find in a typical eukaryote. *Giardia intestinalis* is an example of a eukaryotic protist that lacks mitochondria, peroxisomes, and has a very underdeveloped endomembrane system. *Giardia* generates the ATP energy it needs through an anaerobic metabolic pathway, located in the cytoplasm of the cell that relies on the use of enzymes that contain iron–sulfur (Fe–S) cofactors. These *metalloenzymes* are involved in redox reactions and electron transport pathways that lead to ATP synthesis. Biosynthesis of the Fe–S cofactor in turn, requires the activity of a highly conserved class of proteins known as **Isc proteins**.

Giardia is a medically relevant parasitic protozoan that is responsible for diarrheal diseases that affect hundreds of thousands of people and animals worldwide. Ingestion of food or water contaminated by *Giardia* cysts leads to the infection of host organisms. Once inside a host, the cyst opens up to release one or two **trophozoites**, the free-swimming, cellular phase of the *Giardia* life cycle. The trophozoites populate the intestine of the host, causing a variety of unpleasant symptoms and producing more cysts that are released to continue the cycle of infection and disease.

- Isc proteins are highly conserved from bacteria to humans. What are the implications of this statement?
- Drugs used to treat *Giardia* infections work by entering the cytoplasm of the trophozoite and disrupting ATP production. Why would such a drug be safe for use in humans and animals?

METHODS

Subcellular fractionation of *giardia*

G. intestinalis trophozoites were lysed by mechanical disruption under isotonic conditions to release their cellular contents. The trophozoite cell lysate was centrifuged at a low speed (1000*g*) for 10 min to sediment cells that had not lysed. The supernatant, containing material from lysed cells, was then centrifuged at a high speed (108,000*g*) for 30 min. The resulting supernatant and pellet were collected. Proteins from each of the samples, low-speed sediment, high-speed sediment, and supernatant (cytosol), were separated using SDS-PAGE and then transferred to nitrocellulose membranes. The nitrocellulose membranes were then reacted with antibodies specific to either IscU or IscS, and antibody binding was detected using an enzyme-linked secondary antibody.

Transmission electron immunomicroscopy

Trophozoites were preserved or *fixed* using the chemicals glutaraldehyde and paraformaldehyde. Some of these fixed cells were frozen and cryosectioned into 50–100 nm slices. The slices were reacted with the IscS or IscU antibodies under conditions that would allow the antibody to bind to any of the respective Isc proteins present in the region of the cell exposed in the section. Antibody binding was detected using a secondary antibody linked to an 8 nm (IscU) or 12 nm (IscS) gold particle. Gold particles showed up as black dots when the sections were viewed using a transmission electron microscope (TEM). Other fixed cells were placed in an epoxy plastic resin, sectioned without freezing, and viewed using the TEM.

- Why was the low-speed centrifugation step important?
- Suggest why it would be necessary to chemically preserve cells before sectioning them.
- Why was it necessary to use gold-labeled antibodies instead of fluorescently labeled antibodies in this work?

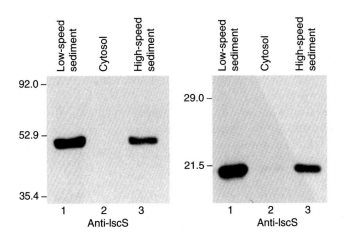

FIGURE 2.2.1 Distribution of Isc proteins in *Giardia* cell fractions.
The distribution of Isc proteins was examined following differential centrifugation of *Giardia* lysate. Trophozoite extracts were analyzed by SDS-PAGE and immunoblotting with antibodies specific to IscS and IscU. Size markers are given in kilodaltons.

RESULTS

- Name the cellular components you would find in the three samples in Figure 2.2.1?
- What is the approximate size (in kDa) of the IscS protein?
- Predict how IscU is different from IscS based on the data in Figure 2.2.1.
- Where in the trophozoite cell are the Isc proteins located?
- Why was Isc protein detected in both the low-speed and high-speed sediment lanes?
- Draw a circle around the 8 nm gold particles and a triangle round the 12 nm gold particles in parts (c) and (d) of Figure 2.2.2. What do you observe?
- How are the images in Figure 2.2.2e–h different from Figure 2.2.2a–d?
- What conclusion is supported by the images in Figure 2.2.2a–d?
- What conclusion is supported by the images in Figure 2.2.2e–h?
- What conclusion is supported by the data in the graph in Figure 2.2.2i?
- What other cellular structures can be seen in Figure 2.2.2f?
- Use the scale bar in Figure 2.2.2h to estimate the dimensions of the structure indicated by the arrow and arrowheads.
- Explain the significance of the double membranes found associated with these structures?
- The authors propose that *Giardia* contain a mitochondrial remnant they call a **mitosome** whose function is to assemble the Fe-S proteins required for anaerobic energy production. Based on this statement and the data in this chapter, how does a mitosome differ from a mitochondrion?
- The biosynthesis of Fe–S cofactors is inhibited by the presence of O_2. However, this process occurs in the matrix of your mitochondria. How can this be true?

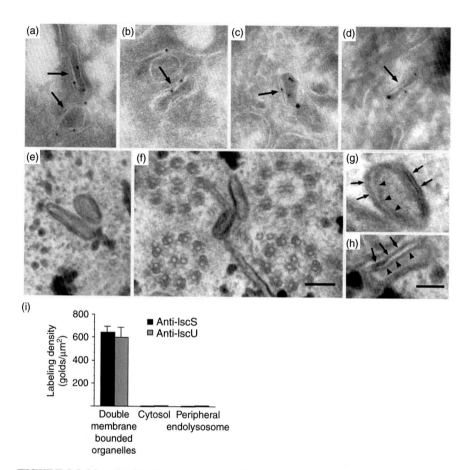

FIGURE 2.2.2 Localization of IscS and IscU in *Giardia* trophozoites by transmission electron immunomicroscopy.

a–d. Thawed frozen cryosections of glutaraldehyde-fixed *Giardia* were labeled with anti-IscS (a) and anti-IscU antibodies (b) or double-labeled (c, d) with anti-IscU (8 nm gold particles) followed by anti-IscS antibodies (12 nm gold particles). e–h. Epoxy resin sections of glutaraldehyde-fixed trophozoites. f. Double membrane–bound structures are shown in close proximity to flagellar axonemes. g, h. High-magnification images showing organelle double membranes. Arrows point toward outer membranes, while arrowheads identify inner membranes. i. Densities of labeling from typical labeling experiments were obtained from scanned micrographs ($n = 8$ for IscS and $n = 7$ for IscU). Scale bars, 100 nm (a–f) and 50 nm (g, h).

References

[1] Stomp M, Huisman J, deJongh F, Veraart AJ, Gerla D, Rijkeboer M, Ibelings BW, Wollenzien UIA, Stal LJ. Adaptive divergence in pigment composition promotes phytoplankton biodiversity. Nature 2004;432:104–7.

[2] Tovar J, León-Avila G, Sánchez LB, Sutak R, Tachezy J, van der Giezen M, Hernández M, Müller M, Lucocq JM. Mitochondrial remnant organelles of Giardia function in iron–sulphur protein maturation. Nature 2003;426:172–6.

CHAPTER 3

Proteins

Bite of the Brown Recluse Spider: An Introduction to Protein Gel Electrophoresis [1]

INTRODUCTION

Proteins are involved in every aspect of cellular life. Proteins function to transport materials across membranes, catalyze chemical reactions, organize DNA, support the movement of materials within a cell, and even drive movement of the entire cell. The technique of **sodium dodecyl sulfate polyacrylamide gel electrophoresis** or **SDS-PAGE** is a standard technique for the analysis of the protein composition of a sample. The nature of the sample will vary depending on the specific experiment.

Proteins differ in the number and type of amino acids that assemble to form a polypeptide chain. The chemistry of the amino acid side chains (R groups) determines protein folding and the overall charge of the protein. SDS-PAGE works by separating proteins based on their relative size. In order to do this, protein folding and the differences in charge of the various proteins must be eliminated. Protein gel samples are treated with the nonionic detergent, **sodium dodecyl sulfate (SDS)**. SDS coats the proteins with a net negative charge, masking any charges normally present on the protein's surface. The sample is also treated to disrupt protein folding by the addition of chemical reagents such as **dithiothreitol (DTT)** or **beta mercaptoethanol (BME)** and heat. DTT and BME function to disrupt any disulfide bonds within the protein. The combination of SDS, DTT/BME, and heat will **denature** or unfold the protein.

23

The protein sample is **loaded** into a **well** or space that is formed at the top of the polyacrylamide gel. Polyacrylamide is chemically crosslinked to form a Jello-like matrix or gel. The density of the gel can be varied based on the percent composition of polyacrylamide. High percent acrylamide (dense) gels are useful for examining small proteins while lower percent acrylamide (thin) gels are best for examining larger proteins. Gels with an acrylamide concentration of around 8–10% are standard for most research applications.

The power of SDS-PAGE comes from its ability to separate a mixture of proteins into individual **bands**. Proteins migrate out from the well and through the polyacrylamide matrix in response to an electrical current. Recall that all the proteins in a sample have been treated with SDS and so are now negatively charged. Negatively charged proteins will migrate away from a anode and toward an cathode in an electrical circuit. The speed with which the various proteins migrate will be influenced by the degree of crosslinking of the acrylamide gel. Smaller proteins will have an easier time moving through the crosslinked matrix while larger proteins will move much more slowly. The net effect is the separation of proteins based on size, with smaller proteins at the bottom of a gel and larger proteins nearer the top. The size of a protein can be estimated by comparing the distance a protein has migrated relative to the movement of **molecular weight standards**. These standards are a mix of proteins of known molecular weight (in units of **kilodaltons** or **kDa**).

Protein gels must be stained to visualize the individual protein bands before they can be analyzed. Various staining protocols exist; however, one very common method is the use of a stain called **Coomassie Blue**. The Coomassie dye has the advantage of binding to proteins in a 1:1 ratio. This means that the staining intensity of a protein band on the gel is a measure of the concentration of that particular protein in the sample.

- Name the amino acids that would contribute a positive charge to the surface of a protein.
- Which amino acids are involved in the formation of a disulfide bond?
- Conduct a search for images of a polyacrylamide gel and the apparatus used to run the gel.
- Explain why a small protein migrates faster than a larger protein in a gel.

BACKGROUND

Red blood cells (RBCs) are specialized to carry oxygen to all parts of an organism. RBCs are unusual as they lack most of the organelles you would expect to find in a typical eukaryotic cell. These "cells" are really nothing more than a plasma membrane surrounding a cytoplasm full of the oxygen-binding protein, hemoglobin. Breaking open the RBCs (lysis) releases hemoglobin and leaves behind a pure preparation of RBC plasma membranes or **RBC ghosts**.

Because these cells must be able to move through small capillaries, their plasma membrane is reinforced by a special collection of peripheral membrane proteins. The proteins **α/β spectrin, ankryin, actin**, and **protein 4.1** bind to one another, forming a cage-like scaffold or **membrane cytoskeleton** on the cytoplasmic surface of the RBC plasma membrane. The membrane cytoskeleton is anchored to the plasma membrane through its attachment to the many **Band 3** anion transporter proteins that exist as transmembrane proteins in the RBC plasma membrane.

Spiders belonging to the genus *Loxoceles* are commonly known as brown recluse spiders. Despite their reputation, there are few fatalities linked to brown recluse spider bites (**envenomation**); however, their venom can cause serious health problems. The **cutaneous** form of the bite is localized to the skin and appears as a painful sore that can be followed by necrosis or the loss of tissue. If the venom is carried into the bloodstream, it can cause RBCs to lyse, resulting in the **viscerocutanous** or **systemic** form of the bite, which causes anemia, jaundice, and fever. Spider venom is a mixture of many proteins, some of which function as **proteases**. A protease is a type of enzyme that catalyzes the breakdown of proteins based on its ability to recognize specific short amino acid sequences. This case study illustrates how the technique of SDS-PAGE was used to examine the effect of spider venom on blood cells.

- Name the cellular components typically associated with eukaryotic cells.
- Describe the difference between an integral and peripheral membrane protein.
- Research/review the structure of the RBC membrane cytoskeleton. Note the interactions between the various protein components.
- Anemia results when the number of available RBCs is too low. Predict some likely symptoms that would be associated with anemia.
- Suggest a reason why RBCs lacking hemoglobin are called "ghosts."
- Proteases catalyze the reverse reaction that occurs during translation. Explain how this reaction would cause a protein to fall apart.

METHODS

Red blood cell ghosts

Blood samples were collected from four patients who had previously presented with the viscerocutaneous form of *Loxoceles* envenomation, four patients who had presented the cutaneous form of *Loxoceles* envenomation, and four individuals who had not been envenomated. RBCs were washed three times with saline at 4°C and lysed using 50 mM phosphate buffer (pH 8.3) at 4°C. Membrane ghosts were washed by repeated centrifugation at 25,000g until a colorless solution of ghosts was obtained. Membrane ghosts were stored at −70°C until use.

Venom digestion assays

Loxoceles venom was obtained from spiders by electrostimulation of the fangs. 80 µg of RBC membrane ghosts were incubated with 2.5 µg crude *Loxoceles* venom in 50 mM phosphate buffer at 37°C for 0, 30, 60, or 120 min. Other digests used 100 µg of membrane ghosts incubated for 60 min with increasing concentrations of *Loxoceles* venom. Protease inhibitors, N-ethylmaleimide (NEM), paramethylsulfonyl fluoride (PMSF), or ethylenediaminetetra-acetic acid (EDTA) were added to 80 µg of normal RBC ghosts in the presence of 2.5 µg *Loxoceles* venom and incubated for 120 min.

Protein gel electrophoresis

Samples were solubilized by incubation in gel sample buffer containing 1% SDS, 10% glycerol, 80 mM DTT, and 1 mM EDTA in Tris buffer. Samples were boiled for 10 min and then applied to an SDS-polyacrylamide gel. Electrophoresis was carried out at 25 V for 17 h. Gels were stained with 0.05% Coomassie Blue R250 and scanned using a densitometer.

- Conduct a search to find the chemical composition (recipe) of physiological saline. Explain why treatment of the RBCs with 50 mM phosphate buffer would cause them to lyse.
- What color would you see after lysis of the RBCs?
- Why was it necessary to remove all the hemoglobin from the membrane ghost preparation?

RESULTS

- Use the information in Figure 3.1.1 to estimate the molecular weight (in kDa) for β spectrin, Band 3, and actin.
- Predict how the proteins α and β spectrin are similar and how they are different.
- Compare the pattern of protein migration in the gel shown in Figure 3.1.2 with that shown in Figure 3.1.1. What might account for the difference?
- Which of the protein bands in Figure 3.1.2 correspond to venom proteins? Justify your answer.
- What accounts for the difference in staining intensity of the major venom protein band between lane 2 and lane 5 of Figure 3.1.2?
- Propose an explanation for the change in the appearance of the protein band indicated with the arrowhead in Figure 3.1.2.
- Summarize the results of the experiment shown in Figure 3.1.2.
- Why are the protein bands associated with the *Loxoceles* venom not visible on the gel in Figure 3.1.3?

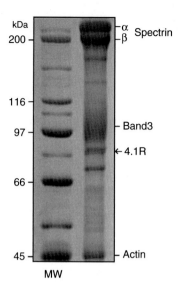

FIGURE 3.1.1 Protein composition of the membrane cytoskeleton of a human red blood cell (RBC).
SDS-PAGE of purified human RBC plasma membrane. Major cytoskeletal proteins are labeled. Protein
molecular weight standards (MW) are indicated. Samples were run on a 7.5% polyacrylamide gel and
stained with Coomassie Blue. *Image from [2].*

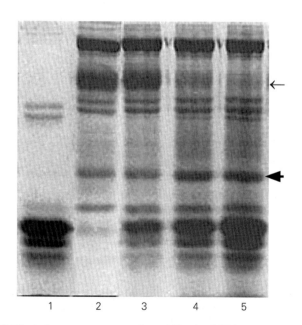

**FIGURE 3.1.2 Effect of venom exposure on the red blood cell (RBC) membrane cytoskeleton of
normal patients.**
RBC ghosts isolated from normal patients were treated with increasing concentrations of *Loxosceles*
spider venom. Lane 1, 100 μg venom; lane 2, 100 μg normal ghosts; lane 3, 100 μg normal ghosts
incubated with 25 μg venom; lane 4, 100 μg normal ghosts incubated with 50 μg venom; lane 5,
100 μg normal ghosts incubated with 100 μg venom. Arrow indicates the position of Band 3 protein.

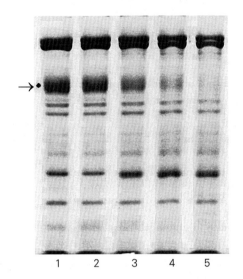

FIGURE 3.1.3 Effect of venom exposure on the red blood cell (RBC) membrane cytoskeleton of envenomated patients.

RBC membrane ghosts were isolated from patients who had previously presented with the viscerocutaneous symptoms of *Loxosceles* spider bite. Lane 1, membrane ghosts without incubation with venom. Lane 2, membrane ghosts mixed with 2.5 μg venom, but not incubated. Lane 3, membrane ghosts incubated with 2.5 μg venom for 30 min at 37°C. Membrane ghosts incubated with 2.5 μg venom for 60 min (lane 4) and 120 min (lane 5). Arrow points to the position of Band 3 protein.

- Describe the effect of prolonged exposure to *Loxoceles* venom on the RBC membrane cytoskeleton as illustrated in Figure 3.1.3. Are other proteins affected?
- How do the data in Table 3.1.1 relate to the data in Figure 3.1.3?
- Discuss whether the differences between values reported in Table 3.1.1 should be considered significant?
- Is there long-term damage to the RBC membrane cytoskeleton of patients with a history of *Loxoceles* spider bite? Use the data in Figure 3.1.3 and Table 3.1.1 to justify your answer.
- Which of the conditions tested in the experiment shown in Figure 3.1.4 resulted in inhibition of the degradation of Band 3?
- Design an experiment that would determine which of the venom protein bands is responsible for the loss of Band 3.
- The researchers conclude that the venom protein responsible for the loss of Band 3 is a *metalloprotease*. Conduct a search to learn more about this class of enzymes. Discuss whether the claim is supported by the data presented in this case study.

Table 3.1.1 Hydrolysis of Red Blood Cell (RBC) Ghost Band 3 From Patients With the Viscerocutaneous and the Cutaneous Forms of Loxoscelism Caused by Venom From the *Loxosceles gaucho* Spider

	0 min	30 min	60 min	120 min
Controls (N = 4)	29.2 ± 2.2	20.7 ± 3.9	15.2 ± 3.2	6.5 ± 3.1
Viscerocutaneous form (N = 4)	30.0 ± 2.1	20.2 ± 3.4	17.7 ± 3.8	7.5 ± 3.4
Cutaneous form (N = 4)	29.7 ± 1.7	21.5 ± 2.6	11.0 ± 2.2	7.5 ± 3.4

The ghosts (80 μg) were incubated with 2.5 μg venom in 100 μL of 5 mM phosphate buffer, pH 7.4, at 37°C. After SDS-PAGE (Figure 3.1.1), the protein content of Band 3 was determined by densitometry. Data are reported as percentage of total applied protein. RBC ghosts were isolated from patients who had previously suffered from the viscerocutaneous or cutaneous forms of Loxoceles envenomation; RBC ghosts from individuals who had not been bitten were used as a control; membranes were incubated with 2.5 μg venom for the indicated times; staining intensity of Band 3 on the protein gel was measured and reported as a percentage of the total protein present in a lane.

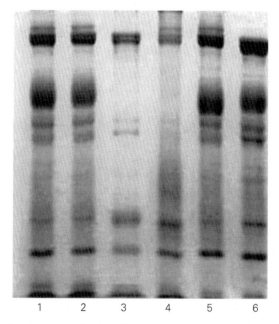

FIGURE 3.1.4 Action of protease inhibitors on the ability of venom to degrade Band 3. RBC membrane ghosts isolated from normal patients were incubated with venom in the absence or presence of protease inhibitors. Lanes 1 and 6, normal ghosts without venom; lane 2, normal ghosts mixed with 2.5 μg venom without incubation; lane 3, normal ghosts incubated for 120 min with 2.5 μg venom and NEM, a compound that can inhibit enzyme activity by binding to thiol groups; lane 4, normal ghosts incubated for 120 min with 2.5 μg venom and the compound PMSF, a serine protease inhibitor; lane 5, normal ghosts incubated for 120 min with 2.5 μg venom and the calcium chelator EDTA.

- Predict how the loss of Band 3 might affect the RBC membrane cytoskeleton. How might these changes translate into symptoms associated with envenomation?
- Discuss how well the experimental system used in this study matches the conditions that would exist under normal physiological conditions? What variables might be influenced by the design of these experiments?
- What are the medical implications of this study? What would you suggest is the "next step" for research on this topic?

Where to Start? A Case of Too Many AUGs [3]

INTRODUCTION

Information for the construction of proteins is stored within the DNA of a eukaryotic organism in units known as **genes**. Before a protein can be assembled, the genetic instructions or gene sequence must be copied or **transcribed** from the DNA into a molecule of **messenger RNA** (mRNA). Messenger RNAs travel from the nucleus into the cytoplasm where they interact with ribosomal subunits in preparation for the **translation** or synthesis of the protein. The sequence of the mRNA is "read" in units of three nucleotides called **codons**. Most of the 64 possible codons "code" for an amino acid; however, three of the codons, the stop codons, serve as signals for the termination of protein synthesis. One codon, the **AUG start codon**, signals the point where protein synthesis begins. The start codon is significant because it establishes the **reading frame** for the ribosome.

- Outline the steps involved in transcription of a gene.
- Research/review mRNA processing from the primary transcript to the mature mRNA.
- Outline the steps involved in translation of an mRNA.
- What amino acid does the AUG codon code for? What conclusion can you make about ALL proteins?
- Demonstrate the concept of a reading frame using a simple sentence.

BACKGROUND

A cell requires a vast constellation of proteins in order to function. Many of these are enzymes, proteins that catalyze a reaction. One example of a critically important enzyme is **DNA ligase**. As its name implies, a DNA ligase functions to join two pieces of DNA. DNA ligase I specifically functions to ligate short strands of DNA, known as Okazaki fragments, which are formed during the process of DNA replication. In the yeast, *Saccharomyces cerevisiae*, DNA ligase I is encoded by the gene *CDC9*. The protein product of this gene, $CDC9_p$, is synthesized in the cytoplasm before entering the nucleus where it can interact with the cell's DNA.

The nucleus is not the only place in a yeast cell where DNA can be found. Yeast have mitochondria and mitochondria have their own DNA. Mitochondrial DNA must replicate in order to maintain the genome of mitochondria as they undergo binary fission and are segregated into daughter cells following yeast mitotic cell division. Analysis of the genome of *S. cerevisiae* suggests that *CDC9* is the only gene for DNA ligase I. How can a single gene produce two proteins that maintain the same function but are located in two very different organelles? Several mitochondrial proteins are encoded by genes located in the nucleus and are synthesized in the cell's cytoplasm. Mitochondrial **presequences** direct these proteins to be imported into the mitochondria. Once the protein enters the mitochondria, a mitochondrial presequence peptidase (MPP), binds to the consensus amino acid sequence **ARFFT** and cleaves off the presequence.

- Brainstorm a list of other proteins and their functions in a cell.
- Research/review DNA replication. Identify the role of Okazaki fragments.
- Predict whether DNA replication could proceed in the absence of DNA ligase I.
- How can a protein move from the cytoplasm into the nucleus?
- Mitochondria possess a circular molecule of DNA that replicates in the same manner as the linear DNA found in the eukaryotic nucleus. Discuss the evolutionary significance of these two observations.
- Compare and contrast the process of binary fission and mitosis.
- Speculate on the consequences to the cell of the loss of mitochondrial DNA replication.
- Compare and contrast the import of proteins into mitochondria with the import of proteins into the rough endoplasmic reticulum (RER).

METHODS

Yeast culture conditions
All yeast strains used in this study were derived from a single wild-type strain, W303. Yeast cultures were grown on YPD media (2% peptone, 1% yeast extract, 2% glucose) supplemented with appropriate amino acids. Mitochondrial function was assessed by plating cells on YPEG media (2% peptone, 1% yeast extract, 3% glycerol, 3% ethanol).

CDC9–GFP fusions, CDC9 mutagenesis, and HA tagging
Plasmids containing various lengths of the amino terminus of the *CDC9* gene were fused in-frame with the gene for green fluorescence protein (GFP). Site-directed mutagenesis was used to alter AUG sequences in the CDC9 gene. The first AUG site was converted to UAG and the second was changed to GCG. Changes to the DNA sequence were confirmed by sequencing. A human

influenza hemagglutinin (HA) epitope tag was added to the *CDC9* gene using transposon mutagenesis. A triple HA tag was inserted at position 75 in the amino acid sequence of CDC9 (CDC9–HA). It was determined that the insert did not affect the growth rates of the cells.

Subcellular fractionation and protease K digestion

Yeast were harvested by centrifugation (5 min at 3000*g*), washed in distilled water and resuspended in buffer and incubated at 30°C for 10 min. Zymolase 5000 (5 mg/g of cell, wet weight) was added to the suspension and incubated at 30°C with gentle shaking. Yeast spheroplasts were harvested by centrifugation (5 min at 3000 rpm). Spheroplasts were resuspended in buffer and disrupted using a Dounce homogenizer on ice. The cell homogenate was centrifuged (5 min at 3500 rpm) and the supernatant was saved. A crude mitochondrial fraction was collected by centrifugation of the supernatant at 9000 rpm for 10 min at 4°C. The mitochondrial pellet was washed twice and the postmitochondrial supernatant was saved.

Yeast nuclei were prepared from spheroplasts that were homogenized as described earlier. The cell homogenate was centrifuged at 1000*g* for 10 min at 4°C to remove unbroken cells and cell debris. The supernatant was transferred to a fresh tube and the nuclei were pelleted by centrifugation at 12,000*g* for 25 min at 4°C. The pellet was resuspended in a small amount of buffer and layered on top of a sucrose gradient. The sample was centrifuged using a swing bucket rotor at 25,000 rpm for 60 min at 4°C. Intact nuclei pelleted, leaving other cellular material in the overlying sucrose layers of the gradient.

Proteinase K treatment of isolated mitochondria or nuclei was carried out by adding an equal volume of 500 μg/mL proteinase K. Triton X-100 was added to a final concentration of 0.5%. After incubation at 0°C for 30 min, proteins were precipitated by incubation with TCA.

Immunoblotting

Protein samples were analyzed by SDS-PAGE followed by immunoblotting. Antiserum directed against yeast Hsp60 was used at a 1:10,000 dilution, anti-NpI3 antiserum at 1:3,000 dilution, anti-GFP antiserum at 1:1,000 dilution, and anti-HA monoclonal antibody at 1:1,000 dilution. Secondary HRP-conjugated antirabbit and antimouse antibodies were used at 1:1000 dilution.

- What is a "wild-type" cell?
- Create a flow chart that summarizes the steps used in the subcellular fractionation of yeast cells.
- What was the final concentration of proteinase K used in the digestion assays?

(a)
```
*
MRRLLTGCLLSSARPLKSRLPLLMSSSLPSSAGKKPKQATLARFFTSMKNKPTEGTPSPK  60
                  *
KSSKHMLEDRMDNVSGEEEYATKKLKQTAVTHTVAAPSSMGSNFSSIPSSAPSSGVADSP 120
QQSQRLVGEVEDALSSNNNDHYSSNIPYSEVCEVFNKIEAISSRLEIIRICSDFFIKIMK 180
QSSKNLIPTTYLFINRLGPDYEAGLELGLGENLLMKTISETCGKSMSQIKLKYKDIGDLG 240
EIAMGARNVQPTMFKPKPLTVGEVFKNLRAIAKTQGKDSQLKKMKLIKRMLTACKGIEAK 300
FLIRSLESKLRIGLAEKTVLISLSKALLLHDENREDSPDKDVPMDVLESAQQKIRDAFCQ 360
VPNYEIVINSCLEHGIMNLDKYCTLRPGIPLKPMLAKPTKAINEVLDRFQGETFTSEYKY 420
DGERAQVHLLNDGTMRIYSRNGENMTERYPEINITDFIQDLDTTKNLILDCEAVAWDKDQ 480
GKILPFQVLSTRKRKDVELNDVKVKVCLFAFDILCYNDERLINKSLKERREYLTKVTKVV 540
PGEFQYATQITTNNLDELQKFLDESVNHSCEGLMVKMLEGPESHYEPSKRSRNWLKLKKD 600
YLEGVGDSLDLCVLGAYYGRGKRTGTYGGFLLGCYNQDTGEFETCCKIGTGFSDEMLQLL 660
HDRLTPTIIDGPKATFVFDSSAEPDVWFEPTTLFEVLTADLSLSPIYKAGSATFDKGVSL 720
RFPRFLRIREDKGVEDATSSDQIVELYENQSHMQN  755
```

(b)
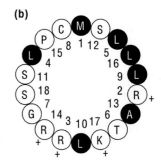

FIGURE 3.2.1 Amino acid sequence of CDC9.

a. Predicted sequence of the CDC9 protein. Analysis of the sequence of CDC9$_p$ revealed the presence of four, in-frame AUG codons (asterisks) located within the first 70 amino acids of the protein. Three potential nuclear localization sequences (underlined) were also identified. b. Alpha helical structure of the first 18 amino acids (shown in single letter code). Black circles represent hydrophobic amino acids and (+) indicates positively charged amino acids.

RESULTS

■ Label the methionine amino acids representing the AUG codons in Figure 3.2.1a AUG1, AUG2, AUG3, and AUG4. Count the number of amino acids between AUG1 and AUG2. Calculate the number of nucleotide base pairs that would have made up that region of the original mRNA.

■ Predict how the protein products would differ if translation initiated at AUG3 instead of AUG1 *in vivo*.

■ Which of the following contains a second, **in-frame** AUG sequence?
  ■ <u>AUG</u>CGUCGGAAUGAGACUGGA
  ■ <u>AUG</u>GAGCGGAAUCUUAUGGUU

■ Study the diagram in Figure 3.2.1b. What do you observe about the numbering and chemical properties of the amino acids? Attempt to match this diagram with an illustration of alpha helical structure.

■ Look for an MPP consensus sequence in the overall CDC9 sequence (Figure 3.2.1a). Confirm your findings based on the information in Figure 3.2.2a.

■ Estimate the molecular weight (size; kDa) of each of the bands shown in Figure 3.2.2b. Match the bands to the fusion protein constructs shown in Figure 3.2.2a.

■ Use the data presented in Figure 3.2.2a and b to develop an argument in support of the statement, "the 30.3 kDa form in pMW197-containing cells cannot be the product of initiation at AUG3, but rather it must be a proteolytic product of the full-length polypeptide expressed from AUG1."

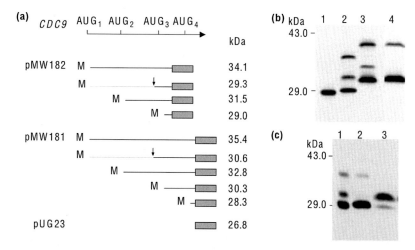

FIGURE 3.2.2 CDC9–GFP fusion protein constructs are translated from multiple start codons and proteolytically processed by mitochondria.

a. The 5′ end of the CDC9 reading frame is shown schematically with the positions of AUG 1–4 indicated to scale. The structure of the GFP fusion proteins expressed from AUG1–3 using the pMW182 plasmid vector and the AUG1–4 inpMW181 are shown together with their predicted molecular weights. M, initiating methionine; gray box, GFP; arrow, predicted mitochondrial presequence cleavage site. b. Immunoblot of total cell extracts from yeast expressing the CDC9–GFP constructs and probed with anti-GFP antibody. Lane 1, pUG23 (GFP only, no insert); lane 2, pMW182 (AUG1–3); lane 3, pMW181 (AUG1–4); lane 4, pMW197 (AUG1, 4; AUG 2 and 3 sequences changed to GCG). c. Subcellular fractionation of yeast expressing the CDC9–GFP fusion proteins encoded by pMW182 (AUG1–3). Lane 1, total extract; lane 2, mitochondrial fraction; lane 3, postmitochondrial supernatant.

- What evidence is used in Figure 3.2.2c to support the conclusion that CDC9–GFP protein is imported into mitochondria?
- Based on the data presented in Figure 3.2.3, has the addition of an HA tag altered the behavior of the CDC9 protein?
- Which organelle, mitochondria or nuclei, would travel farther into a sucrose gradient during sucrose gradient density centrifugation? Explain your answer.
- Propose an explanation for the presence of antibody staining in lanes 2–4 in Figure 3.2.3b.
- Which of the lanes in Figure 3.2.4a supports the conclusion that CDC9–HA is inside the mitochondria? Explain your answer.
- Create a drawing to illustrate how Figure 3.2.4a would appear if CDC9–HA were attached to the outside (cytoplasmic side) of the mitochondria.
- Does the presence of Hsp60 staining in Figure 3.2.4e diminish the strength of the conclusion that CDC9–HA is located inside nuclei? Justify your answer.

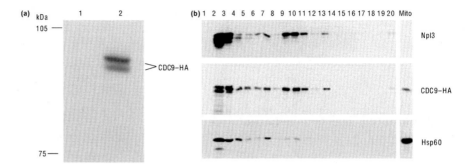

FIGURE 3.2.3 Expression of an HA-tagged CDC9 in yeast.
a. Immunoblot of total cell extracts from wild-type yeast (lane 1) or cells expressing the HA-tagged CDC9 protein (lane 2) using an anti-HA antibody. b. CDC9–HA is found in nuclear and mitochondrial fractions following subcellular fractionation. Intact nuclei were prepared from CDC9–HA expressing cells using a sucrose step gradient. Fractions collected from the gradient were immunoblotted with antibodies to the nuclear marker protein, Npl3, HA, or the mitochondrial marker protein, Hsp60. Lane 1, top of gradient; lane 20, bottom of gradient. Mitochondria (Mito) were purified from the yeast and analyzed by immunoblotting with the same set of antibodies.

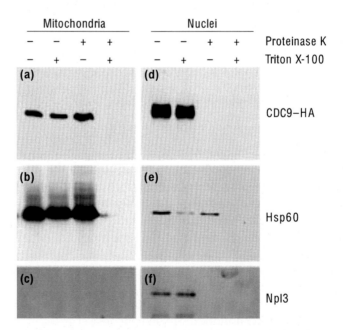

FIGURE 3.2.4 CDC9–HA is protected against protease digestion in the mitochondrial fraction.
Nuclei and mitochondria were purified from yeast cells expressing CDC9–HA and treated with proteinase K in the absence or presence of the detergent, Triton X-100 as indicated. Proteins were precipitated by incubation with tricarboxylic acid (TCA) and analyzed by immunoblotting with antibodies to HA, the nuclear marker protein Npl3, or the mitochondrial marker protein, Hsp60. Both the mitochondrial fraction (a–c) and nuclear fraction (d–f) were subjected to four different combinations of protease and detergent.

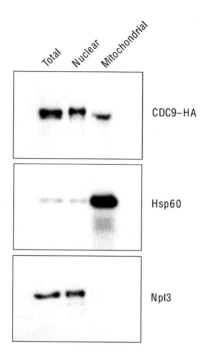

FIGURE 3.2.5 The forms of CDC9–HA found in nuclei and mitochondria are different.
Immunoblot of total cell extract (total), nuclear and mitochondrial preparations from CDC9–HA expressing cells probed with antibodies to HA, Hsp60, or Npl3.

- Propose an explanation for the results shown in Figure 3.2.4d and f.
- Explain why the mitochondrial form of CDC9–HA is smaller than the nuclear form as shown in Figure 3.2.5. Use data from Figure 3.2.1a to support your explanation.
- Refer back to Figure 3.2.1a. CDC9 has three potential nuclear localization sequences. Design an experiment to establish which of the sequences is being used to target import of CDC9 into the nucleus.
- Why would mitochondrial function not be a factor affecting cell growth on glucose?
- Compare the growth of wild-type cell culture on glycerol and glucose shown in Figure 3.2.6. What do you observe? Propose an explanation for your observation.
- How would the mutations used in the experiment shown in Figure 3.2.6 (AUG1 changed to UAG; AUG2 changed to GCG) alter the CDC9 protein?
- Immunoblot analysis of the mutant cells shown in Figure 3.2.6 yielded the following results: (1) the upper band of the CDC9–HA doublet was

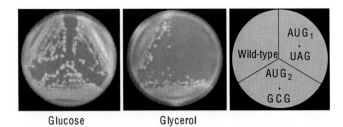

Glucose   Glycerol

FIGURE 3.2.6 Functional distinctions between CDC9 encoded from AUG1 and AUG2.
Yeast cells rely on their mitochondria in order to grow on glycerol, a nonfermentable carbon source. Mutations that impact mitochondrial function will inhibit cell growth on glycerol. Wild-type and CDC9 mutant cell growth on glucose and glycerol was compared. AUG1 was mutated to UAG and AUG2 was mutated to GCG as indicated.

missing from the AUG2–GCG mutant cells. (2) The lower band of the CDC9–HA doublet was missing from the AUG1–UAG mutant cells.

- Develop an argument for or against the conclusion that synthesis of the mitochondrial and nuclear forms of CDC9 is initiated at two different AUG sequences.

- The experiment described was presented as "data not shown" in the original paper. Discuss whether the absence of an image of the blot influences your confidence in the conclusion.

- Yeast cells were able to undergo cellular division in the absence of a functional AUG2 start codon (AUG2–GCG; Figure 3.2.6). Develop a hypothesis to account for the ability of yeast cells to progress through multiple rounds of the cell cycle in the absence of AUG2.

Cotranslational Translocation: Gatekeepers of the RER [4]

INTRODUCTION

Proteins that are destined to be inserted into the plasma membrane, exported from the cell, or imported into the lysosome must be synthesized on the RER. While these proteins serve very diverse functions, one thing they have in common is the presence of a sequence of amino acids that acts as a **signal peptide**, targeting synthesis of the protein to the surface of the RER. The specific amino acid composition of the signal peptide can vary between proteins; however, all signal peptides contain a hydrophobic core.

Initiation of translation begins in the cytoplasm. As elongation proceeds, the newly synthesized polypeptide (**nascent chain**) begins to emerge from the ribosome. If the polypeptide carries a signal sequence it will be recognized and bound by the **signal recognition particle (SRP)**, temporarily pausing elongation. The SRP escorts the complex of ribosome, mRNA, and nascent chain to the RER membrane. There, the SRP binds to the SRP receptor and the ribosome attaches to a channel protein in the RER membrane called a **translocon**. The SRP is released from the signal sequence, elongation resumes, and the growing polypeptide chain moves through the translocon and into the lumen of the RER in a process known as **cotranslational translocation**. **Signal peptidases**, enzymes located in the lumen of the RER, catalyze the removal of the signal peptide from the protein.

The steps leading up to attachment of a ribosome to a translocon are well known. What is less well known is what happens as the signal sequence moves through the translocon channel. What proteins are involved, in what order, and does the signal sequence have any role in movement of the nascent chain through the channel?

- Research some specific examples of proteins that would be synthesized on the RER and transported either out of the cell, into a lysosome, or inserted into the plasma membrane.
- Using your own words, explain what is meant by cotranslational translocation.

- Conduct an online search for a database of signal peptide sequences. Compare the signal peptide sequences from different proteins or different organisms.
- Review the steps involved in translation of mRNA.

BACKGROUND

RER membrane proteins play a critical role in the synthesis and transport of proteins. In addition to the SRP receptor, there is a collection of proteins that make up the protein-conducting channel known as the translocon. The **Sec61 complex**, in combination with **TRAM** protein, makes up the walls of the translocon. Studies investigating the movement of proteins across the RER membrane have taken advantage of two experimental protocols. In the first, isolated RER membranes are purified, allowed to form into small vesicles (**microsomes**), and used in assays that test the ability of a newly synthesized protein to bind to or move through the membrane. Because these RER microsomal membranes (**PK-RM**) are isolated from cells, they possess all the proteins that you would find associated with the RER. In the second experimental approach, researchers use artificial membranes known as **proteoliposomes**. Since these membranes are assembled in the lab, the protein composition can be manipulated.

In vitro assays of cotranslational translocation make use of well-characterized, secretory proteins. In the work described here, that protein is **prolactin**, a hormone associated with lactation in mammals. When the mRNA for prolactin is translated in the absence of RER membranes the protein that is produced is slightly larger than when the same reaction occurs in the presence of RER microsomal membranes. The larger version of the protein is called **preprolactin (pPL)**.

- Explain why a biological membrane can form a vesicle.
- Propose an explanation for the difference in size of prolactin synthesized in the presence and absence of RER membranes.
- Conduct a search to find diagrams that illustrate the structure of the translocon. Locate Sec61α and TRAM.

METHODS

In vitro translation

mRNAs coding for full-length preprolactin (pPL), pPL fragments, or signal peptide mutant pPL were translated for 20 min at 26°C in a wheat germ translation system in the presence of a ^{35}S labeling mixture of methionine and cysteine. Reactions also included 40 nM SRP and, where indicated, salt-washed RER microsomal membranes (PK-RM). For crosslinking experiments, a photoreactive derivative of the amino acid lysine was also included during translation. The compound puromycin was included in some reactions to trigger release of the nascent chain from the ribosome.

Crosslinking assay

Radioactively labeled pPL fragments synthesized in the presence of a photoreactive form of the amino acid lysine were incubated as described, followed by UV irradiation. Proteins that are in close proximity to modified lysine amino acids will become chemically crosslinked to the pPL fragment. Crosslinked partner proteins are detected as radiolabeled bands on an SDS-PAGE gel due to ^{35}S methionine/cysteine labeling of smaller pPL fragments. Immunoprecipitation was used to identify crosslinking between RER membrane proteins, Sec61α, and TRAM.

Flotation assay

Centrifugation through a sucrose cushion was used to separate membrane bound from cytosolic pPL. Samples were placed in 2 M sucrose and layered underneath a 100 μL cushion containing 1.8 M sucrose in high salt. 50 μL of 250 mM sucrose was layered on top of the cushion. The samples were then centrifuged at 100,000 rpm for 1 h at 4°C. Membranes floated through the high salt, sucrose cushion and collected in the top layer.

Protease resistance assay

Samples of pPL nascent chains in the presence or absence of microsomal membranes or proteoliposomes were incubated on ice for 30 min in the presence of 0.5 mg/mL proteinase K. The reaction was stopped by the addition of excess protese inhibitor. Some membrane samples were incubated with the detergent TritonX-100 prior to protease digestion.

- Conduct an online search to learn more about wheat germ *in vitro* translation.
- Predict what might happen if the SRP was not included in the wheat germ translation system.
- What was the purpose of the flotation assay?
- How would a membrane be affected by the presence of a detergent?

RESULTS

- Annotate the schematic diagram in Figure 3.3.1a to indicate the number of amino acids available to interact with cytoplasmic or membrane proteins.
- What is the minimum fragment size required for SRP binding (Figure 3.3.1b)?

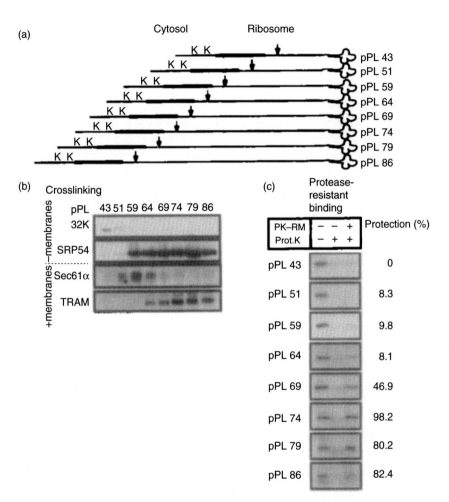

FIGURE 3.3.1 The length of the amino acid chain influences interactions between the nascent protein and cytosolic and RER membrane components required for cotranslational translocation.
a. Schematic drawing of the different pPL fragments used in this study. The total number of amino acids in each fragment is indicated on the right. It is estimated that the ribosome covers approximately 30 amino acids of the nascent chain. The hydrophobic core of the signal peptide is drawn as a thick line, and arrows indicate the signal peptide cleavage site. Lysine amino acids (K) within the signal peptide were replaced by a photoreactive lysine derivative for use in crosslinking studies. b. Crosslinking partners of nascent pPL fragments in the absence or presence of RER membranes. pPL was synthesized in a wheat germ translation system in the presence of [^{35}S]methionine and [^{35}S]cysteine, SRP, and a modified lysine–tRNA carrying a photoreactive group. Aliquots of each translation reaction were irradiated in the absence of RER membranes (minus membranes) and analyzed using SDS-PAGE. Parallel samples were incubated with RER membranes (plus membrane) prior to irradiation and then immunoprecipitated with antibodies specific to Sec61α or TRAM. c. Protease resistance of pPL. After *in vitro* translation, each of the samples was divided into three equal portions. One portion was analyzed immediately (lane 1). The other two were incubated either with buffer (lane 2) or with RER microsomal membranes (PK-RM) (lane 3) and treated with the enzyme proteinase K (Prot. K). Percent protection reflects the amount of radioactive protein in the samples shown in lane 3 compared with those in lane 1.

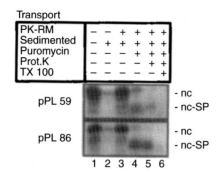

Transport

PK-RM	−	−	+	+	+	+
Sedimented	−	+	+	+	+	+
Puromycin	−	−	+	+	+	+
Prot.K	−	−	−	−	+	+
TX 100	−	−	−	−	−	+

pPL 59
- nc
- nc-SP

pPL 86
- nc
- nc-SP

1 2 3 4 5 6

FIGURE 3.3.2 Transport of pPL fragments containing 59 or 86 amino acids.
Nascent chains of the indicated size were synthesized in the presence of an SRP and incubated in the absence or presence of PK-RMs. Membranes were sedimented through a sucrose cushion containing high salt. Aliquots of resuspended membrane were treated with puromycin, and translocation of the nascent chain was tested by treatment with proteinase K in the absence or presence of the detergent, Triton X-100 (TX 100). The nascent pPL fragment (nc) and the signal peptide cleaved form (nc-SP) are indicated.

- Summarize the results of crosslinking reactions between pPL fragments and RER membranes (Figure 3.3.1b).
- Use the information from Figure 3.3.1b to outline the sequence of interactions that occur between the newly synthesized signal peptide and the cytoplasmic and membrane proteins it encounters during cotranslational translocation.
- Explain why protein bands corresponding to all the pPL fragments are seen in lane 1 of Figure 3.3.1c and no protein bands are visible in lane 2.
- Use your knowledge of membrane structure to explain why proteins inside a microsome are protected from digestion by proteases.
- Propose a model to explain why shorter pPL fragments are not protected from protease digestion by microsomal membranes.
- Relate the information presented in the box at the top of Figure 3.3.2 to the design of the experiment.
- Why are the protein bands in lane 1 of Figure 3.3.2 missing in lane 2?
- Why have the protein bands in lane 3 of Figure 3.3.2 changed size in lane 4?
- Why are the protein bands in lane 4 of Figure 3.3.2 missing in lane 6?
- Discuss whether there is an effect of the size of the pPL fragment on translocation based on the experiment shown in Figure 3.3.2.
- Using Figure 3.3.2 as a model, design an experiment to demonstrate that an SRP is required for the translocation of pPL into RER microsomal membranes (PK-RM).
- Examine the chemistry of the R groups for the amino acids shown in bold in Figure 3.3.3a. What can you conclude?

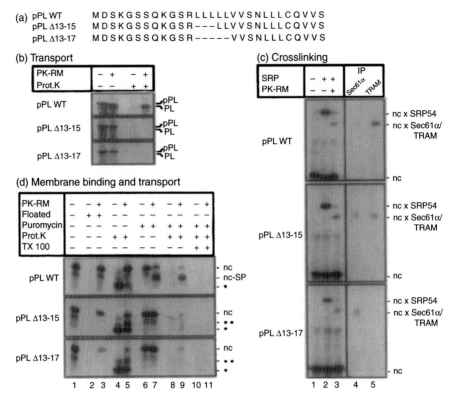

(a)

pPL WT	M D S K G S S Q K G S R L L L L L V V S N L L L C Q V V S
pPL Δ13-15	M D S K G S S Q K G S R – – – L L V V S N L L L C Q V V S
pPL Δ13-17	M D S K G S S Q K G S R – – – – – V V S N L L L C Q V V S

(b) Transport

(c) Crosslinking

(d) Membrane binding and transport

FIGURE 3.3.3 Transport and membrane binding of signal peptide mutants of pPL.
a. Structures of the signal peptides of pPL and two deletion mutants. Amino acids located in the hydrophobic core are in bold. b. Transport of wild-type and mutant pPL into microsomes. Full length protein carrying the wild-type signal peptide (pPL-wt) or a signal peptide deletion (pPLΔ13-15 or pPLΔ13-17) were synthesized in the presence of SRP and microsomes (where indicated). After translation, the samples were split into two aliquots, one of which was treated with proteinase K (Prot.K). The positions of preprolactin (pPL) and mature prolactin (PL) are indicated. c. Crosslinking of wild-type and mutant pPL fragments. Eighty-six amino acid fragments of pPL carrying the wild type or mutant signal peptides were synthesized in the presence of the photoreactive lysine derivative. Aliquots of the samples were incubated with SRP and microsomes as indicated, and irradiated. Crosslinks were analyzed directly (lanes 1–3) or after immunoprecipitation with antibodies to Sec61α (lane 4) or TRAM (lane 5). The position of the nascent chains (nc) and their crosslinked products with SRP54 and the membrane proteins are indicated. d. Membrane binding and translocation of wild-type and mutant pPL fragments. Ribosome-bound fragments of wild-type or mutant pPL were incubated with or without microsomes as indicated (PK-RM). The samples were split into three aliquots. One aliquot was centrifuged through a sucrose gradient containing high salt (floated) to disrupt weak binding between the ribosome and the membranes. The second aliquot was incubated with proteinase K to test for resistance to protease digestion. The third aliquot was treated with puromycin to inhibit translation. After puromycin treatment the samples were treated with proteinase K in the presence or absence of Triton X-100. The position of the nascent chain (nc), signal cleaved fragments (nc-SP), and protease-protected fragments of 30 amino acids (*) or 50–60 amino acids (**) are indicated.

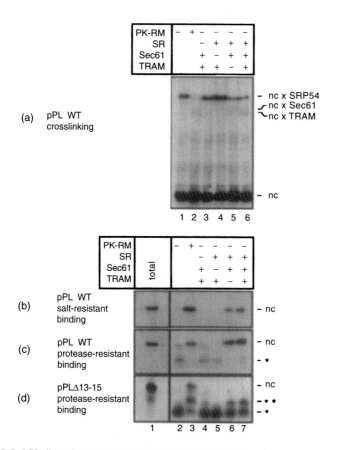

PK-RM	–	+	–	–	–	–
SR			–	+	+	+
Sec61			+	–	+	+
TRAM			+	+	–	+

(a) pPL WT crosslinking

– nc x SRP54
⌐ nc x Sec61
⌐ nc x TRAM

– nc

1 2 3 4 5 6

PK-RM		–	+	–	–	–	–
SR				–	+	+	+
Sec61	total			+	–	+	+
TRAM				+	+	–	+

(b) pPL WT salt-resistant binding

– nc

(c) pPL WT protease-resistant binding

– nc
– *

(d) pPLΔ13-15 protease-resistant binding

– nc
– * *
– *

1   2 3 4 5 6 7

FIGURE 3.3.4 Binding of nascent pPL to reconstituted proteoliposomes.

a. Photocrosslinking with 86 amino acid pPL fragments carrying the wild-type signal peptide. Fragments were synthesized in the presence of SRP and a photoreactive lysine. Aliquots were incubated in the absence or presence of microsome (PK-RM) or proteoliposomes containing the SRP receptor, Sec61p complex (Sec61), or TRAM protein in the indicated combinations. After irradiation, the samples were analyzed by SDS-PAGE. The positions of the nascent chain (NC) and crosslinked products are indicated. b. High salt-resistant binding of wild-type pPL86 to microsomes or proteoliposomes. After synthesis in the presence of SRP, equal aliquots were incubated with microsomes or proteoliposomes as indicated, and subjected to centrifugation in a sucrose gradient at a high salt concentration. The sample in lane 1 shows the starting material. Lanes 2–7 show the floated material. c. Protease-resistant binding of wild-type pPL86 to microsomes or proteoliposomes. After synthesis of pPL86 in the presence of SRP, equal aliquots were incubated with microsomes or proteoliposomes as indicated followed by treatment with proteinase K. The position of the nascent chain (nc) and the ribosome-protected fragment (*) are indicated. Lane 1 is the starting material. Lanes 2–7 are digests. d. Protease-resistant binding of the signal peptide mutant pPL fragment to microsomes or proteoliposomes. The experiment was the same as described for (c), however the pPL fragment that was synthesized was lacking the three lysine amino acids in the hydrophobic core of the signal peptide (pPLΔ13-15). The position of the 50–60 amino acid protease protected fragment is indicated (**).

- Summarize the conclusion that is supported by the results shown in Figure 3.3.3b.
- Use the data in Figure 3.3.3c to formulate an argument in favor or against the following statement: SRP binding to the signal peptide is highly specific.
- Develop a model that describes the role of the signal peptide during translocation across the RER membrane based on the data in Figure 3.3.3c.
- Which of the lanes in Figure 3.3.3d is testing the strength of the binding between the ribosome and the RER membrane? Explain your choice.
- Explain the significance of the data shown in Figure 3.3.3d, lane 4.
- Describe what has occurred to each of the pPL fragments in Figure 3.3.3d, lane 5.
- Are mutant pPL fragments capable of crossing the RER membrane? Justify your answer.
- Predict the outcome of the following experiment for pPL WT and pPLΔ13-17 based on your model:

PK-RM	−	+	+
Puromycin	+	+	+
Protease K	+	+	+
TX100	−	−	+

- Explain the value of including lanes 1 and 2 in Figure 3.3.4a.
- What is the minimal combination of proteins required for translocation of pPL across the proteoliposome membrane (Figure 3.3.4b and c)?
- Compare the protease resistance of wild-type and mutant pPL fragments (Figure 3.3.4d). Develop a hypothesis to account for the difference.

From Sequence to Function [5]

INTRODUCTION

A key function common to the plasma membranes that surround all cells is the ability to regulate the movement of molecules into and out of the cell. **Selective permeability** of a membrane is determined both by the chemistry of the phospholipid bilayer and by the types of proteins located within that bilayer. **Transport proteins** are a class of integral membrane proteins that function to facilitate the movement of specific molecules across a membrane. Integral membrane proteins must be able to reside within the hydrophobic environment that exists in the interior of a phospholipid bilayer. However, since the molecules that move through transport proteins are hydrophilic, these proteins must also possess an aqueous channel through which the molecules can move.

Protein function depends upon the proper folding of a chain of amino acids into a specific three-dimensional structure. How a protein folds is determined by its amino acid sequence. Amino acids are characterized based on their side chains or **R groups** that extend from their central carbon. The size and chemical properties of the various R groups control the ability of the amino acid chain to form the intramolecular bonds required to stabilize the protein's structure.

Protein structure is categorized into four levels. **Primary structure** represents the simple, linear sequence of amino acids. **Secondary structure** occurs when short segments of the amino acid sequence take on the shape of a coil (**alpha helix**) or a pleated fold (**β-pleated sheet**). Complete three-dimensional folding of the protein represents its **tertiary structure**, while the interaction of two or more folding proteins is characterized as the **quaternary structure**. This case study will illustrate the relationship between the amino acid sequence of a protein and its function as a transport protein.

- Create a diagram of a biological membrane. Label your drawing to indicate a phospholipid, its polar head group and its fatty acid tails, a molecule of cholesterol, an integral membrane protein, and a peripheral membrane protein.
- Define the concept of selective permeability using your own words.
- Explain why a phospholipid bilayer can act as a barrier to the movement of an ion but cannot control the movement of small nonpolar molecules.

- Research/review the different types of transport proteins found in a cell.
- Is membrane transport limited to the plasma membrane? Justify your answer.
- Research/review the structure of an amino acid. Compare the R groups of the amino acids cysteine, glutamate, tryptophan, and glycine.
- Name the chemical bonds that are found at each level of protein folding.

BACKGROUND

Aquaporins (AQPs) are a class of channel protein that are found in the plasma membrane of bacteria, plant, and animal cells. Like all proteins, aquaporins are made up of a linear sequence of amino acids. Proteins that insert into a membrane usually do so by folding into an alpha helix structure at the point where the protein crosses the phospholipid bilayer. These alpha helical regions are called **membrane-spanning domains**. AQP has six membrane-spanning domains and two domains that partially insert into the membrane (M1–M8). All of these domains are predicted to form alpha helical regions within the protein.

As the AQP1 protein inserts into the membrane the alpha helical domains wrap around, forming an amino acid-lined pore down the center of the protein. Four individual AQP1 **monomers** assemble into a unit called a **homotetramer**. One of the properties of AQP1 is the ability to facilitate the bidirectional transport of water molecules, but not ions, including hydrogen ions.

- Describe the levels of protein structure using aquaporin as your example.
- What does the word "domain" mean in terms of protein structure?
- Provide a definition, using your own words, for the terms monomer and homotetramer.
- Where in a cell would the AQP1 protein be synthesized?
- Predict how the chemical nature of the amino acids that interact with the fatty acid tails of phospholipids in the membrane-spanning domain of aquaporin might be different from those that line the channel?
- Explain what causes water molecules to move across a membrane.

METHODS

Isolation and structural analysis of bovine AQP1

AQP1 was isolated from bovine red blood cell (RBC) plasma membranes. Purified AQP1 proteins were crystallized and examined by X-ray crystallography. X-ray crystallography is a technique that provides information about protein structure based on **diffraction patterns** generated by collisions between the X-ray beam and the protein. Structural domains such as alpha helices and beta pleated sheets generate unique diffraction patterns. Diffraction patterns were analyzed using specialized software.

Data processing

A range of computer programs exist that aid in the analysis of patterns and prediction of protein structure. AQPs and glycerol facilitator protein, GlpF, gene sequences were compared using a software program called CLUSTAL W5. This program allows for comparison between two or more amino acid sequences. The program makes it possible to identify homologies between proteins and to recognize sequences that are likely to fold in specific ways. Using this program the researchers were able to identify the eight alpha helical domains of AQP1. The programs MOLSCRIPT and Raster3D were used to convert data from X-ray crystallography into three-dimensional models of the folded AQP1 protein. HOLE is the program that was used to analyze the diameter of the pore in AQP1. Kyte–Dolittle analysis of amino acid sequences makes it possible to look for hydrophilic and hydrophobic regions within a protein, facilitating identification of possible membrane-spanning domains.

- Research/review the structure of an RBC. How is an RBC different from other types of eukaryotic cells?
- Discuss the role of the scientist when data analysis is based on computer programs.

RESULTS

- Study the amino acid sequence alignment between human and bovine AQP1 shown in Figure 3.4.1. What do you observe?
- Expand your analysis of amino acid sequences to the other transport proteins shown in Figure 3.4.1. Are the aquaporin sequences conserved? Explain your answer.
- Is the glycerol transport protein, GlpF, similar to an aquaporin? Justify your answer.
- Create a table that characterizes the chemical nature of the R groups of the amino acids found lining the pore (red italicized).
- The beginning of the sequence (upper left corner) corresponds to the amino terminus of the protein. The amino terminus of AQP1 is known to be located on the cytoplasmic side of the plasma membrane. Working from this fact, construct a diagram to illustrate how AQP1 inserts into the membrane. Use rectangles to represent alpha helical domains and a connecting line representing the intervening linear amino acid sequence.
- Propose an explanation for why some of the membrane-spanning domains in the AQP1 sequence in Figure 3.4.1 did not include any red italicized amino acids.
- Compare and contrast the type of information you can gather from a ribbon diagram and a space-filling model of a protein structure. Which model do you prefer and why?

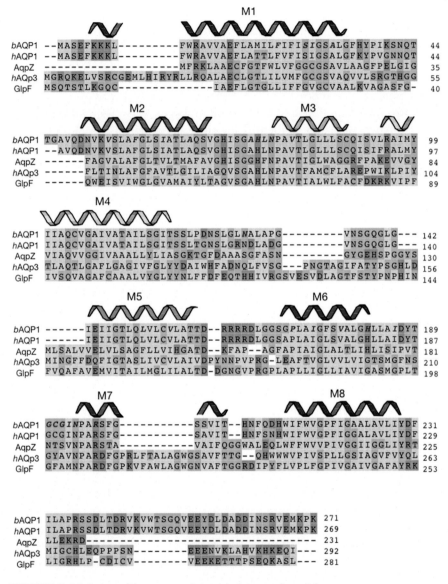

FIGURE 3.4.1 Amino acid sequence alignment of aquaporin proteins.
Bovine AQP1, human AQP1, *E. coli* AqpZ, human AQP3, and *E. coli* GlpF (a glycerol transport protein)
protein sequences are compared. AQP1 membrane-spanning domains are color coded and labeled.
M3 and M7 are membrane-inserted, nonmembrane-spanning helices. Red italicized letters indicate the
amino acids in bovine AQP1 that are oriented toward the center of the channel's pore.

- Locate the pore in the space-filling model of an AQP1 monomer in Figure 3.4.2b (lower left corner).
- Connect the alpha helical domains shown in Figure 3.4.2a with their sequences in Figure 3.4.1. Label the helices that participate in forming the pore.

(a)

(b)

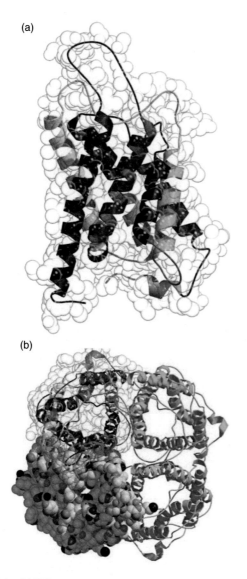

FIGURE 3.4.2 Models of AQP1 structure.
a. Combined ribbon and space-filling model of an AQP1 monomer viewed parallel to the plane of the membrane. Colors of the alpha helical domains correspond to those in the sequence alignment in Figure 3.4.1. b. AQP1 homotetramer viewed from above the plane of the membrane. Two of the monomers also include a space-filling model to show the pore entrance.

- The same color-coding in Figure 3.4.2a is used to represent the alpha helices in the upper left monomer in Figure 3.4.2b. Refer back to your labeling of the helices involved in forming the pore. What do you observe?
- Describe what is meant by the axis labels **effective pore diameter** and **relative position** using your own words.
- Both AQP1 and GlpF are approximately 60Å long. Approximately how much of the overall length of these proteins is buried in the phospholipid bilayer?
- Convert the length of an AQP1 protein into meters.
- Use the data in Figure 3.4.3 to create a diagram of the shape of the pore in AQP1. Include the estimated diameters of the extracellular, membrane-spanning, and cytoplasmic regions of the pore.

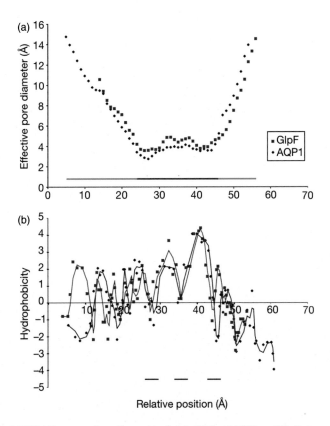

FIGURE 3.4.3 Effective pore diameter and hydrophobicity of AQP1 and GlpF channels.
a. Effective pore diameter relative to the long axis (relative position Å) of the protein was determined for AQP1 (blue) and GlpF (red) transport proteins. The light blue bar indicates the extracellular domain of the proteins. The red bar indicates the membrane-spanning domains, and the green bar represents the cytoplasmic domains of the proteins. b. Hydropathy profile showing the regions of hydrophobic (positive numbers) and hydrophilic (negative numbers) relative to the long axis of AQP1 (blue line) and GlpF (red line). Three black lines indicated hydrophilic nodes.

- Predict how the shape of the pore might influence the function of the channel protein.
- All of the aquaporins studied to date have the amino acids arginine, histidine, and phenylalanine located in the narrow pore region of the protein. Predict how these three amino acids might contribute to movement of a water molecule through the channel.
- Imagine that you have sequenced a newly discovered protein that contains multiple alpha helical domains, some of which share sequence homology with AQP1. Develop an argument either in favor or against the conclusion that this new protein is a water channel.
- Refer back to the amino acid sequences of AQP1 and GlpF. Discuss how proteins with nonidentical amino acid sequences could form identically shaped pores.
- Look again at your diagram of AQP1. Use the information in Figure 3.4.3b to label the hydrophilic and hydrophobic domains along the pore. Do these data match your expectations? Explain your answer.

Visualizing Protein Conformation [6]

INTRODUCTION

Protein function depends on the shape or **conformation** of the protein. Receptor–ligand binding and enzyme–substrate binding are two examples of how the conformation of a protein determines that protein's function. Not only do these proteins need to adopt the proper three-dimensional shape in order to interact with their ligand or substrate, but the binding sites must also be surrounded by amino acids with side chains that will promote binding.

Conformation changes occur when proteins bind to other proteins or molecules. For example, oxygen binding by one subunit of the protein hemoglobin triggers conformational changes that affect oxygen binding to the other subunits. Regulation of protein function in a cell is very often linked to addition of negatively charged phosphate groups to a protein in a process known as phosphorylation. In other cases, conformational changes occur due to the release of energy. Proteins that catalyze the breakdown of ATP into ADP, **ATPases**, undergo conformational changes linked to hydrolysis of the covalent bond between phosphate groups.

- Explain the relationship between the amino acid sequence of a protein and its conformation.
- Research/review cooperative binding in hemoglobin. Describe the process in terms of conformational changes.
- Conduct a search for animations of protein conformation changes.

BACKGROUND

Movement of materials between membrane compartments in the cell relies on the ability of transport vesicles to "dock" and "fuse" with the correct target membrane. Docking between vesicle and target membrane is mediated by a class of membrane proteins known as **SNAREs**. There are two types of SNAREs. **v-SNAREs** are found in the membranes of small transport vesicles. **t-SNAREs** are located in the membranes of the target membrane. SNARE proteins have long cytoplasmic domains that "tangle" together, bringing the two membranes into close proximity in advance of fusing. After membrane fusion has occurred, the SNARE protein complex remains intact until a soluble protein **oligomer** made up of multiple subunits of the protein **NSF** (N-ethylmaleimide sensitive factor) binds to the complex and catalyzes a change in the conformation of the SNAREs; effectively "untangling" them. NSF requires energy from the hydrolysis of ATP to disassemble the SNARE complex.

The role played by NSF and the conformational changes involved in docking and fusion is explored in this case study using the technique of **quick freeze/deep etch electron microscopy**. In this method samples are frozen at −269°C and then broken open or fractured before being "warmed" to −100°C under vacuum causing sublimation of some of the frozen material and leaving an etched surface. The sample is then coated with platinum to preserve the contours and structures revealed by the deep etch procedure. It is this platinum **replica** that is viewed using a transmission electron microscope. The resolution of this method is such that it is possible to visualize individual proteins.

- What is an oligomer? What level of protein structure does an oligomer represent?
- Explain why it makes sense that NSF activity would require ATP energy.
- Biological specimens are almost always in an aqueous environment. Suggest an advantage of rapidly freezing a biological sample.

METHODS

Protein expression and purification
Wild-type NSF protein and NSF protein fragments were expressed using the sequences from a mammalian cell line inserted into plasmids and expressed in bacteria. Sequences for rat v-SNARE and t-SNARE were also inserted into plasmids and expressed in bacteria. Proteins were isolated following lysis of the bacterial culture. Proteins were purified by column chromatography. For some experiments the sequence for maltose binding protein (MBP) was added to either the amino or carboxyl terminus of t-SNARE. MBP folds into a distinctive spherical shape, making it possible to identify the amino or carboxyl ends of the SNARE protein.

Electron microscopy
Purified proteins (~10 µg) were adsorbed to finely ground mica flakes for 30 s. The mica flakes were pelleted by gentle centrifugation and washed with buffer. Quick freezing was accomplished by dropping the samples onto a liquid helium–cooled cooper block. Samples were deep etched for 4 min at −100°C and coated with a 2 nm layer of platinum. Shadow casting was prepared by applying platinum from one direction at an angle of 15° above horizontal. Platinum replicas were cleaned by floating them overnight in

a solution of concentrated hydrofluoric acid, washed in water, and finally placed onto a formvar-coated microscope grid. Replicas were viewed using a transmission electron microscope

SNARE complex labeling

Purified SNARE complexes containing MBP-tagged t-SNARE were combined with specific antibodies (IgG) as indicted. After a brief incubation (30–60 min), these mixtures were adsorbed to mica chips at a final concentration of 15 µg/mL and 4 µg/mL IgG. Samples were subjected to the quick freeze, deep etch protocol described earlier.

- Discuss whether the use of mammalian proteins purified from a bacterial expression system would affect the results of these experiments.
- What are the advantages of using a bacterial expression system?
- Predict what would happen to the original protein samples after exposure to concentrated hydrofluoric acid.

RESULTS

- Describe the conformation of an NSF oligomer as shown in Figure 3.5.1.
- Compare the appearance of NSF with the other examples of ATPases in Figure 3.5.1. Speculate on the significance of your observations.
- Explain the relationships between the schematic diagram and the various gene constructs illustrated in Figure 3.5.2a.
- Which domain of NSF includes the carboxyl terminus of the protein? Explain your answer.
- Why are the protein bands in the lanes labeled "D2" and "N" about the same size?
- What is the minimal domain required for oligomerization of NSF?
- The height of a protein structure can be estimated by the "shadow" that it casts in this technique. What can you conclude about the height of the three forms of NSF shown in Figure 3.5.2c?
- Compare the images of NSF from the top row of Figure 3.5.3a with those in Figure 3.5.1b. How are these images the same? How are they different?
- Conduct a search for images of ATP·γS. How does this molecule differ from ATP?
- What effect does inhibition of ATP hydrolysis have on conformation of an NSF oligomer?
- Refer back to Figure 3.5.2a. Which domain of NSF is involved in ATP hydrolysis? Propose a model to explain the difference in conformation based on ATP binding.
- Discuss the evolutionary significance of the conformation of mammalian and yeast NSF oligomers.

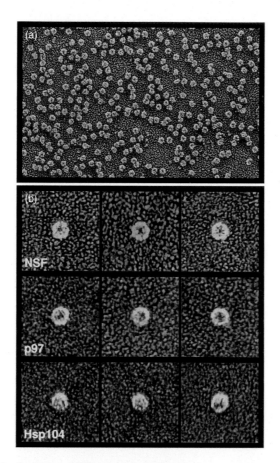

FIGURE 3.5.1 Conformation of NSF and related ATPases.

a. Survey view of NSF oligomers formed in the presence of 1 mM ATP × 115,000. b. Comparative views of NSF, p97, and Hsp104 oligomers. All three proteins are known to have ATPase activity and, in the case of Hsp104, are known to disassemble protein aggregates. All images are at ×300,000.

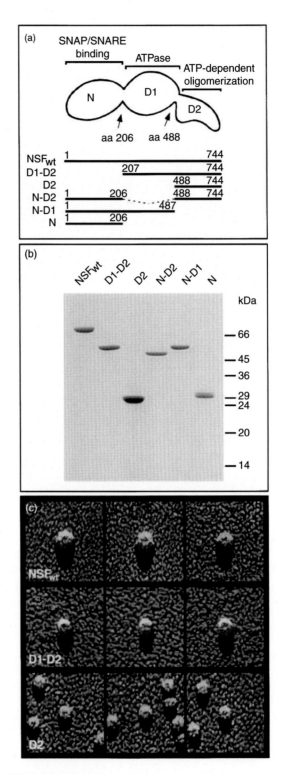

FIGURE 3.5.2 NSF domain structure.
a. Schematic diagram of the domains of NSF and gene constructs used to assess the role of each domain in NSF protein folding. b. Coomassie stained SDS-PAGE of mutant NSF proteins diagrammed in (a).
c. Quick freeze/deep etch images of wild-type and mutant NSF proteins.

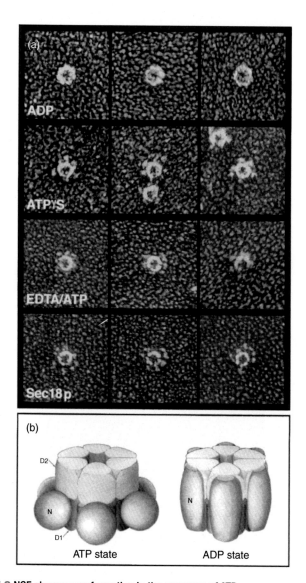

FIGURE 3.5.3 NSF changes conformation in the presence of ATP.

a. Top row: NSF prepared in the presence of ADP. Second row: NSF prepared in the presence of the nonhydrolyzable analog of ATP, ATP-γS. Third row: NSF prepared in the presence of ATP and the divalent chelator, EDTA. Fourth row: yeast NSF homolog, Sec18p. b. Diagram of the conformation of NSF in the presence (ATP state) and absence (ADP state) of ATP.

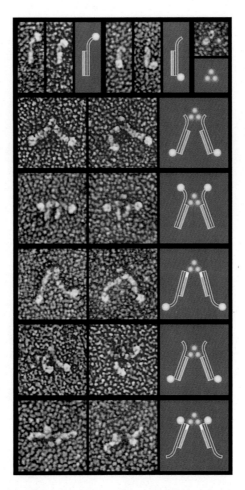

FIGURE 3.5.4 Demonstration of the arrangement of v-SNAREs and t-SNAREs.
The ability of NSF to bind to the SNARE complex is influenced by orientation of the interaction between the v-SNARE and the t-SNARE. Two possible models have been proposed: parallel interactions or antiparallel interactions. These two models were tested using the quick freeze/deep etch technique combined with labeling of the amino or carboxyl terminus of a t-SNARE (curved line) and a v-SNARE (gray rectangle). The ends of the t-SNARE were indicated by genetically engineering the addition of the sequence for maltose binding protein (MBP; yellow spheres) to the sequence of the SNARE. Antibody proteins (IgG) specific to the amino or carboxyl termini were also used. IgG proteins have a distinctive appearance (red spheres). For each experiment two images are shown next to a drawing. Top left: t-SNARE with labeled amino terminus. Top center: t-SNARE with labeled carboxyl terminus. Top right: IgG proteins alone. Second row: SNARE complex with MBP labeling of t-SNARE amino terminus and antibody reacting to t-SNARE carboxyl terminus. Third row: SNARE complex with MBP labeling of t-SNARE amino terminus and antibody labeling of v-SNARE amino terminus. Fourth row: SNARE complex with MBP labeling of t-SNARE amino terminus and antibody labeling of v-SNARE carboxyl terminus. Fifth row: SNARE complexes with MBP labeling of t-SNARE carboxyl terminus and antibody binding to v-SNARE amino terminus. Sixth row: SNARE complex with MBP labeling of t-SNARE carboxyl terminal domain and antibody binding to v-SNARE carboxyl terminal domain.

- Summarize the main conclusion supported by the data presented in Figure 3.5.4.
- Critique the use of colors to highlight the labeling of proteins in the images in Figure 3.5.4.
- v-SNAREs and t-SNAREs insert into their respective membranes at their carboxyl terminal domains. Using the data from Figure 3.5.4 create a schematic diagram of how these two proteins must interact during vesicle docking.
- Based on your model, how would an NSF oligomer interact with the SNARE complex?

Second Chance Chaperones: How Misfolded Proteins get Refolded [7]

INTRODUCTION

The amino acid sequence of a protein determines how that protein will fold. Protein folding influences the overall shape of the protein, and protein shape, in turn, influences protein function. Any variable that alters the shape of a protein has the potential to reduce or eliminate that protein's function, potentially impacting survival of the cell.

Proteins can **misfold** either during translation or afterwards, as a consequence of some change in the environment of the cell. **Molecular chaperones** are a special class of proteins that bind to regions of a protein and help it fold properly. Imagine the case of a protein that forms a hairpin structure through the interaction between two hydrophobic domains located at its amino and carboxyl terminal ends of the polypeptide. As the protein is emerging from the ribosome, the chaperone protein, **heat shock protein 70 (Hsp70)**, binds to the amino terminal hydrophobic domain, stabilizing it until the carboxyl domain has been synthesized. Chaperones also facilitate the assembly of multisubunit protein complexes. Similarly, chaperones can contribute to the breakup of protein aggregates formed as a consequence of elevated temperatures (heat shock) or exposure to toxic compounds. Chaperone proteins are found in eukaryotic and prokaryotic cells. One of the best studied examples in bacteria is the protein **GroEL chaperonin** which forms a hollow chamber that traps misfolded proteins and creates an environment that favors proper folding.

- Name the levels of protein structure and the types of chemical bonds associated with each.
- Describe an example of how protein structure influences protein function.
- In the example described earlier, why would the chaperone attach to the amino terminus of the protein before the carboxyl terminus?
- Explain why elevated temperature has an adverse affect on protein folding.

BACKGROUND

Hsp70 is not the only molecular chaperone functioning in cells. In the yeast, *S. cerevisiae*, the heat shock protein Hsp104 plays an important role in protecting cells from the effects of elevated temperature or high concentrations of ethanol. Exposure to high temperatures causes proteins to misfold and form large aggregates in the yeast's cytoplasm. Upon return to normal temperatures, these protein aggregates disappear from the cytoplasm of wild-type cells, but are retained in cells that lack the *hsp104* gene. Does Hsp104 contribute to the elimination of protein aggregates by acting on proteins prior to unfolding or after unfolding and protein aggregation has occurred?

Assays for chaperone function have been developed using the firefly protein **luciferase**. Luciferase is a 61 kDa protein that generates a short burst of light in the presence of ATP. Treatment of luciferase with urea causes the protein to denature, eliminating light production; however, unfolding is reversible and light production can be recovered. The intensity of the light produced can be measured allowing for quantitative analysis of protein refolding. A similar light-based assay has been developed for the protein β-galactosidase.

The technique of column chromatography features prominently in this case study. Column chromatography is a method that allows for purification of a specific protein out of a mixture of proteins, as you would find in the cytoplasm of a cell, based on its physical properties. A column is filled with a matrix or "beads." The physical and/or chemical nature of the matrix determines how the column will work. Some columns separate proteins based on their size, a method known as **gel filtration**. The beads in these columns have holes that exclude large proteins and protein aggregates, but allow small proteins to enter. As the column is washed with buffer solution the large proteins **elute** or exit the column before the smaller proteins. Soluble material leaving the column is collected in units known as **fractions**. Fractions contain a set volume of **elutants** and are numbered to reflect the sequence in which they left the column. Other types of columns make use of hydrophobic or ionic interactions with proteins, while others take advantage of the binding between antibodies and antigens, receptors, and ligands, or other specific interactions to select for a particular protein.

- Speculate on whether multicellular organisms have heat shock proteins. Explain your thinking.
- How is "disaggregating" different from "refolding" of a protein?
- Evaluate the relative costs to the cell of refolding versus destroying a misfolded protein.
- Conduct a search to find images of the structure of firefly luciferase. Characterize the structural motifs in the folded protein.
- Why does a small protein elute more slowly than a large protein during gel filtration chromatography?

METHODS

Purification of Hsp104

Wild-type *S. cerevisiae* were resuspended in a homogenization buffer (buffer A) containing protease inhibitors and 0.45 mm glass beads. The cell mixture was blended 12 times for 30 s in an ice water bath. The cell homogenate was centrifuged at 12,000 rpm for 20 min and the supernatant was loaded onto a 20 mL Affi-Gel-Blue column (Bio-Rad). The column was washed with homogenization buffer A and then with buffer A containing 1 M KCl to elute bound proteins. The eluted proteins were dialyzed against buffer A to remove the KCl and then loaded onto a 15 mL DEAE column. This column was washed with buffer A. Proteins were eluted by washing the column with increasing concentrations of KCl (linear gradient from 0 mM to 500 mM) in buffer A. The majority of Hsp104 eluted in fractions corresponding to a KCl concentration between 70 mM and 140 mM. Peak fractions were pooled and dialyzed against phosphate buffer B. The sample was then loaded onto a 11 mL hydroxylapatite column, washed with buffer B, and eluted with a linear gradient of buffer B containing 50–400 mM potassium phosphate, pH 6.8. Hsp104 elutes between 136 mM and 180 mM potassium phosphate. Fractions were pooled and Hsp104 was precipitated with ammonium sulfate added to saturation. The precipitate was collected by centrifugation, resuspended into buffer A, and loaded onto a 20 mL DEAE column and eluted with a linear gradient of 50–300 mM KCl in buffer A. Hsp104 elutes at a salt concentration of 105–150 mM. Purified Hsp104 was finally dialyzed against buffer A containing 10% glycerol and was stored at −80°C.

Purification of Hsp40 and Hsp70

Plasmids containing the sequences for yeast Hsp40 (Ydj1) and Hsp70 (SSA1) were overexpressed in *Escherichia coli*. Cells were lysed and the soluble protein was purified following protocols similar to those described for the isolation of Hsp104.

Preparation of yeast lysates

Cells from stationary phase yeast cultures were disrupted with glass beads in buffer supplemented with protease inhibitors. The lysate was clarified by centrifugation at 100,000g for 30 min at 4°C. For analysis of Hsp104 function in refolding, the yeast strain BJ5457 carrying a disruption of *HSP104* was used. Lysates deficient in Hsp40, Hsp70, or both were generated using yeast strains known to be lacking the genes for these proteins.

Turbidity assay

Luciferase (30 μM) was unfolded in 4 M urea for 30 min at 30°C and diluted 50-fold into refolding buffer at room temperature with the specified additions. Turbidity was measured using a spectrophotometer reading absorbance at a wavelength of 320 nm.

Refolding assays

Luciferase was unfolded by incubation with 4–8 M urea for 30 min at 30°C. For refolding in lysates, the unfolded protein was diluted 100-fold into refolding buffer consisting of, unless otherwise indicated, 100 mg lysate, 5 mM ATP, and an ATP-regenerating system. Light emission over 5 s intervals was measured using a luminometer. Control measurements were made as described but without incubation in urea. β-galactosidase activity was measured in the same manner but light emissions were collected for 10 s.

Gel filtration

Luciferase (50 μM) was unfolded in 4 M urea for 30 min at 30°C and diluted 50-fold into ice-cold refolding buffer or at 25°C into the same buffer with the indicated additions. β-galactosidase was unfolded in 8 M urea for 30 min at 30°C and diluted 100-fold into cold buffer. After 30 min incubation, reactions were cooled on ice and filtered through a precooled 0.22 μm filter. 0.25 mL of soluble protein aggregate was added to a Superose 6 gel filtration column equilibrated in refolding buffer. Fractions (0.4 mL) were collected beginning 15 min after addition of the sample. 25 μL of each fraction was analyzed by SDS-PAGE. Gel filtration standards were blue dextran (2000 kDa), thyroglobin (669 kDa), ferritin (440 kDa), aldolase (158 kDa), and bovine serum albumin (66 kDa).

- Outline the protocol used to isolate purified Hsp104 protein from yeast.
- Conduct a search to learn about the properties of the following types of column chromatography:
  - Affi-Gel-Blue
  - DEAE
  - Hydroxylapatite
  - Superose 6
- Explain the purpose of the various dialysis steps described in the isolation of Hsp104.
- Suggest a reason why glass beads are used when lysing yeast cells.
- What volume of unfolded luciferase would you add to a total reaction volume of 50 μL to make a 100-fold dilution?
- Why was it necessary to include an ATP-regenerating system in the refolding assays?

RESULTS

- Describe the relationship between turbidity and protein aggregation.
- Using the data in Figure 3.6.1, estimate the time required for unfolded luciferase to form an aggregate.

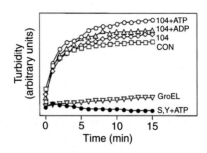

FIGURE 3.6.1 Hsp104 does not prevent aggregation.
Unfolded luciferase was diluted to 0.6 μM in buffer alone (open squares), buffer with 3 μM Hsp104 but no ATP (diamonds), or Hsp104 with 5 mM ATP (open circles) or 5 mM ADP (triangles). Unfolded luciferase was also diluted into 0.6 μm *E. coli* GroEL (inverted triangles) and 3 μM each of yeast chaperones Hsp70 (S) and Hsp40 (Y) with 5 mM ATP (filled circles).

- Which of the conditions described in the figure legend for Figure 3.6.1 is the control condition?
- Discuss whether differences in the shape of the curves for GroEL and S,Y + ATP in Figure 3.6.1 are significant. Justify your conclusion.
- What conclusion is supported by the data shown in Figure 3.6.2a?
- How might the data in Figure 3.6.1 relate to the data in Figure 3.6.2a?
- Define the term "endogenous." Explain the results in Figure 3.6.2b based on your definition.
- Approximately how long does it take to refold luciferase (Figure 3.6.2b)?
- Why might it make sense that protein refolding requires ATP?
- How might the lysine-to-threonine change in the amino acids of the nucleotide-binding domain of Hsp104 influence its function? Is your prediction supported by the data in Figure 3.6.2c?
- Compare the data for Figure 3.6.2a and c. Are the results consistent between these two experiments?
- Which chaperone proteins are present in yeast cell lysate? Which of these chaperones interact with Hsp104 (Figure 3.6.3a)?
- Create a drawing to illustrate how the technique of affinity chromatography produced the result shown in Figure 3.6.3a.
- Compare the refolding activity of wild-type and mutant lysates alone shown in Figure 3.6.3c. What can you conclude?
- How do the data in Figure 3.6.3c support the conclusion that Hsp40 is necessary for protein refolding by Hsp104?
- Describe the relationship between the size of a protein (or protein aggregate) and its behavior on a gel filtration column using Figure 3.6.4a as your example.

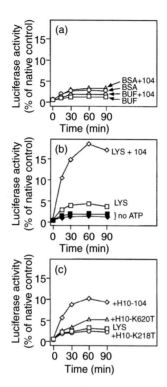

FIGURE 3.6.2 Hsp104-mediated protein refolding requires other cytosolic factors and ATP.
a. Unfolded luciferase was diluted to 20 nM in buffers containing 5 mM ATP and an ATP-regenerating system without Hsp104 (open squares), with 5 µg Hsp104 (diamonds), with 100 µg bovine serum albumin (BSA; open circles), or with 100 µg BSA and 5 µg Hsp104 (triangles). Refolding was measured as enzyme activity detected in aliquots collected at the indicated times. b. Unfolded luciferase was diluted as in (a) into buffer containing 100 µg of yeast cell lysate lacking endogenous Hsp104 (open squares) and into the same amount of lysate supplemented with 5 µg Hsp104 (diamonds). ATP requirement for refolding was demonstrated by omitting ATP from identical reactions (filled symbols). c. Unfolded luciferase was diluted as described for (a) and was incubated with Hsp104-deficient yeast cell lysate (open squares), or lysate supplemented with a His-tagged form of wild-type Hsp104 (+H10-104; diamonds), with nucleotide-binding mutants lysine to threonine (+H10-K218T; open circles, or +H10-K620T; triangles). All reactions contained 5 mM ATP and an ATP-regenerating system.

- Notice the *Y*-axis label for the graphs in Figure 3.6.4. How are these different from previous figures reporting luciferase activity? Suggest a reason the axis was changed.
- Propose an explanation for why the peak of luciferase activity is high in fractions 3–5 and absent in fractions 28–30.
- What can you conclude about the ability of Hsp40 and Hsp70 to refold luciferase in the absence of Hsp104?

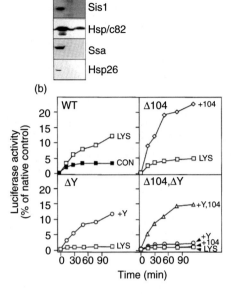

FIGURE 3.6.3 Yeast Hsp40 physically interacts with Hsp104 and is necessary for Hsp104-mediated refolding.

a. Affinity chromatography using His-tagged Hsp104 immobilized on Ni^{2+} resin and incubated with yeast cell lysate as described in "Methods." Aliquots of total lysate (T) and fractions eluted from the affinity column with (+) or without (−) bound Hsp104 were analyzed by SDS-PAGE and immunoblotted with antibodies specific to the indicated chaperone proteins. b. Unfolded luciferase (20 nM) was refolded in buffer containing 100 μg BSA (CON; filled squares), or 100 μg lysates from yeast cells lacking Hsp104 (Δ104) or Hsp40 (ΔY) (open squares), lysates supplemented with 2.5 μg Hsp104 (+104; diamonds), 5 μg Hsp40 (+Y; open circles), or both Hsp104 and Hsp40 (+104, +Y; triangles). All reactions contained 5 mM ATP and an ATP-regenerating system.

FIGURE 3.6.4 Refolding of preformed aggregates of denatured luciferase requires Hsp40, Hsp70, and Hsp104. ▶

a. Denatured luciferase was diluted to 1 μM in ice-cold buffer. After 30 min, particulate material was removed by centrifugation and the remaining aggregated luciferase was fractionated by size using gel filtration at 4°C. The elution profile of luciferase was determined by immunoblotting with antifirefly luciferase antibody (α-LUC). For refolding, ice-cold column fractions were supplemented with 5 mM ATP without chaperones (open squares), with 1 μM Hsp40 and Hsp70 (S,Y; filled diamonds) or with 1 μM Hsp40, Hsp70, and Hsp104 (S,Y,104; open circles), then incubated at 25°C. Luciferase activity was determined after 90 min of refolding. Inset displays refolding for no chaperone control and Hsp 40 and Hsp70 (S,Y) on an expanded axis. Arrowheads indicate elution positions of molecular size standards. b. Luciferase (1 μM) was denatured in the presence of 5 mM ATP and 5 μM each of Hsp40 (Y) and Hsp70 (S). After 30 min the reaction was cooled on ice, particulates were removed by filtration, and the remaining aggregates were fractionated by gel filtration. The elution profile of Hsp40 (Ydj1) was detected by Coomassie Blue staining (CB). Hsp70 (α-Hsp70) and luciferase (α-LUC) were detected by immunoblotting. For refolding, ice-cold column fractions were supplemented with 5 mM ATP without chaperones (open squares) with 1 μM Hsp40 and Hsp70 (S, Y; filled diamonds) or with 1 μM Hsp40, Hsp70, and Hsp104 (S, Y, 104; open circles). Luciferase activity was determined after 90 min of refolding at 25°C. c. The experiment as described in (b) was repeated with the inclusion of Hsp104 (5 μM) in the initial reaction. The elution profile of Hsp104 was determined by CB staining. All other proteins were detected as described in (b).

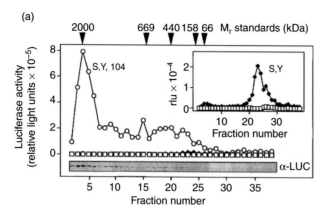

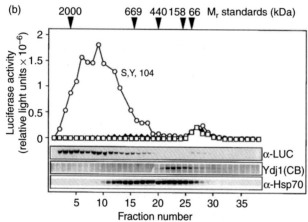

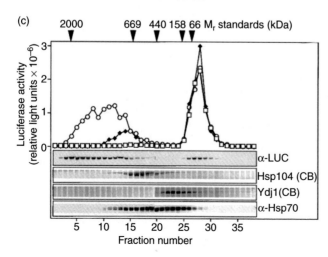

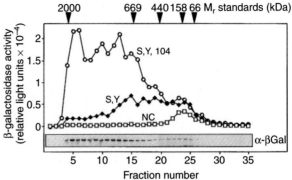

FIGURE 3.6.5 Refolding of previously aggregated β-galactosidase.
Denatured β-galactosidase was diluted to 0.8 μM (monomer concentration) in ice-cold buffer. After 30 min, particulate material was removed by filtration and the remaining aggregated β-galactosidase was fractionated by gel filtration at 4°C. The elution profile of β-galactosidase was determined by immunoblotting (α-βGal). For refolding, ice-cold column fractions were supplemented with 5 mM ATP without chaperones (open squares), with 1 μM Hsp40 and Hsp70 (S, Y; diamonds) or with 1 μM Hsp40, Hsp70, and Hsp104 (S, Y, 104; open circles), then incubated at 25°C. β-galactosidase activity was determined after 90 min of refolding.

- Does the degree of aggregation affect the ability of Hsp104 to restore luciferase activity?
- Describe the pattern of elution of luciferase, Hsp40, and Hsp70 shown in Figure 3.6.4b.
- Did the presence of Hsp40 and Hsp70 influence the aggregation of unfolded luciferase? Justify your answer.
- Speculate on the significance that all three chaperones must be present in order to generate the large peak of luciferase activity in Figure 3.6.4b.
- How is the peak of luciferase activity corresponding to fractions 6–10 different from the peak corresponding to fractions 27–29?
- Propose an explanation for the presence of the two peaks in Figure 3.6.4c.
- Did the presence of all three chaperones prevent the formation of protein aggregates from denatured luciferase (Figure 3.6.4)?
- Describe the correlation between the amount of protein seen on the blot and the measure of β-galactosidase activity in Figure 3.6.5. What can you conclude about the protein found in fractions 23–26?
- Explain the significance of the result shown in Figure 3.6.5.
- The original research question asked whether Hsp104 acts on unfolded proteins before or after aggregation. Use the data presented in this case study to provide an answer to that question.

References

[1] Barretto OC de O, Satake M, Nonoyama K, Cardoso JLC. The calcium-dependent protease of *Loxosceles gaucho* venom acts preferentially upon red cell band 3 transmembrane protein. Braz J Med Biol Res 2003;36:309–13.

[2] Orlacchio et al. Neuroacanthocytosis associated with a defect of the 4.1R membrane protein. BMC Neurol 2007;7:4.

[3] Willer M, Rainey M, Pullen RT, Stirling CJ. The yeast *CDC9* gene encodes both a nuclear and a mitochondrial form of DNA ligase I. Curr Biol 1999;9:1085–94.

[4] Jungnickel B, Rapoport T. A posttargeting signal sequence recognition event in the endoplasmic reticulum membrane. Cell 1995;82:261–70.

[5] Sui H, Han B-G, Lee JK, Walian P, Jap BK. Structural basis of water-specific transport through the AQP1 water channel. Nature 2001;414:872–8.

[6] Hanson PI, Roth R, Morisaki H, Jahn R, Heuser JE. Structure and conformational changes in NSF and its membrane receptor complexes visualized by quick-freeze/deep-etch electron microscopy. Cell 1997;90:523–35.

[7] Glover JR, Lindquist S. Hsp 104, Hsp 70 and Hsp 40: a novel chaperone system that rescues previously aggregated proteins. 1998. Cell;94:73–82.

CHAPTER 4

The Nucleus

Nuclear Pore Complex Assembly is a House of Cards [1]

INTRODUCTION

A eukaryotic cell can be thought of as consisting of two major compartments: the **nucleoplasm** and the cytoplasm. The double membrane of the **nuclear envelope** defines the boundary between these two compartments. The nuclear envelope acts as a physical barrier, however, cellular function depends on the ability of proteins, RNAs, and ribosomes to travel between the cytoplasm and nucleoplasm. Transport between the compartments occurs through **nuclear pores**, specialized openings along the surface of the nuclear envelope.

A nuclear pore is a structure that is made up of a collection of 30 different proteins called **nucleoporins** that assemble to form the complete **nuclear pore complex** (NPC). The NPC spans the membranes of the nuclear envelope creating a central aqueous channel that accommodates the movement of material between the nucleoplasm and cytoplasm. Repeated clusters of the amino acids phenylalanine and glycine extend from the nucleoporins that line the channel, helping to create a "filter" that prevents the passive diffusion of large molecules. In addition to the proteins that form the channel, the nuclear pore complex includes nucleoporins that stabilize the entire structure within the membrane. Cytoplasmic filaments and the nuclear basket extend from their respective surfaces on the NPC and function in the recognition of proteins and RNAs that must transport through the pore. **Nuclear**

73

localization signals (NLS) are conserved amino acid sequences that target proteins for import into the nucleus.

- Create a Venn diagram listing the contents you would expect to find in the cytoplasm and nucleoplasm of a cell. What cellular contents do these compartments have in common?
- Describe how the membrane structure of the nuclear envelope contributes to its ability to create distinct compartments within a cell.
- Connect the structure of the nuclear envelope to the organization and structure of the endoplasmic reticulum.
- Research the chemistry of the amino acids phenylalanine and glycine. How would the presence of repeated clusters of these amino acids lining the central channel influence transport through a nuclear pore?

BACKGROUND

Disassembly and reassembly of the nuclear envelope is a normal part of the cell cycle. The ability of a fully functional nuclear envelope to form around the chromosomes at the end of mitosis can be replicated *in vitro* using a **nuclear reconstitution assay**. In this assay, eggs from the frog, *X. laevis*, are lysed and the cytoplasm is separated into a fraction containing membranes and a fraction containing soluble proteins. These two fractions are then incubated in the presence of sperm chromatin. Under control conditions, not only does a nuclear envelope form around the chromatin, but this artificial nucleus is also capable of importing nuclear proteins and supporting DNA replication.

The nuclear pore complex assembles through the interaction of multiple subcomplexes of nucleoporin proteins. The proteins gp210 and pom121 are both transmembrane proteins that help anchor the NCP to the membrane of the nuclear envelope. Nucleoporins Nup62, 58, 54, and 45 assemble to form the central channel complex of the NPC. The nucleoporins Nup153 and Nup98 play a role in mRNA export and the protein TPR is associated with the nuclear basket. A collection of six nucleoporins, known as Nup107-160, forms a ring-like structure on either side of the nuclear pore. The formation of a functional nuclear pore complex presumably requires that all of the components are present in their proper orientation. This case study tests that presumption through the use of the technique of immunodepletion to remove the nucleoporin protein, Nup85.

- Research/review the stages of mitosis. Identify the stages when the nuclear envelope disassembles and when it reassembles.
- Suggest a reason why frog eggs would be a good source for the proteins and membranes required to build a nucleus.
- Speculate on the significance of the numbers used to identify the various nucleoporin proteins.

■ Design an experiment that would test some aspect of the formation of a nuclear envelope based on the information provided about the nuclear reconstitution assay.

METHODS

Nuclear reconstitution assay

Cytosolic and membrane fractions were isolated from *Xenopus* egg extracts. Eggs were washed in buffer and then lysed by centrifugation at 12,000g for 15 min. The supernatant containing the crude soluble fraction was collected and centrifuged for 200,000g for 90 min in a swinging bucket rotor. Cellular contents of the crude soluble fraction separate into a layer of cytosol, a layer of membranes, and a pellet containing mitochondria, glycogen, and ribosomes. The membrane fraction was washed with 500 mM KCl. The cytosol and membrane fractions were used immediately or frozen for future use.

Sperm were collected from the testes of a mature *Xenopus* male. Sperm were rinsed in buffer and treated with a compound that removes the plasma and nuclear membranes surrounding the sperm DNA while leaving the chromatin in a condensed state. Demembranated sperms were mixed with cytosol and membranes, and incubated at 22°C for 1 h to allow the assembly of the nuclear envelope and nuclear pore complexes.

Immunodepletion

For immunodepletion reactions 300 μg of anti-Nup85, 600 μg of anti-Nup133, or equivalent volumes of preimmune serum were bound to Sepharose beads. 30 μL of preimmune or antibody-coated beads were mixed with 200 μL of egg cytosol and incubated with agitation for 1 h at 4°C. Sepharose beads and any attached protein sediment, leaving a supernatant of depleted cytosol.

Nuclear import

Fluorescently labeled human serum albumin carrying a nuclear localization signal (NLS–HSA) was added to the nuclear reconstitution assay 60 min after the start of nuclear envelope assembly. Nuclei were collected 15–20 min later, rinsed with buffer, and visualized using fluorescence microscopy.

Immunoblotting

Protein samples were prepared for separation by SDS polyacrylamide gel electrophoresis (SDS-PAGE). Following separation, the proteins were electrophoretically transferred to a nitrocellulose membrane. The membrane was blocked with 5% nonfat dry milk and incubated with dilute primary antibodies to the various nucleoporin proteins for 1 h at room temperature.

After incubation with the primary antibody, the membranes were washed and then incubated with secondary antibody conjugated to the reporter molecule, horseradish peroxidase.

DNA replication assay

Demembranated sperm chromatin was mixed with cytosol and membranes to initiate nuclear envelope assembly. Two microcuries of $[\alpha^{32}P]$ dCTP were added to 40 μL of the reaction mixture and allowed to incubate for a total of 5 h. At each time point a sample of the mixture was collected. Samples were run on a 0.8% agarose gel. The intensity of the radioactively labeled bands was measured using a phosphoimager. Band intensity for the 5 h mock depletion sample was defined as 100% DNA replication. Percent DNA replication for all other samples was based on their relative intensity.

Scanning electron microscopy

Mock and anti-Nup133 depleted nuclei were assembled for 60 min and then prepared for visualization using scanning electron microscopy. The samples were critical point dried and coated with 3.4 nm chromium.

■ Create a flow chart that outlines the steps involved in the preparation of the *Xenopus* cytosol and membrane fractions used in the nuclear reconstitution assay.
■ What effect would the 500 mM KCl wash have on membranes isolated from *Xenopus* eggs?
■ Research/review the amino acid sequence most commonly associated with a NLS.
■ Calculate the relative percent DNA replication for each of these samples based on the following data set:

Sample	Time (h)	Band Intensity (Arbitrary Units)
Mock	1	174
Mock	2	755
Mock	4	988
Mock	5	1163
Δ133	3	93
Δ133	5	128

RESULTS

■ Conduct a search for an image of the structure of a metazoan nuclear pore complex. Locate the proteins mentioned in Figure 4.1.1.
■ What is "preimmune serum?" Why is it used as a control in Figure 4.1.1?

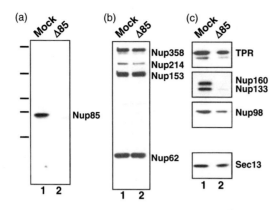

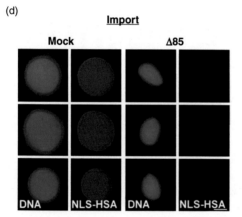

FIGURE 4.1.1 Depletion of Nup85 results in small, import-deficient nuclei.
a. Incubation of *Xenopus* egg cytosol with anti-Nup85 antibodies (lane 2; Δ85) effectively removed the Nup85 protein as compared to immunodepletion using preimmune serum (lane 1; Mock).
b. Immunodepletion using the anti-Nup85 antibody had no effect on nucleoporin proteins Nup62, 153, 214, or 358. c. Nucleoporin proteins Nup133 and 160 were codepleted by the anti-Nup85 antibody. The amount of proteins Sec 13 and TPR were slightly reduced following immunodepletion. d. Nuclear reconstitution assays were performed following immunodepletion of the *Xenopus* cytosol using preimmune serum (Mock) or anti-Nup85 antibodies (Δ85). Transport into these nuclei was assayed using fluorescently labeled human serum albumin engineered to carry a nuclear localization signal (NLS-HSA). Size of the nuclei was determined using a DNA-specific stain.

- Propose an explanation for why immunodepletion of Nup85 results in the loss of Nup133 and Nup160.
- Develop a hypothesis to account for the apparent lack of import of the NLS-HSA protein in Δ85-treated nuclei.
- Suggest an explanation for why Figure 4.1.2a shows two different loading volumes of the same samples.

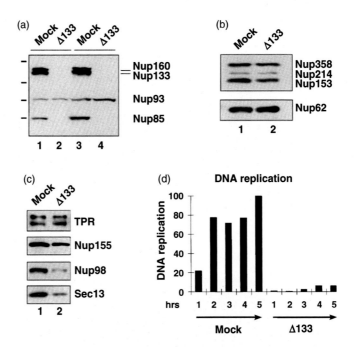

FIGURE 4.1.2 Immunodepletion of the Nup107-160 complex inhibits DNA replication.
a. *Xenopus* cytosol was incubated with either preimmune serum (lanes 1 and 3; mock) or anti-Nup133 antibodies (lanes 2 and 4; Δ133). Lanes 1 and 2 were loaded with 0.5 μL of sample. Lanes 3 and 4 were loaded with 1 μL of sample. b. Cytosol from (a) was immunoblotted with an antibody that recognizes Nup62, 153, 214, and 358. c. Cytosol from (a) was immunoblotted with an antibody that recognizes nuclear pore proteins Nup98, Nup153, Sec13, and TPR. d. DNA replication was measured for nuclei reconstituted using either mock or Nup133 immunodepleted cytosol. Percent replication was normalized to 100% of the replication found in mock-depleted nuclei after 5 h.

- What effect does immunodepletion with an antibody to Nup133 have on *Xenopus* cytosol (Figure 4.1.2)?
- Refer back to the image of the nuclear pore. How does the data from Figure 4.1.2 add to your understanding of the interactions that occur within the NPC?
- How does immunodepletion of Nup133 influence DNA replication in Δ133 nuclei?
- Connect the result shown in Figure 4.1.2d with the result from Figure 4.1.1d.
- Calculate an average diameter of the NPC in the mock-depleted nuclear membrane shown in Figure 4.1.3a.

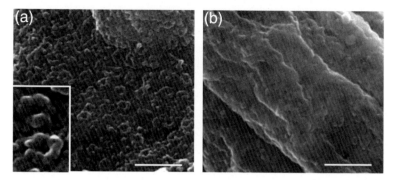

FIGURE 4.1.3 Nuclei formed in the absence of the Nup107-160 complex lack NPCs.
Nuclei prepared from mock-depleted (a) and anti-Nup133 depleted (b) cytosol were examined using scanning electron microscopy. Typical views are shown. a. Abundant NPCs with eightfold symmetries can be seen in the mock-depleted nuclei. Inset image is a 3× higher magnification. b. NPCs are not visible in the Δ133 membranes. Scale bars represent 500 nm.

- Would you expect to find the transmembrane proteins gp210 and pom121 in the membrane shown in Figure 4.1.3b? Explain your answer.
- The researchers conclude that the loss of the Nup107-160 complex inhibits the ability of the NPC to assemble. Design an experiment that would test this idea.

Importin β: The Important Importin [2]

INTRODUCTION

The nuclear envelope divides a eukaryotic cell into two compartments; the cytoplasm and the nucleoplasm. While these compartments serve very different functions in the life of a cell, transport between the two is vital. Nuclear pores, complex protein structures that span the double membrane of the nuclear envelope, serve as selective channels for the movement of macromolecules in and out of the nucleus.

Proteins that function inside the nucleoplasm must travel from the cytoplasm, where they are synthesized, to the nuclear envelope and through a nuclear pore. Targeting a protein for import into the nucleus is made possible by the presence of an nuclear localization signal (NLS). An NLS is typically a short amino acid sequence that contains a domain of positively charged amino acids. Cytosolic proteins known as **importins** act like receptors for the NLS domain, binding the protein and escorting it to the nuclear pore complex. Importin α binds to NLS-containing proteins. Importin β binds to importin α, forming a heterodimer. From there, the importin heterodimer with the attached nuclear protein moves through the nuclear pore and into the nucleoplasm. Once inside the nucleoplasm, the protein **Ran-GTP** binds to importin β, triggering disassembly of the complex and release of the protein. Ran-GTP then functions to return importin β through the nuclear pore and into the cytoplasm. Importin α is also exported back into the cytoplasm and the process of nuclear import repeats.

- Make a list of cellular activities that occur only in the cytoplasm or only in the nucleoplasm.
- Research/review the chemical nature of the various amino acids. Which amino acids would you find in the NLS of a protein?
- Using your own words, write a definition for the term, "heterodimer."
- Predict what would happen if an NLS was genetically engineered onto a cytoplasmic protein.

BACKGROUND

The earliest studies about how macromolecules are imported into the nucleus from the cytoplasm used crude cytoplasmic and nucleoplasmic extracts to assay the movement of proteins and protein-coated gold particles.

Over time, the specific, individual components responsible for each step in the process have been identified. At the time that the research described in this case study was conducted, scientists knew that import into the nucleus involved two distinct steps: NLS-dependent targeting of the nuclear pore complex and transport through the nuclear pore and into the nucleoplasm. It was also known that cytoplasmic proteins were required in order for these steps to occur. The cytoplasmic protein, **importin 60**, now known as **importin α**, had been found to be involved in the first step of nuclear import. This case study tells the story of the discovery of **importin 90**, the protein we now call **importin β**.

METHODS

Preparation of a high-speed supernatant from *Xenopus* eggs

Xenopus eggs were collected, dejellied, and activated by the addition of $CaCl_2$. Activated eggs were washed with buffer, homogenized, and centrifuged using a swinging bucket rotor to pellet large particles. High-speed supernatant (HSS) was collected from the middle of the centrifuge tube. HSS fractions were frozen in liquid nitrogen until needed. All HSS fractions were centrifuged at 100,000g for 20 min prior to use, to remove residual large particles and protein aggregates.

Immunodepletion of importin from cytosol

A sample of 10 mL of HSS was passed through a 2 mL column consisting of immobilized anti-importin 60 antibody, Ab1. Fractions (0.5 mL) were collected and analyzed by immunoblotting. Protein-containing fractions that lacked importin 60 were combined and frozen. Proteins bound to the immobilized Ab1 antibody were released by washing in the presence of the antigenic peptide and 1 M NaCl. A similar process was used for protein binding to the anti-importin 60 antibody, Ab2. Ab2-bound proteins were released using 4% SDS.

Nuclear protein import assay

Import of proteins into the nucleus was assayed using permeabilized tissue culture cells. HeLa cells were grown in culture and sedimented at 500g for 5 min. The cell pellet was washed repeatedly with buffer and finally resuspended into cold, permeabilization buffer. The detergent, digatonin, was added in increasing concentrations until it was determined that the plasma membrane of the cell had been removed, but the nuclear envelope remained intact. The permeabilization reaction was stopped and the cells were pelleted and transferred into fresh permeabilization buffer lacking digitonin.

The import assay used various combinations of HSS, energy-regenerating systems, and fluorescent import targets (NLS-BSA or nucleoplasmin). For some assays recombinant importin 60 purified from *Escherichia coli* was used to supplement the HSS. Assays started with the addition of permeabilized cells (intact nuclei) to the HSS solution and were allowed to proceed for 60 min at room temperature. The cells were then fixed, transferred to a coverslip, and visualized using a confocal microscope. Confocal microscopy allows optical sectioning, limiting the field of view to a discrete plane, and thereby improving the resolution of the fluorescent image. All images were collected using the same conditions including exposure time, and all photographic processes were identical for all samples in a given experiment.

Nuclear localization signal binding assay

NLS peptides and reverse sequence NLS peptides were conjugated to bovine serum albumin (BSA). Biotin was coupled to the conjugated proteins. 2 mg/mL of reverse sequence NLS-BSA was added to 100 μL of HSS followed by the addition of 40 μg/mL biotinylated NLS-BSA. After 1 h incubation on ice, the solution was centrifuged to remove large aggregates and the supernatant was mixed with 50 μL of streptavidin agarose beads and incubated for another hour. The beads were washed by repeated changes of buffer and bound proteins were eluted using 100 μL of 1 M $MgCl_2$. Proteins were precipitated using 90% ethanol and analyzed by SDS-PAGE.

Protein gel electrophoresis and immunoblotting

Protein samples were analyzed by SDS-PAGE using a 10% polyacrylamide gel. Gels were stained with Coomassie Blue to visualize the protein bands. Sample volumes were adjusted to insure equal concentrations of total protein were loaded onto the gels. Proteins were transferred to nitrocellulose by standard blotting techniques. Blots were incubated with antibodies raised against the amino terminal peptide of importin 60 (Ab1), the entire importin 60 protein (Ab2), or an internal peptide sequence of importin 90 (αimp90).

- Exposure of frog eggs to $CaCl_2$ triggers cellular changes leading to mitotic cell division; a process known as "activation." What might be the relationship between egg activation and increased nuclear import?
- How many fractions would you plan on collecting to be sure that all of the HSS solution has passed through the immobilized Ab1 column?
- HSS, importin 60, importin 90, and nucleoplasmin are all derived from the frog, *Xenopus*; however, the nuclear import assay is based on the ability of protein to enter a human nucleus. Speculate on why frog protein can function in a human cell.

RESULTS

- Explain the difference between the information in Figure 4.2.1a and Figure 4.2.1b.

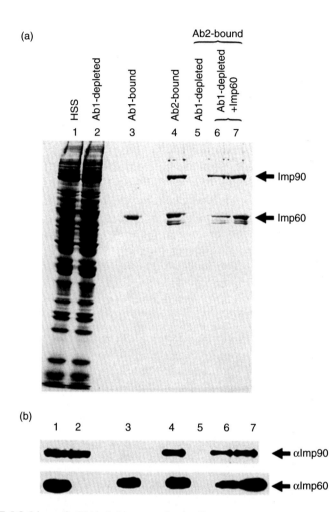

FIGURE 4.2.1 Importin 60 binds to a second subunit.
a. Coomassie Blue-stained SDS-PAGE of *Xenopus* egg HSS before (lane 1) and after (lane 2) immunoprecipitation using the anti-importin 60 antibody Ab1. Importin 60 binds to the Ab1 antibody (lane 3). However, immunoprecipitation with the anti-importin 60 antibody Ab2, results in the binding of an additional 90 kDa protein (Imp90; lane 4). Pretreatment of the HSS with the Ab1 antibody eliminated the ability of the Ab2 antibody to immunoprecipitate any proteins (lane 5). Addition of 20 μg/mL (lane 6) or 40 μg/mL (lane 7) recombinant importin 60 proteins to the Ab1-depleted supernatant restored the ability of the Ab2 antibody to immunoprecipitate the 90 kDa protein. b. Samples from the gel shown in (a) were immunoblotted using a mixture of the Ab1 and Ab2 antibodies (αImp60) or an antibody to the 90 kDa protein (αImp90).

- Is the 90 kDa protein present in the *Xenopus* high-speed supernatant? Explain your answer.
- Why is there no staining with the αImp60 antibody mix in lane 2 of Figure 4.2.1b?
- Develop an argument to defend the statement, "the anti-importin 60 antibody, Ab1, is specific to importin 60."
- Develop an argument to defend the statement, "the anti-importin 60 antibody, Ab2, is *also* specific to importin 60."
- Suggest a reason why the Ab1 and Ab2 antibodies react differently.
- Describe the significance of the results shown in Figure 4.2.1a, lanes 6 and 7.
- Characterize the pattern of fluorescent staining you observe in Figure 4.2.2.
- Can importin 90 stimulate binding to the nuclear envelope on its own? Describe the data that supports your answer (Figure 4.2.2).
- Explain why the image of the nuclei in the panel labeled "cytosol" in Figure 4.2.3 looks different from the confocal images shown in Figure 4.2.2.
- What are the minimal proteins required for import of nucleoplasmin into the nucleus?
- Predict what would happen if the nuclear import assays shown in Figure 4.2.3 were conducted in the absence of an energy-regenerating system.
- Summarize the evidence from Figures 4.2.2 and 4.2.3 that supports the researchers' conclusion that importin 60 and importin 90 are working collaboratively?
- Explain the significance of the biotinylated NLS-BSA experiment shown in Figure 4.2.4a.
- How would your explanation be changed if either of the importin antibodies had reacted in the lane labeled, "reversed-NLS bound"?
- How can the researchers be sure that importin 60 and 90 are the only cytosolic proteins that function as NLS receptors?
- Predict which importin is most likely to interact directly with the NLS sequence. Justify your answer.
- Conduct a search for a diagram of the current model of nuclear import. Connect the conclusions supported by this study with the information provided in the diagram. Recall that importin 60 = importin α and importin 90 = importin β.
- Assume that you have access to an antibody that blocks the binding of importin α to importin β. Design an experiment to test the importance of the interaction between importin α and β using one of the assays described in this case study.
- Predict what the results of your experiment would look like for the following two possible outcomes: (1) importin α and β binding is required for nuclear import, (2) importin α and β binding is *not* required for nuclear import.

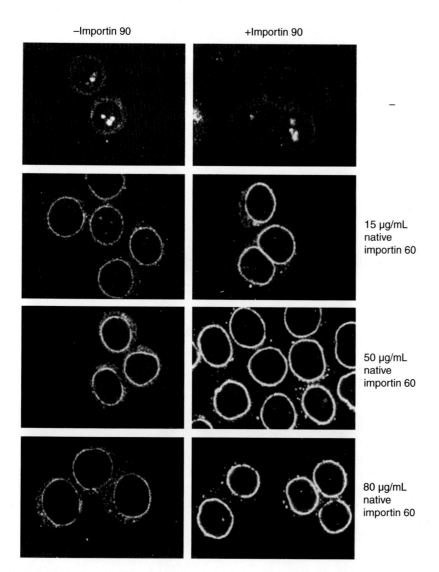

−Importin 90   +Importin 90

15 µg/mL
native
importin 60

50 µg/mL
native
importin 60

80 µg/mL
native
importin 60

FIGURE 4.2.2 Importin 60 and Importin 90 work cooperatively to bind proteins to the NE.
Nuclear transport assays were conducted using fluorescently labeled NLS-BSA in the absence (−importin 90) or presence (+importin 90) of importin 90. HeLa cells lacking Ran protein were incubated in the presence or absence of importin 60 and 90. The ability of the fluorescent NLS-BSA to bind to the nuclear envelope was determined by confocal microscopy sections through the nuclei of HeLa cells. Background levels of NLS-BSA bound to the nuclear envelope in the absence of importin 60 were detected (−). NLS-BSA staining of the nuclear envelope increases with increasing amounts of importin 60. The intensity of the NLS-BSA fluorescent signal is enhanced by the presence of importin 90. All panels were scanned and photographed under identical conditions.

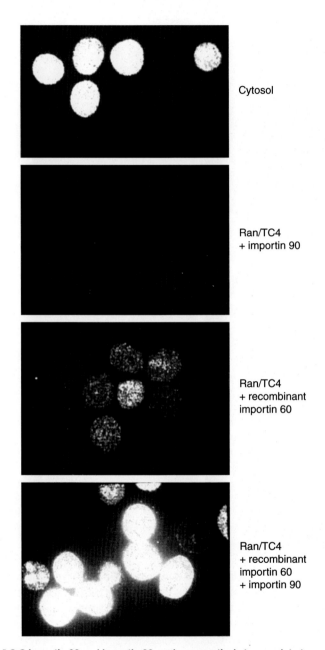

Cytosol

Ran/TC4
+ importin 90

Ran/TC4
+ recombinant
importin 60

Ran/TC4
+ recombinant
importin 60
+ importin 90

FIGURE 4.2.3 Importin 60 and Importin 90 work cooperatively to complete transport into the nucleus.

Nuclear import assays were conducted using fluorescently labeled nucleoplasmin protein in the presence of an energy-regenerating system. Permeabilized cells were treated with *Xenopus* HSS (cytosol), or a combination of Ran-GTP and importin 90, a recombinant form of importin 60. Confocal sections of nuclei were photographed and printed under identical settings.

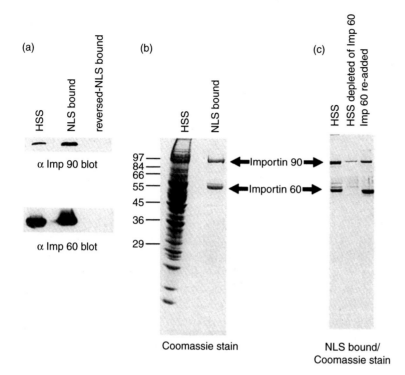

Coomassie stain

NLS bound/
Coomassie stain

FIGURE 4.2.4 The Importin 60/90 complex constitutes a cytosolic nuclear localization signal (NLS) receptor.

Xenopus HSS was incubated with biotinylated NLS-BSA, a nuclear import competent substrate, or with biotinylated reverse sequence NLS-BSA, a construct that cannot be imported into the nucleus but which carries the same amino acids as present in the NLS sequence. Biotinylated BSA was collected from the HSS by binding to streptavidin agarose, and any proteins that were bound to the NLS were released by incubation with 1 M $MgCl_2$, precipitated with 90% ethanol and analyzed by SDS-PAGE and immunoblotting. a. Importin proteins 60 and 90 were present in the HSS starting material, and both proteins were bound to the NLS sequence of the biotinylated NLS-BSA, but not to the reverse NLS sequence. b. Coomassie Blue-stained SDS-PAGE of the HSS starting material and NLS bound proteins. c. Immunodepletion of HSS with the anti-importin 60 antibody Ab1 inhibits the ability of the importin complex to bind to NLS-BSA. Binding is recovered by the addition of recombinant importin 60.

A Tale of tRNA Transport [3]

INTRODUCTION

Communication between the cytoplasm and interior of the nucleus, or the nucleoplasm, flows in both directions. Proteins must be able to move from the cytoplasm into the nucleoplasm, where they contribute to the assembly of ribosomes or bind to DNA. Binding of proteins that function as transcription factors to the DNA, alters gene expression resulting in the transcription of molecules of RNA. Although some types of RNA such as small nuclear RNAs (snRNAs) remain in the nucleus, others such as messenger RNAs (mRNAs) and transfer RNAs (tRNAs) must move from the nucleoplasm to the cytoplasm where they participate in the process of translation.

Proteins that are imported from the cytoplasm to the nucleoplasm carry a short amino acid sequence that acts as an nuclear localization signal (NLS). Movement of these proteins into the nucleus relies on interactions between the NLS and receptor proteins (importins) that bind to the signal and facilitate transport of the protein through a nuclear pore and into the nucleoplasm. **RanGTPase** proteins also play a critical role in this process. Ran proteins are present in the cytoplasm and the nucleus. However, the concentrations of **RanGTP** are higher in the nucleus where they are involved in catalyzing the dissociation of importin β from imported proteins. GTP is eventually catalyzed to GDP through the activity of the Ran protein. The concentrations of **RanGDP** are highest in the cytoplasm. The conversion of RanGDP back to RanGTP is catalyzed by a guanine nucleotide exchange factor or GEF protein. The activity of Ran changes depending on which guanine nucleotide is bound to it.

- List the different types of RNA molecules that are exported out of the nucleus.
- Research/review the steps involved in transcription.
- How is a molecule of tRNA different from a molecule of mRNA?
- Propose an explanation for why the activity of the Ran protein would change depending on whether it was bound to GTP or GDP.
- What prevents an NLS protein from moving back into the cytoplasm?

BACKGROUND

RNA molecules exit the nucleus as **ribonucleoproteins** (RNPs), a complex of RNA and RNA-binding proteins. A leucine-rich amino acid sequence in an RNA-binding protein acts as a **nuclear export signal** (NES), targeting the protein and the attached RNA at the cytoplasm. Just as with import into the nucleus, export requires an NES receptor protein that functions to facilitate movement of the RNP through the nuclear pore. Export of RNAs has also been shown to be dependent on the presence of nuclear RanGTP.

Analysis of yeast mutants unable to export tRNAs led to the discovery of NES receptor proteins. The yeast NES receptor, Los1p, shares sequence homology with importin β. Knowing the gene sequence for Los1p allowed researchers to look for homologous genes in other eukaryotic organisms, including humans, by searching for similar sequences in genomic databases. Using these techniques, **Exportin(tRNA)** (XPO-1), a Los1p-related protein, was identified in humans.

The discovery of two proteins with similar amino acid sequences does not guarantee that they serve the same function in a cell. Before a protein can be designated as an NES receptor it must behave like known NES receptors. NES receptors must be capable of moving into and out of the nucleus. NES receptors should bind to their target proteins in the presence of RanGTP, leading to the export of RNA molecules. Finally, the rate of RNA export should be influenced by the amount or availability of the receptor. The ability of human Exportin(tRNA) to meet the criteria for an NES receptor will be examined in this case study.

- . Research/review examples of proteins that bind to RNA.
- Examine the chemical structure of the amino acid leucine. How could these amino acids act as a signal?
- Predict whether a small nuclear RNA would interact with an Exportin-type protein. Explain your answer.
- Human Exportin(tRNA) is 21% similar to yeast Los1p. Discuss what it means to say that two proteins are "similar"? What accounts for the differences between these two proteins?
- Describe the relationship between a protein's amino acid sequence and its function.

METHODS

In vitro translation and ^{35}S-labeling

The full-length cDNA sequences of the various proteins used in this study were cloned into expression plasmids. The type of plasmid used varied depending on whether the protein of interest needed to include a polyhistidine tag (His$_6$-tag) or a *Staphylococcus aureus* z domain sequence (a sequence that binds to IgG proteins). Messenger RNAs for Exportin(tRNA) or the NLS protein CBP80 were translated *in vitro* using a rabbit reticulocyte lysate system. Reticulocytes are a type of immature red blood cell (RBC), lacking a nucleus, but containing large amounts of mRNAs for the synthesis of hemoglobin. Since the primary function of a reticulocyte is to synthesize protein, it contains all the cellular components required for efficient translation of eukaryotic mRNAs. The reticulocytes are lysed and their cytoplasmic contents are preserved for use in *in vitro* protein synthesis assays. Translation was conducted in the presence of ^{35}S-methionine. The radioactive amino acid becomes incorporated into the growing polypeptide chain causing the final protein to become **radiolabeled**. The presence of the protein can then be detected by autoradiography and the low-level emission of radioactive particles by exposure to X-ray film.

Injection into *Xenopus* frog eggs

RNAs and proteins were microinjected into individual *Xenopus* eggs. Solutions were mixed with dextran blue to allow visual tracking of the injected cells. Contents of the cytoplasm and nucleus were separated by microdissection at the time points indicated.

In vitro binding assays

His$_6$-tagged RanQ69LGTP or RanT24NGDP were preincubated with Ni-NTA agarose beads. His$_6$-tagged proteins have been genetically engineered to carry a sequence of six histidine amino acids at either the amino or carboxy terminus. The histidine-repeat sequence binds with high affinity to nickel, allowing selective immobilization of proteins carrying a His$_6$ tag onto the surface of agarose beads. The bead plus immobilized protein is then incubated in a solution of other proteins. Proteins that are able to bind to the immobilized protein will remain with the beads, while unbound proteins are removed by repeated buffer washes. The populations of bound proteins were finally released from the beads by addition of an elution buffer.

RanQ69LGTP or importin β were also fused to a *S. aureus* z domain, allowing the protein to be immobilized on a Sepharose bead coated with IgG protein. His-tagged Exportin(tRNA) was incubated with the z-tagged RanQ69LGTP or the z tag alone. Exportin protein that bound to the beads was separated from unbound protein and analyzed by SDS-PAGE. In another assay, z-tagged Exportin(tRNA) or z-tagged importin β were immobilized on IgG–Sepharose beads and incubated with ^{32}P-labeled RNAs. ^{32}P-labeled transcripts were synthesized from cDNA templates for the proteins dihydrofolate reductase (DHFR) and histone H4, the small nuclear RNAs U1ΔSm and U6Δss, and the transfer RNAs, tRNA^{Met} and tRNA^{Phe}.

Gel electrophoresis and immunoblotting

Protein samples were analyzed using SDS-PAGE. SDS-PAGE separates proteins based on their relative size. Sample preparation for SDS-PAGE includes the reagent dithiothreitol (DTT), a reducing agent that denatures (unfolds) the proteins. The SDS component of the sample preparation coats the protein with a negative charge, eliminating any inherent charge properties that the protein would have. The gels were run, dried, and then exposed to X-ray film to detect the presence of radiolabeled proteins.

Native gels are a type of nondenaturing polyacrylamide gel electrophoresis. In this method the size, shape, and inherent charge of a protein is retained. Sample preparation omits the SDS and DTT. The movement of proteins in the gel depends on their shape and charge, with small, compact, and/or negatively charged proteins moving faster (lower in the gel) than larger, more positively charged proteins.

Immunoblotting involves the transfer of proteins from an SDS-PAGE gel onto a substrate such as nitrocellulose blotting paper. The nitrocellulose blot is incubated with a solution of dry milk, dissolved in buffer to block the areas of the blot, that do not already have proteins bound. The blot is then incubated with a primary antibody, specific to the protein of interest, in this case Exportin(tRNA). Excess primary antibody is removed by buffer washes, followed by incubation with a secondary antibody. The secondary antibody recognizes and binds to the primary antibody and acts as a "reporter." The secondary antibody used in this work was linked to the enzyme horseradish peroxidase. Reaction of the blot with the substrate for this enzyme results in the formation of a black precipitate on the surface of the blot.

RNA samples can be analyzed using a denaturing polyacrylamide gel combined with urea. The urea helps unfold the secondary structure of the RNA, making it easier for it to move through the matrix of the gel.

- List the cellular contents you would expect to find in a sample of reticulocyte lysate.
- Why is ^{35}S-Meth a good choice for an amino acid that would label any protein?
- Create a labeled diagram that illustrates the following *in vitro* binding assays:
  - Radiolabeled Exportin(tRNA) + His$_6$-tagged RanQ69LGTP
  - Exportin(tRNA) + z-tagged RanQ69LGTP
  - Exportin(tRNA) + z-tagged alone
- Compare and contrast various methods of gel electrophoresis used in this case study. What are the advantages and applications of each method?

RESULTS

- What part of a protein is labeled with ^{35}S?
- Summarize the conclusions supported by the data shown in Figure 4.3.1.
- Explain why the protein CBP80 cannot leave the nucleus.
- Why would it be important for a NES receptor to reenter the nucleus?
- Create a diagram to illustrate how the presence or absence of GTP influences the interaction between the Ran protein and Exportin(tRNA) (Figure 4.3.2a).
- What was the purpose of the experiment shown in lanes 6–9 in Figure 4.3.2a?
- Use the data in Figure 4.3.2 to explain the difference between SDS-PAGE and nondenaturing native gel electrophoresis.
- Propose an explanation for why the RNaseA treatment eliminated band III from the gel shown in Figure 4.3.2b.
- What part of a nucleic acid is labeled by ^{32}P?
- Use the data presented in Figure 4.3.3 to justify the following statements:
  - Exportin specifically binds to tRNAs.
  - Exportin binding requires RanGTP.
  - Importin β is not involved in tRNA export.
- Discuss the significance of the appearance of a band corresponding to U1ΔSm in multiple lanes in Figure 4.3.3.
- Predict the outcome of an experiment, like that shown in Figure 4.3.3, that uses tRNA^{Ala} and rRNA.
- Speculate on whether human Exportin(tRNA) could bind to *Xenopus* tRNAs.

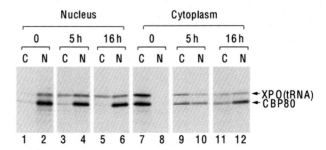

FIGURE 4.3.1 Exportin(tRNA) shuttles between the nucleus and the cytoplasm.
^{35}S-labeled Exportin (XPO(tRNA)) and the NLS-containing protein CBP80, were mixed and microinjected into the nucleus (lanes 1–6) or cytoplasm (lanes 7–12) of a *Xenopus* frog egg. Eggs were dissected into a cytoplasmic fraction (C) or a nuclear fraction (N) at the indicated times after microinjection. Proteins were isolated and analyzed by SDS-PAGE. Radiolabeled proteins were visualized using X-ray film.

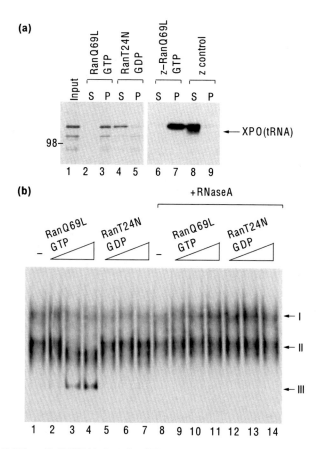

FIGURE 4.3.2 Exportin(tRNA) binds to RanGTP.

a. ^{35}S-labeled Exportin (XPO(tRNA)) was incubated with RanQ69L, a mutant form of Ran that cannot hydrolyze GTP, or RanT24N, a protein that cannot bind GTP. Starting material included the XPO(tRNA) protein (lane 1). Mutant Ran proteins, carrying a His$_6$ tag, were immobilized on Ni-NTA agarose beads. Bound (P) and unbound (S) proteins were examined by SDS-PAGE and autoradiography (lanes 2–5). Unlabeled recombinant Exportin was incubated with RanQ69LGTP fused with the IgG-binding z domain of *S. aureus* protein A or the z domain alone (lanes 6–9). Samples were incubated with IgG–Sepharose beads. Bound (P) and unbound (S) proteins were analyzed by SDS-PAGE and immunoblotting using an antibody that recognized the recombinant Exportin(tRNA). b. ^{35}S-labeled Exportin (lanes 1 and 8) was incubated with increasing concentrations of RanQ69LGTP (lanes 2–4) or RanT24NGDP (lanes 5–7) and analyzed using native gel electrophoresis. Exportin(tRNA) resolves as two bands (I and II). A highly mobile form of Exportin (III) appears in the presence of 120 nM and 400 nM RanQ69LGTP. Pretreatment of the mixture with RNaseA eliminates band III.

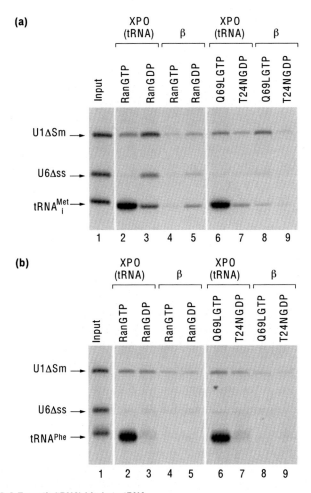

FIGURE 4.3.3 Exportin(tRNA) binds to tRNA.
Recombinant Exportin(tRNA) (XPO(tRNA)) or importin β were bound to Sepharose beads and then incubated with a mixture of ^{32}P-labeled small nuclear RNAs (U1ΔSm and U6Δss) and tRNA^{Met} (part a) or tRNA^{Phe} (part b). Incubations were conducted in the presence of RanGTP, RanGDP, RanQ69LGTP, or RanT24NGDP. Bound RNAs were extracted and analyzed using denaturing polyacrylamide–urea gel electrophoresis. The RNA starting material is shown in lane 1.

- Approximately how long does it take for Exportin(tRNA) to transport tRNAs out of the nucleus (Figure 4.3.4)?
- Why is tRNA^{Met} showing up in the cytoplasm of the control after 2.5 h?
- What other RNAs are exiting the nucleus? Is their transport dependent on Exportin? Explain your answer.
- Refer back to the list of characteristics expected of an NES receptor protein. Match each of the characteristics to the data presented in this case study that supports the conclusion that Exportin(tRNA) is an NES receptor.

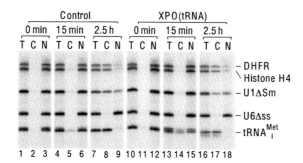

FIGURE 4.3.4 Exportin(tRNA) stimulates transport of tRNAs out of the nucleus.
A collection of RNAs were injected into the nucleus of a *Xenopus* frog egg either alone (lanes 1–9) or mixed with recombinant Exportin(tRNA) (lanes 10–18). The RNA collection included mRNAs for the proteins DHFR and histone H4, small nuclear RNAs U1ΔSm and U6Δss, and tRNA^{Met}. RNA export was measured immediately or after 15 min, or 2.5 h after injection by extracting RNA from the total egg (T), the cytoplasm (C), or the nucleus (N).

- Reflect on the function of an NES receptor protein. Are there any other properties that you would expect such a protein to possess? Write a hypothesis that addresses your idea(s).
- Draw on the methodologies used in this case study to design an experiment that would test one of the further questions:
  - Does Exportin(tRNA) bind preferentially to one type of tRNA?
  - What other nuclear proteins bind to RanGTP?
- Does Exportin(tRNA) interact with importin β?

When It Comes to the Nucleus – Size Matters [4]

INTRODUCTION

A eukaryotic cell is defined by the presence of membrane-enclosed organelles, the most prominent of which is the nucleus. The **nuclear envelope (NE)** forms a barrier between the cytoplasm of the cell and the interior of the nucleus, or nucleoplasm. The nuclear envelope is made up of two distinct membranes. The outer membrane, which faces the cytoplasm, is continuous with the endoplasmic reticulum of the cell, and the intermembrane space that separates the membranes of the nuclear envelope is continuous with the lumen or interior space of the endoplasmic reticulum. The inner membrane, which faces the nucleoplasm, is supported by a matrix of intermediate filament proteins known as the **nuclear lamina**. **Nuclear pore complexes** cross both membranes of the NE, allowing a bidirectional transport of materials between the nucleoplasm and the cytoplasm.

The physical size of a nucleus varies across the domain Eukarya. For example, in the yeast *Saccharomyces cerevisiae* the nuclear volume is 3 μm^3 while in the frog *Xenopus laevis* the volume is 300 μm^3. One explanation for this phenomenon is that the size of the nucleus depends on the amount of DNA inside it. Another idea is that the size of the nucleus scales with the size of the cell. Do bigger cells mean bigger nuclei?

Nuclei go through cycles of assembly and disassembly as part of the cell cycle. Construction of a new nucleus requires the reformation of an NE complete with functional nuclear pore proteins and supported by nuclear lamins. Nuclear lamin proteins must be transported out of the cytoplasm, through the nuclear pore complexes, and into the interior of the nucleus. Nuclear proteins are targeted for import by the presence of an nuclear localization signal (NLS) located in their amino acid sequence. The protein, **importin α**, binds to NLS sequences. A second protein, **importin β**, binds to importin α, and all three proteins (NLS protein, importin α, and importin β) are then transported through the nuclear pore complex. Once inside the nucleus, importin β binds to the protein **RanGTP** triggering the release of the NLS protein into the nucleoplasm. Importin β recycles back to the cytoplasm, still attached to RanGTP. The Ran protein eventually hydrolyzes GTP to GDP, freeing importin β to bind to another importin α protein. The RanGDP protein is carried from the cytoplasm back into the nucleus by the protein **Ntf2,** where it is converted back to RanGTP allowing the cycle to continue.

What then controls the size of a nucleus? Is size dependent upon the efficiency of transport of materials, like the nuclear lamins, into the nucleoplasm? Can nuclear size be altered? This case study looks at the *in vitro* assembly of nuclei using cell-free systems derived from two related species of the frog *Xenopus*.

- Create a labeled diagram that illustrates the structure of the nuclear envelope.
- Explain why nuclear lamin proteins originate in the cytoplasm of a cell.
- Illustrate the process of nuclear import by tracing the pathways of the following proteins: *NLS protein, importin α, importin β, RanGTP, RanGDP, and Ntf2.*
- Research/review the amino acid sequence associated with an NLS.
- Research/review the equation for determining the volume of a sphere. Calculate the volume of the nucleus in the following cell types assuming that a nucleus is a sphere:

Species	Diameter of the Nucleus (μm)	DNA Content (pg)
Drosophila melanogaster (common fruit fly)	5.30	0.30
Gallus gallus (chicken)	7.37	2.68
Homo sapiens (human cancer cell)	8.93	10.68
Homo sapiens (human white blood cell)	7.62	6.24
Mus musculus (mouse L cell)	9.62	13.8
Notophthalmus viridescens (eastern newt)	19.97	95.6

- Evaluate the data you calculated for nuclear volume. What trends do you observe?
- Use the aforementioned data to calculate the area of the NE assuming that the nucleus is a circle.

BACKGROUND

Assembly of a nucleus from its component parts can be replicated *in vitro* using a cell-free assay system. One of the best-studied cell-free systems uses cellular components isolated from eggs of the frog *Xenopus*. An egg stores large quantities of the proteins, phospholipids, and other molecules required to support the rapid mitotic cell divisions that occur following fertilization. Frog eggs can also be collected in large quantities making it possible to isolate enough cellular material for experimental use. *Xenopus* eggs are naturally arrested in metaphase of meiosis II.

X. laevis is the species of frog most often used to generate egg extracts. However, another species of frog *Xenopus tropicalis* can also be a source. *X. laevis* frogs are larger than *X. tropicalis* frogs and so are their cells, eggs, DNA content, and nuclei. Assembly of nuclei using cell-free systems based on the egg extracts of the two frog species makes it possible to determine if the size difference observed *in vivo* should be replicated *in vitro*.

■ Predict whether a cell arrested at metaphase would have an intact nucleus. Explain your answer.
■ How is a cell in meiosis II different from a cell in meiosis I?
■ Generate a hypothesis about the relationship between nuclear size and the cytoplasmic contents in egg extracts.

METHODS

Xenopus egg extracts and nuclear assembly

X. laevis and *X. tropicalis* unfertilized eggs were collected, dejellied, rinsed with buffer, and crushed by centrifugation to generate metaphase-arrested egg extracts. *Xenopus* sperm nuclei were isolated from frog testes and their membranes were removed by incubation with a mild detergent. The sperm nuclei were rinsed with buffer to remove the detergent. *In vitro* nuclear assembly was initiated by incubating 25 μL fresh egg extract with 1000 sperm nuclei per microliter reaction volume. *X. laevis* nuclei were used for all experiments except for the one in Figure 4.4.1d, in which *X. tropicalis* sperm nuclei were used. Reactions were incubated at 19–22°C. Import-competent nuclei were generally formed within 30–40 min.

Immunodepletions and recombinant proteins

Partial depletion of importin α from *X. laevis* egg extract was achieved using a polyclonal antibody to importin α. Protein A-coated beads were incubated with either the anti-importin α antibody or IgG. Unbound antibody was removed by washing, and the beads were incubated with *X. laevis* extract for 45 min on ice. Importin α and mock-depleted extracts were collected and used for nuclear assembly reactions. Depletion of Ntf2 from *Xenopus tropicalis*, followed a similar protocol as aforementioned but using a monoclonal antibody to Ntf2.

Proteins GPF-NLS, importin α–E, and importin α–EΔIBB were expressed in *Escherichia coli* and purified using standard protocols.

Fluorescence microscopy and image analysis

Nuclear assembly was conducted in the presence of various fluorescent markers of nuclear import. GFP-NLS represents a small soluble marker while

IBB-labeled quantum dots (IBB–Qdots) represent large cargo. The total surface area of a NE was estimated as four times the cross-sectional area of thresholded images of a nucleus. A thresholded image is made up of pixels that are either black or white. Fluorescence intensity was measured by acquiring an image of the nuclei and an area of background at the same exposure time. Total fluorescence intensity per unit area was calculated after corrections for background fluorescence.

Nuclei were also stained using immunofluorescence. Nuclei were fixed for 15 min using a buffer containing 15% glycerol and 2.6% paraformaldehyde, layered onto a 5 mL buffer cushion of 200 mM sucrose and 25% glycerol, and centrifuged onto a 12 mm circular coverslip at 1000g for 15 min. Nuclei were postfixed with methanol and incubated with the indicated primary antibody followed by reaction with a fluorescently labeled secondary antibody.

Immunoblots

Egg extract protein concentrations were determined using a Bradford protein assay. Three samples of 25 µg of egg extract protein from X. *laevis* and X. *tropicalis* were analyzed by SDS-PAGE and transferred to a nitrocellulose membrane. Blots were blocked with 5% milk in buffer and reacted with the indicated primary antibodies. Fluorescently labeled secondary antibodies were used to detect primary antibody binding on the blot. The intensity of antibody binding was quantified using an infrared fluorescence imager. Concentrations of importin and Ntf2 proteins were determined by comparing the intensity of bands from the egg extracts with the band for importin or Ntf2-recombinant proteins of known concentration on the same blot.

- How many membranes surround the genetic information stored in a sperm cell?
- Speculate on why it is necessary to demembranate the sperm.
- How would the use of thresholded images contribute to the analysis of nuclear size?
- What is the difference between a monoclonal and polyclonal antibody?
- Explain what "mock depletion" is and why it was used.

RESULTS

- Summarize your observation of the nuclear assembly assay shown in Figure 4.4.1a.
- How do the data in Figure 4.4.1b relate to the images in Figure 4.4.1a?
- Estimate the rate of nuclear growth for the two species based on the change in NE surface area using Figure 4.4.1b.
- What does an R^2 value tell you about a data set?

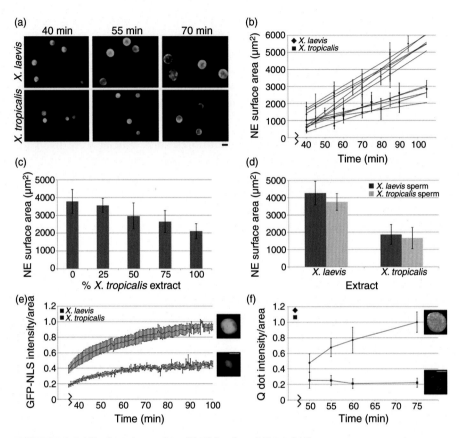

FIGURE 4.4.1 Nuclear size scales with *X. laevis* and *X. tropicalis*.
a. Nuclei were assembled using egg cytosolic extract from *X. laevis* and *X. tropicalis* and stained with an antibody to the nuclear pore complex. Scale bar = 20 μm. b. NE surface area was measured from images like those in (a) at the indicated time points during nuclear assembly using each of the two extracts. Best-fit linear regressions are shown for six *X. laevis* and five *X. tropicalis* egg extracts. R^2 values range from 0.96 to 0.99 for *X. laevis* and from 0.94 to 0.98 for *X. tropicalis.* Error bars represent standard deviation. c. *X. laevis* and *X. tropicalis* extracts were mixed as indicated and nuclear size was measured after 90 min incubation. Error bars represent standard deviation. d. *X. tropicalis* sperm nuclei contain 55% of the DNA content of *X. laevis* sperm nuclei. Nuclei were assembled with the indicated combinations of sperm nuclei and egg extract. Nuclear size was measured after 90 min incubation. Error bars represent standard deviation. e. Nuclei were assembled in the presence of green fluorescence protein fused to a nuclear localization signal (GFP-NLS). Live fluorescent images were captured at 30 s intervals. Nuclear GFP fluorescence intensity per unit area was measured for five nuclei in each extract at each time point. Images show GFP fluorescence after 70 min incubation. f. The ability of nuclear pore complexes to transport large cargo was tested using quantum dots (Qdots) carrying an importin β-binding domain (IBB) sequence. IBB-coated Qdots were added to a nuclei assembly assay. Nuclear Qdot fluorescence intensity per unit area was measured for at least 30 nuclei at the indicated time points and normalized to 1.0 (arbitrary units). Representative images are at 75 min.

- Describe the pattern you observe in the data shown in Figure 4.4.1c. Interpret what the patterns mean in the context of the *in vitro* assembly of a nucleus.
- Does DNA content or egg extract have a greater influence on nuclear size (Figure 4.4.1d)? Explain your answer.
- How are the images in Figure 4.4.1e different from those in Figure 4.4.1a?
- Explain the significance of the data presented in Figure 4.4.1e.
- Summarize the conclusion supported by the data in Figure 4.4.1f.
- Propose an explanation for the difference in the pattern of transport shown for the *X. tropicalis* nuclei in Figure 4.4.1e, f.
- Explain the significance of the statement that each of the lanes in Figure 4.4.2a were loaded with the same amount of total egg extract protein.
- Analyze the results shown in Figure 4.4.2a and b. What can you conclude about the composition of the egg extracts from *X. laevis* and *X. tropicalis*?

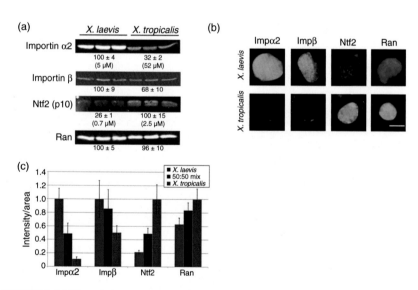

FIGURE 4.4.2 The relative amounts of proteins associated with nuclear import differ in *X. laevis* and *X. tropicalis*.

a. Here, 25 μg of protein from three different samples of egg extracts from *X. laevis* and *X. tropicalis* egg were analyzed by SDS-PAGE and immunoblotting with antibodies specific to the indicated proteins. Values below each set of three lanes represent the relative protein amounts (mean; SD, $n = 3$). Absolute concentrations (μM) were determined by comparing band intensities with known standards on the same blot. b. Nuclei at 80 min were processed for immunofluorescence using the same antibodies as in (a). For a given antibody, images were acquired using the same exposure time. Scale bar = 20 μm. c. Quantification of nuclei displayed in (b). Nuclear fluorescence intensity per unit area was calculated for at least 50 nuclei per condition, averaged, and normalized to 1.0 (arbitrary units). Error bars represent SD.

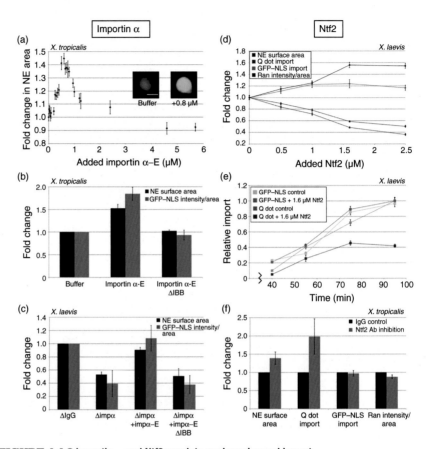

FIGURE 4.4.3 Importin α and Ntf2 regulate nuclear size and import.

a. Nuclei were assembled using *X. tropicalis* egg extract in the presence of GFP-NLS. After 40 min, importin α-E was added at the concentrations indicated. At 80 min, images of at least 50 nuclei per condition were captured and the NE surface area was quantified, averaged, and normalized to the buffer control. Error bars represent standard error (SE). Scale bar = 20 μm. b. Experiments were performed as in (a) with a fixed concentration (0.8 μM) of importin α-E or a mutant version lacking the importin βΔIBB added to the reaction. Nuclear surface area and the intensity of GFP-NLS fluorescence were measured and compared with control (buffer) values. c. Nuclei were assembled in *X. laevis* egg extract that had been depleted of its endogenous importin α (Δimpα) or mock depleted (ΔIgG). Nuclei assembly was also tested using importin α-E-depleted egg extract with importin α-E or the mutant importin α-E (ΔIBB) added back. d. Ntf2 was added back to *X. laevis* egg extract prior to nuclear assembly. GFP-NLS or IBB-coated Qdots were added after 30 min. At 80 min nuclei were processed for immunofluorescence with an antibody to Ran protein. At least 50 images of nuclei per condition were captured. NE surface area and intensity of the GFP-NLS, Qdot, and Ran fluorescence were measured. Error bars represent SE. e. Experiments performed as in (d) using a fixed concentration (1.6 μM) of Ntf2. Qdot or GFP-NLS fluorescence intensities for at least 50 nuclei per time point were averaged and normalized to 1.0 (arbitrary units). Error bars represent SE. f. Nuclei were assembled in *X. tropicalis* egg extract supplemented with anti-Ntf2 or IgG antibodies. At 30 min, Qdots or GFP-NLS were added. At 80 min, immunofluorescence for Ran was performed and nuclear parameters were quantified as in (d). Average fold change from the IgG control and SD are shown.

- Refer back to your illustration of the nuclear import pathway. Modify your illustration to identify the points in the pathway influenced by the experimental results shown in Figure 4.4.2.
- Make a prediction about how one of the differences presented in Figure 4.4.2 would alter nuclear import of an NLS protein.
- Discuss the significance of the results of nuclear import using a 50:50 mix of *X. laevis* and *X. tropicalis* egg extracts (Figure 4.4.2c).
- Predict how supplementing the *X. tropicalis* egg extract with additional importin α–E and GFP-NLS might affect nuclear assembly.
- Describe how the data in Figure 4.4.3a and b fit your prediction.
- "Fold change" is used to represent the difference between two conditions. Use the data in Figure 4.4.3a and b to determine how to calculate fold change.
- Explain the logic behind the experiments shown in Figure 4.4.3c.
- What effect does the addition of supplemental Ntf2 to *X. laevis* egg extract have on nuclear assembly and import (Figure 4.4.3d and e)?
- Ntf2 is known to bind to proteins within the nuclear pore complex. An Ntf2- specific antibody was used to inhibit the activity of Ntf2 in *X. tropicalis* egg extracts. What conclusions regarding the activity of Ntf2 are supported by the data in Figure 4.4.3f?
- Synthesize the experimental results presented in the case study to arrive at a model of nuclear import during nuclear assembly.
- Apply your model to explain how regulation of the import of large cargo, such as nuclear lamins, would influence the size of a nucleus.
- Conduct a search for a scientific video of sea urchin, zebrafish, or frog development. Note the change in cell size with each cell division following fertilization. Develop a hypothesis about nuclear import during early embryogenesis.
- Watch the following video by the authors of this case study to get another perspective on the people and science behind this work: https://www.youtube.com/watch?v=Jbi-nfIa15M

References

[1] Harel A, Orjalo AV, Vincent T, Lachish-Zalait A, et al. Removal of a single pore subcomplex results in vertebrate nuclei devoid of nuclear pores. Mol Cell 2003;11:853–64.

[2] Görlich D, Kostka S, Kraft R, Dingwall C, Laskey RA, Hartmann E, Prehn S. Two different subunits of importin cooperate to recognize nuclear localization signals and bind them to the nuclear envelope. Curr Biol 1995;5:383–92.

[3] Arts G-J, Fornerod M, Mattaj IW. Identification of a nuclear export receptor for tRNA. Curr Biol 1998;8:305–14.

[4] Levy DL, Heald R. Nuclear size is regulated by importin α and Ntf2 in *Xenopus*. Cell 2010;143:288–98.

CHAPTER 5

Membranes and Membrane Transport

Flip This Lipid [1]

INTRODUCTION

All cells are surrounded by a plasma membrane (PM) composed of a phospholipid bilayer that serves to separate the contents of the cell from the external environment. Additionally, eukaryotic cells possess membrane-bound organelles that further delimit compartments within the cell separated from the surrounding cytoplasm. The distinctive organization of a phospholipid bilayer is a consequence of the amphiphatic nature of individual phospholipid molecules. Polar phospho-head groups can interact with the aqueous environment of the cell while the nonpolar fatty acid tails cluster together, away from water. The two sides or **leaflets** of a phospholipid bilayer are not equivalent. The outer leaflet or **exoplasmic** side of the membrane has a high concentration of the phospholipid **phosphotidylcholine** while the inner or **cytosolic** side of the membrane is enriched in the phospholipids phosphotidylethanolamine and phosphatidylserine. Another important feature of biological membranes is the presence of membrane proteins. **Integral** membrane proteins inserted into the bilayer can only be removed using harsh methods such as detergents while **peripheral** membrane proteins associate with the phospho-head groups and can be removed using buffers containing high concentrations of salt.

- Name some of the different membranes and the compartments they create in a eukaryotic cell.
- Define the term "amphipathic" using your own words.

- Create a diagram of a biological membrane that illustrates the asymmetry of the different types of lipids on the two sides of the bilayer. Include an integral and a peripheral membrane protein in your diagram.

BACKGROUND

In eukaryotic cells, phospholipids are synthesized by enzymes located on the cytoplasmic side of membranes of the endoplasmic reticulum (ER). Newly synthesized phospholipids are inserted into the cytosolic leaflet of the ER membrane. There must be a mechanism to transport these phospholipids from the cytosolic to the exoplasmic sides of the membrane in order to achieve the differential distribution of phospholipids found in membranes. Phospholipids cannot simply diffuse from one side of the bilayer to the other. It has been proposed that a membrane protein might facilitate the transfer or "flipping" of a phospholipid across the bilayer. Such a **flippase** would be expected to be located in the membranes of the ER where phospholipids are synthesized.

One way to study the cellular activities that occur on the membrane of the ER is by isolating those membranes in the form of small vesicles called **microsomes**. When a cell is homogenized the membranes of the ER will fragment and reseal upon themselves forming small vesicles. The technique of **differential centrifugation** can be used to isolate the microsomal membrane fraction away from other vesicles and organelles that are present in the cell homogenate. Because the microsomes are derived from ER membrane they contain all the proteins that are present in the ER. If the ER is the site of phospholipid transfer then microsomal membranes should contain a phospholipid flippase. Another experimental approach involves the use of artificial vesicles called **liposomes**. Liposomes form when phospholipids are mixed in the presence of an aqueous solution. Detergent solubilized membrane proteins can become incorporated into the liposome bilayer during **reconstitution** of liposome, forming **proteoliposomes**.

The experiments described in this case study made use of the compound $[^3H]$ **dibutyrylphosphatidylcholine** ($[^3H]diC_4PC$), a water-soluble compound that consists of the phospho-head group of phosphatidylcholine, but lacks the hydrophobic, fatty acid tails. Diffusion will cause the movement of $[^3H]diC_4PC$ from one side of the bilayer; however, given the polar nature of the phospho-head group, transport of the compound can only occur in the presence of a flippase. This case study examines the ability of native microsomes and artificially reconstituted proteoliposomes to transport $[^3H]diC_4PC$ as an indicator of flippase in the ER membrane.

- Predict what would happen if there was no mechanism to transport phospholipids from the cytosolic to the exoplasmic leaflet of a membrane.
- Outline the pathway a newly synthesized phospholipid would take from the point of origin (the ER) and the PM of the cell.

- Describe the nature of the barrier to the passive movement of a phospholipid from one side of the bilayer to the other.
- Why would a fragment of a membrane or mixture of pure phospholipids form a vesicle?
- Debate whether the organization of membrane proteins inserted into a proteolipid is the same as that in the native (unextracted) membrane.
- Debate whether measuring the movement of a phospho-head group across a bilayer can be considered a good assay for phospholipid transport.

METHODS

Microsomes and submicrosomal membrane fractions

Total rough and smooth microsomes were isolated from rat liver. Livers were homogenized in 0.25 M sucrose and then centrifuged at 12,500 rpm for 20 min at 4°C. The supernatant was removed and recentrifuged under the same conditions. Total microsomes were isolated from the second supernatant by centrifugation at 40,000 rpm for 45 min. For separation of rough and smooth microsomal membranes 0.15 mL of 0.1 M $CsCl_2$ was added to each 9.85 mL of second supernatant. The supernatant was then layered onto a 15 mL solution of 1.3 M sucrose containing 15 mM $CsCl_2$, and topped with a solution of 0.25 M sucrose. The layered sample was centrifuged at 25,000 rpm for 160 min in a swinging bucket rotor. Rough microsomes form a pellet and smooth microsomes float at the interface between the 0.25 M sucrose and 1.3 M sucrose solutions. When examined by electron microscopy (data not shown) rough microsomal membranes have bound ribosomes and smooth microsomal membranes do not.

Preparation of salt-washed rough microsomal membranes

A suspension of rough microsomal membranes was diluted with an equal volume of a high salt buffer containing 10 mM EDTA, incubated on ice for 20 min, then centrifuged through a 0.5 M sucrose cushion at 70,000 rpm for 30 min. The resulting salt-washed membranes were resuspended in 0.25 M sucrose and stored at −80°C.

Salt-washed microsomal membranes were extracted with the detergent, Triton X-100, by mixing 20 mg/mL of the vesicle suspension with an equal volume of extraction buffer containing 200 mM NaCl, 2% Triton X-100, and incubating the solution on ice for 30 min before centrifugation at 70,000 rpm for 45 min. The resulting clear supernatant (Triton extract) was carefully removed and stored on ice for further fractionation or reconstitution into proteoliposomes.

Reconstitution of proteoliposomes

Proteoliposomes were reconstituted by adding the Triton X-100 extract containing the membrane proteins and phospholipids from the rough microsomal membrane fraction to a glass tube containing a solution of pure phosphatidylcholine. The mixture was supplemented with [^{3}H]mannose. Radiolabeled mannose was captured inside the proteoliposomes as they formed and was used as an indicator of the size/volume of the liposomes. The resulting vesicles were treated to remove excess detergent and washed several times.

Protease treatment of proteoliposomes

Proteoliposomes reconstituted from a Triton X-100 extract of salt-washed rough microsomal membranes were resuspended to a concentration of 0.35 mg/mL protein. Two aliquots of 300 μL each were analyzed. One aliquot was incubated with proteinase K (33 μL of a 20 mg/mL stock solution) for 40 min at room temperature before addition of 6 μL of 150 mM PMSF, a protease inhibitor. The second aliquot was mock-treated in parallel by incubating the membranes with 33 μL of buffer followed by PMSF.

Transport of [^{3}H]diC$_4$PC into vesicles

An aliquot (20 μL) of vesicle suspension was equilibrated at the assay temperature (30°C) for 1 min before initiating the assay with the addition of 5 μL of [^{3}H]diC$_4$PC. The assay was terminated by adding 600 μL of ice-cold buffer and filtering the mixture through a precooled filter. The assay tube was rinsed with an additional 600 μL of ice-cold buffer and filtered again through the same filter. The filter was rinsed with 5 mL of ice-cold buffer and transferred to a scintillation vial and agitated briefly with 100 μL of 0.1% SDS before addition of the scintillation cocktail. Radioactivity adhered to the filter was measured using liquid scintillation counting in which the energy from the particles emitted from the radioactive sample is used to excite compounds in the scintillation cocktail that will then emit light which is detected and quantified.

The intravesicular space (volume) of the proteoliposomes was determined by liquid scintillation counting of [^{3}H]mannose-treated membranes without prior incubation with [^{3}H]diC$_4$PC.

- Create a flow chart that outlines the procedure for isolating the three microsomal membrane fractions.
- Predict what cellular material you would find in the pellet following the 12,500 rpm and 40,000 rpm centrifugation steps.
- Calculate the final concentration of CsCl$_2$ in the second supernatant prior to the centrifugation step to separate rough and smooth microsomal membranes.
- What effect does incubation in a high salt buffer have on biological membranes like rough microsomal membranes?

- What effect does a detergent like Triton X-100 have on a membrane?
- What was the final concentration of protease K used in the protease digestion experiment?
- What temperature should be used for the mock protease treatment?
- Why was "ice-cold buffer" used to stop the transport assays?

RESULTS

- Ribophorin 1 is a transmembrane glycoprotein that is involved in the transport of newly synthesized proteins across the membrane of the rough endoplasmic reticulum. What conclusion can you draw based on the pattern of ribophorin 1 antibody staining of the three microsomal fractions in Figure 5.1.1a?
- DPPIV and α2,6-ST are marker proteins for the PM and Golgi complex, respectively. How do you interpret the pattern of antibody staining for these two proteins in Figure 5.1.1a?

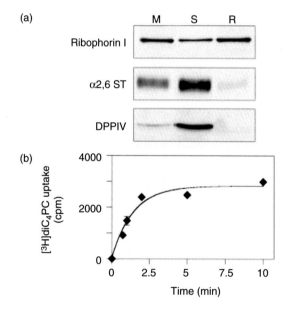

FIGURE 5.1.1 Characterization of a submicrosomal membrane fraction.
Rat liver microsomes (M) were further purified into a smooth ER fraction (S) and a rough ER fraction (R) based on their migration during sucrose gradient centrifugation. a. Each of the microsomal membrane fractions were analyzed by SDS-PAGE and immunoblotting using antibodies to marker proteins for the ER (ribophorin 1), Golgi (α2,6-ST), and the PM (dipeptidylpeptidase IV, DPPIV). b. Rough ER microsomal membranes were treated with high salt and EDTA, washed, and equilibrated to 30°C for 1 min. Transport was initiated by the addition of 5 μL of radioactively labeled [^{3}H]diC$_4$PC. Reactions were stopped at the indicated time and the salt-washed microsomal membranes were separated from free [^{3}H]diC$_4$PC. Transport of [^{3}H]diC$_4$PC into the lumen of the microsomal membranes was measured.

- What is the purpose of pretreating the R microsomes with high salt and EDTA?
- Suggest a reason the transport of $[^3H]diC_4PC$ reaches equilibrium after 3 min (Figure 5.1.1b).
- $[^3H]$mannose was used as a measure of the intravesicular volume of proteoliposomes. The Y-axis values for $[^3H]diC_4PC$ uptake by proteoliposomes is expressed as a "percentage of mannose space." Explain what this means (Figure 5.1.2a).
- Estimate the difference in $[^3H]diC_4PC$ uptake following protease K treatment shown in Figure 5.1.2a. How does your estimate match the data shown in Figure 5.1.2b?
- Propose an explanation for the continued transport of $[^3H]diC_4PC$ after protease K treatment (Figure 5.1.2a).
- Does treatment with protease K alter the integrity of the proteoliposome bilayer (Figure 5.1.2c)? Justify your answer.
- Explain how gradient centrifugation separates proteins using Figure 5.1.3a as an example.
- In which pool of gradient fractions would you find ribophorin I (Figure 5.1.3a)?
- Which of the gradient pools shown in Figure 5.1.3b has the highest concentration of phospholipid? Which has the highest concentration of protein (Figure 5.1.3c)?
- Figure 5.1.3d shows $[^3H]diC_4PC$ transport activity of the individual protein pools reconstituted into proteoliposomes relative to transport in Triton X-100 extracted, but unfractionated proteoliposomes. Summarize the conclusion that is supported by these data.
- Data presented in Figure 5.1.3e supports the conclusion that most $[^3H]diC_4PC$ transport occurs in proteoliposomes reconstituted using proteins from pool B. Explain why the results from Figure 5.1.3d and e are different.
- What additional experimental approaches might have been used to identify the protein or proteins responsible for the transport of $[^3H]diC_4PC$?
- Transport of $[^3H]diC_4PC$ is predicted to be bidirectional. Outline an experimental protocol to test for bidirectional transport using microsomes or proteoliposomes.
- All the microsomal membrane fractions (M, S, and R) were found to be able to transport $[^3H]diC_4PC$. Design an experiment to determine whether flippase proteins are present in the membranes of the Golgi complex and PM.
- The researchers state, "the specific ER proteins implicated by our data are the flippases responsible for the membrane translocation of natural phospholipids." Develop an argument in favor or against this statement.

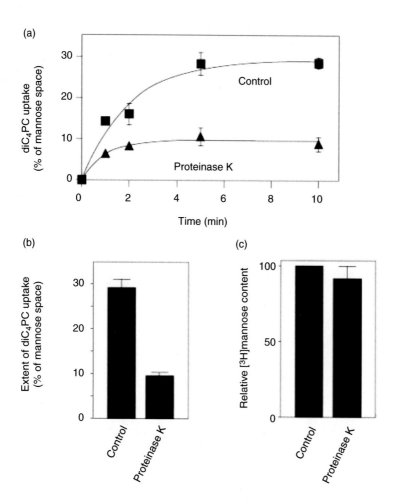

FIGURE 5.1.2 Protease treatment blocks phospholipid transport.
Proteoliposomes composed of lipids and proteins reconstituted from the R microsomal fraction were divided into two aliquots. One aliquot was treated with protease K for 40 min. Protease digestion was stopped by the addition of the protease inhibitor, PMSF. The other aliquot was treated with PMSF alone for 40 min. a. The ability to transport [³H]diC$_4$PC was assayed for both proteoliposomes. [³H]mannose was used as a marker for intravesicular volume and the integrity of the liposome bilayer. b. The extent of [³H]diC$_4$PC transport at equilibrium as a percentage of mannose space. c. Integrity of the liposome membrane as measured by the [³H]mannose content.

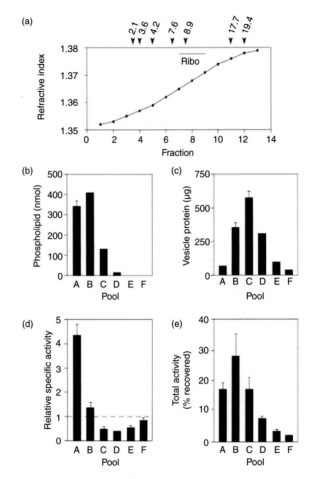

FIGURE 5.1.3 Analysis of phospholipid transport by fractions of detergent-extracted, salt-washed rough microsomal membranes.

Salt-washed rough microsomal membranes were extracted using Triton X-100. Extracts were fractionated by gradient centrifugation. a. Refractive index of gradient fractions from the top of the tube (0) to the bottom of the tube (14). Protein standards used for the gradient included cytochrome c (2.1S), ovalbumin (3.6S), bovine serum albumin (4.2S), alcohol dehydrogenase (7.6S), β-amylase (8.9S), apoferritin (17.7S), and thyroglobin (19.4S). Sedimentation of the microsomal protein ribophorin 1 was at about 11S. Fractions were combined pairwise (starting with fraction 2) to yield six pools representing contents from the top of the gradient (pool A) to the bottom of the gradient (pool F). b. Distribution of Triton X-100 extracted phospholipids in the gradient pools. Phospholipid content was determined by lipid extraction prior to reconstitution of proteoliposomes. c. Distribution of Triton X-100 extracted proteins in the gradient pools. Protein content was determined after reconstitution of proteoliposomes. d. Specific activity of [^{3}H] diC$_4$PC transport in proteoliposomes reconstituted from the indicated gradient pool. Uptake was measured after 12 min incubation at 30°C. Activity is reported relative to the specific activity of proteoliposomes reconstituted from an unfractionated Triton X-100 extract (dashed line). e. Total activity of [^{3}H]diC$_4$PC transport in gradient pools expressed as a percentage of the activity loaded onto the gradient. Total activity recovered in the six pools was 75% of the activity of the sample loaded onto the gradient.

Navigating the Bilayer: Lipid Rafts and Caveolae [2]

INTRODUCTION

A cell's plasma membrane (PM) is more than a homogenous phospholipid bilayer forming a selectively permeable barrier between the intracellular and extracellular environments. Distinct **microdomains** are thought to exist within the plane of the membrane. In some cases, these domains are defined by the presence of specific peripheral membrane proteins, as is the case with clathrin-coated pits. In other examples, microdomains are formed through the clustering of specific lipids such as sphingolipids and cholesterol. Still other microdomains are characterized by a unique combination of lipids and proteins. The segregation of specific types of proteins into distinct microdomains within the PM allows for specialization of function. In order to investigate the properties of the various membrane microdomains it is first necessary to develop a protocol for their isolation. This case study describes a method for the separation of lipid rafts and caveolae from the rest of the PM.

- Create a labeled drawing to illustrate the structure of a biological membrane.
- Research/review the proteins required to form a clathrin-coated pit.
- Conduct a search for images of sphingolipid and cholesterol. Compare the structure of these molecules with that of a generic phospholipid.

BACKGROUND

Lipid rafts are membrane microdomains enriched in cholesterol and sphingolipids. Lipid rafts are also characterized by the presence of glycosylphosphatidylinositol **(GPI)-anchored proteins** and proteins involved in signal transduction. GPI-anchored proteins are peripheral membrane proteins that associate with the bilayer through a covalent attachment to the glycolipid GPI. The 5–25 nm clusters of sphingolipid, cholesterol, and protein that make up a lipid raft are more tightly packed than the phospholipids of the surrounding bilayer, forming distinct microdomains that can freely diffuse in the plane of the membrane. Unlike most of the phospholipid bilayer, lipid raft microdomains are resistant to solubilization by detergents. Lipid rafts are thought to be dynamic, with the ability to rapidly assemble and disassemble.

Caveolae are 50–100 nm "cup-shaped" invaginations of the PM associated with endocytosis, cell signaling, and the entry of pathogens into the cell. Caveolae share the high-cholesterol and sphingolipid concentrations found in lipid rafts, but are distinct from rafts due to the presence of the protein **caveolin-1**. Caveolae are more stable than a lipid raft.

- Conduct a search for images of a lipid raft and caveolae. Identify the similarities and differences in the structure of these two membrane domains.
- Discuss why a caveolae can bend inward while the lipid raft is flat.

METHODS

Plasma membrane purification

Mouse tissue culture 3T3 cells were grown to confluency, washed with buffer, scraped into an isotonic buffer containing protease inhibitors, and homogenized. The cell lysate was centrifuged at 10,000g for 45 min at 4°C and the membrane pellet was resuspended into the same isotonic buffer.

Sucrose density gradient centrifugation

The PM pellet was placed at the bottom of an ultracentrifuge tube and overlaid with 4 mL of 35% sucrose and 4 mL of 5% sucrose in buffer. The sample was centrifuged at 180,000g for 20 h at 4°C in a swinging bucket rotor. The PM fraction accumulates at the interface between the 5% and 35% sucrose layers. Fractions were collected from the top of the gradient.

A modified protocol for the sucrose density gradient centrifugation overlaid the PM sample with 3 mL of 35% sucrose, 4 mL of 21% sucrose, and 1 mL of 5% sucrose. Centrifugation conditions remained the same. Membrane fractions were collected from the interface between the 5% and 21% sucrose steps and between the 21% and 35% sucrose steps.

Extraction of membrane lipid domains

A nondetergent, sodium carbonate solution was used to extract lipid microdomains. Purified PM was incubated with 2 mL of 0.5 M Na_2CO_3 (pH 11.0), sonicated, and mixed with an equal volume of 90% sucrose in buffer containing 0.3 M NaCl. The extract solution was placed at the bottom of an ultracentrifuge tube, overlaid with sucrose, and centrifuged as described before.

Detergent extraction of membrane lipid domains was achieved using 2 mL of ice-cold 0.5% Triton X-100 in buffer. The extract solution was mixed with an equal volume of 80% sucrose and centrifuged as described earlier.

Lipid analysis

Cholesterol was extracted using 1.5 volume chloroform and vortexed for 4 h at 20°C. After centrifugation, the organic phase was collected. The chloroform was evaporated and the specific amount of cholesterol measured using an enzyme-based assay. Glycosphingolipids were analyzed using a dot-blot assay with HRP-conjugated cholera toxin β subunit.

To determine the amount of other lipids, the membranes were diluted in buffer to reduce the concentration of sucrose, pelleted by centrifugation at 150,000g for 4 h. The membrane pellet was extracted with methanol and chloroform and analyzed by thin-layer chromatography. The composition of the lipid microdomains was determined based on the migration pattern of the lipids relative to a set of standards.

Electron microscopy

Membrane samples were diluted to reduce the concentration of sucrose and then pelleted by centrifugation at 150,000g for 4 h. The membrane pellets were fixed in 2.5% glutaraldehyde in buffer, postfixed with 1% osmium tetroxide, and then dehydrated and embedded in epoxy. Samples were sectioned and examined using a transmission electron microscope.

- Define "confluency" using your own words.
- Sketch a diagram to illustrate what occurs during the process of sucrose density gradient centrifugation. Suggest a reason why the membrane fractions move "up" during centrifugation.
- How does a detergent work to extract a membrane?

RESULTS

- Create a schematic diagram of a eukaryotic cell that includes all the membrane compartments and proteins mentioned in the figure legend for Figure 5.2.1a.
- Research/review the concept of buoyant density. Apply your understanding of this concept to explain how the modification illustrated in Figure 5.2.1b would affect the membranes in the PM fraction.
- Which marker proteins are found in the PM fraction from the conventional sucrose gradient centrifugation protocol shown in Figure 5.2.1c? What can you conclude based on their presence?
- How did the use of the modified sucrose gradient alter the pattern of marker protein distribution?
- Use Figure 5.2.1c to characterize which fraction, LLD or HLD, represents the caveolae lipid microdomain and which represents the lipid raft microdomain.
- Refer to the diagram of the conventional gradient in Figure 5.2.1b. Connect the pattern of separation generated by the use of this gradient to the fractionation pattern shown in Figure 5.2.2. Estimate which fraction numbers correspond to the 5:35 sucrose interface.

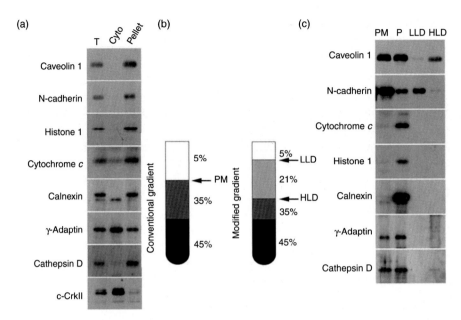

FIGURE 5.2.1 Purification of the plasma membrane (PM) and lipid microdomains from mouse 3T3 cells.

a. Whole cell lysate (T) was centrifuged at 10,000g to produce a supernatant (Cyto) and pellet. Protein markers for the following membrane compartments were detected by immunoblotting. Caveolin-1 and N-cadherin are markers for the PM. Histone H1 is a marker for the nucleus. Cytochrome c is a marker for mitochondria. Calnexin and γ-adaptin are markers for the ER and Golgi complex, respectively. Cathepsin D is a lysosomal marker and c-CrkII is a soluble protein involved in signaling. b. Schematic diagram of conventional and modified sucrose gradients used for purification of the PM and lipid microdomains. Percentages indicate sucrose concentrations (w/v). PMs collected at the 5:35 sucrose interface in the conventional gradient. LLD and HLD indicate the positions in the gradient where noncaveolin (LLD) and caveolin (HLD) membrane fractions collected. c. Conventional sucrose gradient centrifugation of the pellet from (a) resulted in a PM fraction and a membrane pellet (P). Use of the modified sucrose gradient centrifugation protocol resolved two membrane fractions (LLD and HLD) from the original PM fraction. Marker proteins as described were detected by immunoblotting.

- Which method of lipid microdomain solubilization is most effective? Justify your answer.
- Describe the pattern of distribution of cholesterol and protein in the two different centrifugation protocols shown in Figure 5.2.3.
- What is the relationship between Figure 5.2.3a and Figure 5.2.2?
- Approximately what percent of a cell's membrane protein is located in lipid rafts?
- Look closely at the two graphs in Figure 5.2.3. What do you observe? How does this influence your ability to analyze the data?

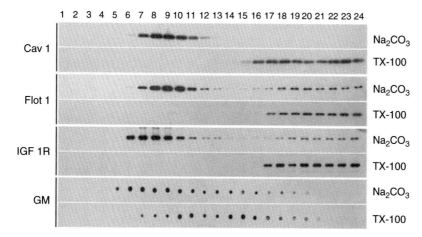

FIGURE 5.2.2 Extraction of lipid microdomains.

PM containing 80 μg of protein was extracted using either a nondetergent (Na_2CO_3) or detergent (Tx-100) based protocol. The PM extract was separated using conventional sucrose density gradient centrifugation. Samples were collected from the top of the gradient (fraction 1) to the bottom of the gradient (fraction 24). The distribution of the caveolar marker proteins Cav 1, Flot 1, and IGF 1R throughout the gradient was examined by immunoblotting. The distribution of the glycosphingolipid GM was determined by binding the HRP-labeled β subunit of cholera toxin.

- Each of the graphs in Figure 5.2.3 is "from one representative experiment." Discuss how this statement affects your interpretation of the results.
- Use the graph in Figure 5.2.3b to identify which fractions in Figure 5.2.3c correspond to the LLD and HLD peak fractions.
- Which of the results presented in Figure 5.2.3c are consistent with our understanding of the structure of lipid rafts and caveolae? Which results are unexpected? Explain you answer.
- Characterize the appearance of the membranes from the LLD and HLD fractions shown in Figure 5.2.3d. Propose an explanation for the differences.
- Does the lipid composition of the membranes from the LLD and HLD fractions in Figure 5.2.4a meet the expected criteria for lipid rafts and caveolae? Explain your answer.
- Which of the lipid microdomains contains the highest concentration of cholesterol and glycosphingolipids based on the data in Figure 5.2.4?
- Predict what the graph in Figure 5.2.4c would look like if the amounts of cholesterol and glycosphingolipids were the same in lipid rafts and caveolae.
- The existence of lipid rafts *in vivo* has been a point of controversy in the cell biology community. How do these experiments contribute to the discussion?

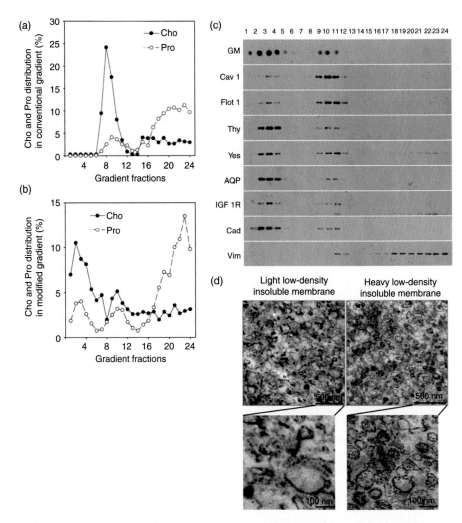

FIGURE 5.2.3 Distribution of cholesterol and protein following sucrose density gradient centrifugation.

a. Fractions from the conventional three-step sucrose density gradient centrifugation of Na_2CO_3 PM extracts (200 μg protein). The amount of cholesterol (Cho) and protein (Pro) in each fraction was determined. The results are from one representative experiment. b. Fractions from modified sucrose density gradient centrifugation of Na_2CO_3 PM extracts. c. Analysis of lipid microdomain-associated proteins and glycosphingolipids in fractions from the modified sucrose density gradient centrifugation protocol. GM glycosphingolipids are the major lipids in lipid microdomains. Marker proteins include the caveolin-specific proteins Cav 1 and Flot 1, the GPI-anchored protein Thy-1 (Thy), and the lipid-tethered protein c-Yes (Yes). Aquaporin-1 (AQP) and IGF-1 (IGF 1R) are transmembrane proteins. N-cadherin (Cad) is a glycoprotein involved in cell adhesion and vimentin (Vim) is an intermediate filament cytoskeletal protein. d. Electron micrographs of the LLD and HLD lipid microdomain fractions.

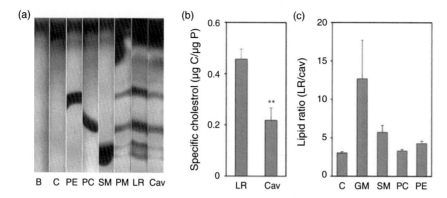

FIGURE 5.2.4 Analysis of the lipids in the LLD and HLD fractions.

a. Thin-layer chromatograph of lipids extracted from the PM LLD lipid microdomain (LR) or HLD lipid microdomain (Cav). Cholesterol (C), phosphatidylethanolamine (PE), phosphatidylcholine (PC), and sphingomyelin (SM) were used as standards. A solvent blank (B) was also included. The samples were normalized to the equivalent of 20 μg membrane protein. b. Measurement of the specific cholesterol present in LLD (LR) and HLD (Cav) fractions. The results are the average of six experiments (**$P < 0.001$). c. The ratio of total lipids in LLD (LR) and HLD (Cav) lipid microdomains.

Shifting Gears: Calcium Transport in Flagella [3]

INTRODUCTION

Movement of ions across a phospholipid bilayer requires the presence of transport proteins. Different types of transport proteins are specific to different ions. While some transport proteins form an open channel for the passive diffusion of ions, others only allow transport to occur under certain circumstances. These **gated channels** undergo a conformation change, transitioning from a "closed" to an "open" state, and back again. Examples of the physiological triggers for gated channels include ligand binding, mechanical deformation, and changes in voltage across the membrane.

Voltage-gated channels are proteins that can respond to small changes in **membrane potential** or the distribution of charge across a phospholipid bilayer. Voltage-gated channels play a vital role in the process of nerve cell communication through their involvement in production of an action potential. In order for a voltage-gated protein to function it must be capable of "sensing" differences in charge across a membrane. Voltage-gated channels are integral membrane proteins with multiple, alpha helical, transmembrane domains. Some of these domains contribute to the formation of a channel that selectively allows for the diffusion of an ion. Other transmembrane domains are able to respond to changes in electrical charge due to the presence of acidic or basic amino acids.

- Explain why an ion must move through a transport protein and cannot simply diffuse through a phospholipid bilayer.
- What determines the direction of movement of an ion?
- Research/review examples of the three types of gated channels mentioned earlier.
- Describe the properties associated with a transmembrane domain.
- Predict how a transmembrane domain of a voltage-gated channel with arginine amino acids near the extracellular side of the membrane would respond to an increase in calcium ions outside the cell.

BACKGROUND

The unicellular green algae *Chlamydomonas reinhardtii* swim through their aqueous environment using flagella. The coordinated movement of these two flagella, in a breaststroke-like pattern, propels the cell forward. *Chlamydomonas* are photosynthetic and so are attracted to light. However, exposure to intense light causes the flagella to change their beat pattern, switching to a more undulating or S-shaped pattern, and causing the cell to reverse direction, a behavior known as **photophobic response**. Alteration of the flagellar waveform requires an influx of extracellular calcium ions into the flagella.

The *Chlamydomonas* mutant, **ppr2**, lacks a photophobic response. The *ppr2* gene shares homology with voltage-gated calcium channels. A search of the *Chlamydomonas* genomic database revealed the presence of eight voltage-gated calcium channels genes that share sequence homology with *ppr2*. The experiments described in this case study investigate the role of voltage-gated calcium channels in the photophobic response of *Chlamydomonas*.

- Are *Chlamydomonas* prokaryotic or eukaryotic cells?
- Speculate on why a photosynthetic cell would avoid intense light.
- What proteins are responsible for the movement of the flagellar axoneme?
- The *Chlamydomonas* eyespot responds to bursts of intense light by altering the cell's membrane potential. How does this fact contribute to your understanding of the photophobic response?

METHODS

Cell cultures
Wild-type and mutant strain *ppr2* of *C. reinhardtii* were used in these experiments. Cells were cultured at 25°C in TAP medium using a 12-h light, 12-h dark light cycle.

Reverse transcriptase – PCR of regenerating flagella
Flagella were detached from the cells using pH shock, and total RNA was extracted from flagellar-regenerating cells at the times indicated. Equal amounts of total RNA were reverse-transcribed into cDNA using Superscript III Reverse Transcriptase. PCR was performed on each of the samples using primers that were specific to a 120–200 base pair fragment of fructose-bisphosphate aldolase 3, beta tubulin, or CAV2. Expression of fructose-bisphosphate aldolase 3 was assumed to be unaffected by flagellar regrowth. Calculation of the amounts of beta tubulin and CAV2 RNA were normalized to fructose-bisphosphate aldolase 3.

RNA interference suppression of CAV2 expression
CAV2 gene expression was knocked down using RNA interference. Genomic DNA sequences corresponding to exons 1–3 of the CAV2 gene was inserted into an RNAi vector and wild-type *Chlamydomonas* were transformed by electroporation. Transformants were evaluated by the level of CAV2 expression and their ability to respond to high-intensity light.

CAV2 antibodies

Polyclonal antibodies specific to CAV2 were developed. For immunoblotting, cells in the logarithmic growth phase were pH-shocked to remove their flagella. Flagella were collected by centrifugation and analyzed by SDS-PAGE followed by blotting onto a nitrocellulose membrane. Membranes were reacted with dilute primary antibody to CAV2, washed, and reacted with an HRP-conjugated secondary antibody. Antibody binding was detected by chemiluminescence.

Localization of CAV2 and alpha tubulin in flagella was achieved by immunofluorescence labeling. Cells were fixed in 1.8% formaldehyde for 20 min and washed with buffer. *Chlamydomonas* were allowed to adhere to the glass slide and then the cells were permeabilized with 0.2% Triton X-100, washed with buffer, and reacted with the CAV2 primary antibody. The samples was washed again and reacted with a FITC-conjugated secondary antibody. Labeled cells were viewed using a fluorescence microscope.

- Why was it necessary to grow *Chlamydomonas* with a light/dark cycle?
- Predict how the RT-PCR experimental results would be altered if fructose-bisphosphate aldolase expression did change during flagellar regeneration.
- Why was a permeabilization step included for the immunofluorescence experiments? How would you change the methods if your goal were to demonstrate that CAV2 is located in the flagellar PM?

RESULTS

- Which of the genes for voltage-gated calcium channels in *Chlamydomonas* is not expressed in the *ppr2* mutant.
- Research/review how RNA interference can suppress expression of a protein.
- Which of the *Chlamydomonas* strains shown in Figure 5.3.1b would you use if you were planning on testing for photophobic response? Explain your answer.
- Suggest a reason why the researchers selected alpha tubulin as the control in Figure 5.3.1b.
- Summarize the significance of the result shown in Figure 5.3.1c?
- Calculate the number of CAV2 knockdown cells that exhibited a photophobic response using the information provided in the figure legend for Figure 5.3.1c.
- Propose an explanation for why the photophobic response in the CAV2 knockdown cells is not 0 (Figure 5.3.1).
- Describe how the pattern of *Chlamydomonas* swimming is different between the control and RNAi treated cells.

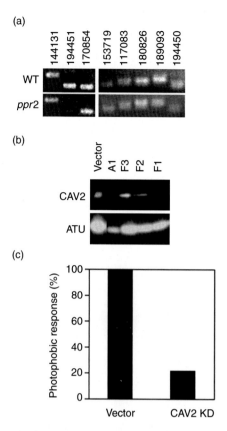

FIGURE 5.3.1 Characterization of the *ppr2* mutant as a voltage-gated calcium channel.
a. RT-PCR of mRNAs for *Chlamydomonas* voltage-gated channels. All mRNAs are present in the wild-type
(WT) cell, but not in the *ppr2* mutant. b. Results from (a) indicated that the *ppr2* mutation affected the
gene for the voltage-gated calcium channel *CAV2*. Expression of *CAV2* was suppressed in WT cells using
RNAi. The presence of CAV2 protein in various *Chlamydomonas* strains was detected by immunoblotting
with an anti-CAV2 antibody. Reaction of the blot to an anti-alpha tubulin antibody is shown as a control
for the amount of protein loaded on the blot. c. Photophobic response in the control (vector; $n = 35$) and
RNAi knockdown (CAV2 knockdown; $n = 64$) *Chlamydomonas* strains.

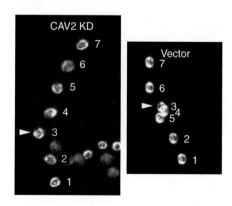

FIGURE 5.3.2 Photophobic response of WT and CAV2 knockdown cells.
The swimming tracks of a control (vector) cell and an RNAi-treated CAV2 knockdown cell following
photostimulation (arrowhead). The interval between consecutive images is 0.2 s.

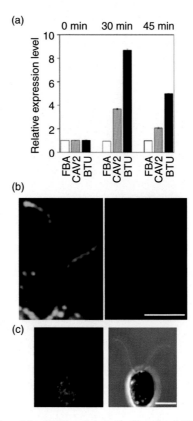

FIGURE 5.3.3 Localization of the CAV2 protein in the flagella of *Chlamydomonas*.
a. RNA was collected for quantitative RT-PCR from WT *Chlamydomonas* cells before (0 min) or after (30 or 45 min) removal of the flagella. *Chlamydomonas* are capable of regrowing flagella following their removal. FBA and beta tubulin (BTU) were used as loading control and positive control, respectively. Expression levels relative to that at 0 min are shown. b. Double-labeled immunofluorescence staining of a WT cell with antibodies to CAV2 (green) and alpha tubulin (red). c. Immunofluorescence staining of a *ppr2* cell with the CAV2 antibody (left panel). A phase contrast image is shown in the right panel.

- About how long does it take the control cell to recover from photostimulation?
- Develop an argument for or against the statement, "knockdown of the CAV2 voltage-gated calcium channel enhanced the swimming speed of *Chlamydomonas*."
- Explain why the results shown in Figure 5.3.3a support the conclusion that CAV2 is a flagella protein.
- What is the function of fructose-bisphosphate aldolase (FBA)? Why does the level of expression for this protein remain the same at each time point?
- Where is the CAV2 protein located in the wild-type cell in Figure 5.3.3b?

- Why is there no staining of the flagella in the *ppr2* cell in Figure 5.3.3c?
- CAV2 antibody staining of the *Chlamydomonas* cell body was dismissed as an artifact. What evidence in Figure 5.3.3 can be used to support this conclusion?
- How does the location of CAV2 protein in Figure 5.3.3b relate to its proposed role in the photophobic response in *Chlamydomonas*?
- Look back at Figure 5.3.1a. Each of the genes amplified in the wild-type cell shared homology with a voltage-gated calcium channel. Design an experiment to investigate whether CAV2 is the only voltage-gated calcium channel required for the photophobic response.

References

[1] Menon AK, Watkins III WE, Hrafnsdóttir S. Specific proteins are required to translocate phosphatidylcholine bidirectionally across the endoplasmic reticulum. Curr Biol 2000;10:241–52.

[2] Yao Y, Hong S, Zhou H, Yuan T, Zeng R, Liao K. The differential protein and lipid compositions of noncaveolar lipid microdomains and caveolae. Cell Res 2009;19:497–506.

[3] Fujiu K, Nakayama Y, Yanagisawa A, Sokabe M, Yoshimura K. *Chlamydomonas* CAV2 encodes a voltage-dependent calcium channel required for the flagellar waveform conversion. Curr Biol 2009;19:133–9.

CHAPTER 6

Cytoskeleton and Intracellular Motility

Plakins: Keeping the Cytoskeleton Safe [1]

INTRODUCTION

The **cytoskeleton** of a cell is made up of three types of protein filaments: **actin microfilaments (MF)**, **intermediate filaments (IF)**, and **microtubules (MT)**. Together, these structural proteins help maintain cell shape and integrity, organize and distribute organelles within the cytoplasm, generate force for cell movement, and control the accurate partitioning of DNA, and cytoplasmic contents during cell division. Although research into the cytoskeleton often focuses on a single filament system, presumably all three filaments function simultaneously in a living cell. To what extent do MFs, IFs, and MTs interact in a cell? What happens when these interactions are disrupted?

■ Research/review the differences in the proteins that make up the three types of cytoskeletal filament.
■ How does the assembly of cytoskeletal filaments from actin or tubulin differ from the assembly of IF proteins?
■ Describe a cellular function that is associated with each of the cytoskeletal filaments.
■ Describe how all three filaments contribute to the process of mitotic cell division.

127

BACKGROUND

Plakins are a class of large cytoskeletal-associated proteins that were originally identified as IF-binding proteins. The carboxy terminal domain of plakins has a conserved IF-binding domain, but the amino terminal end is more variable. Various isoforms of plakin proteins are expressed in different cell types. In fibroblast cells, **plectin** proteins crosslink IFs, actin, and microtubules. In neurons, the protein product of the gene *bullous pemphigold antigen 1* (*BPAG1*) possesses binding domains for actin and IFs, and may play a role in anchoring the IF cytoskeleton to membrane-associated actin. Deletion of the *BPAG1* gene in nerve cells results in the disruption of the IF cytoskeleton and the **neurofilaments** (**NF**), causing swelling in axons and eventual death of the organism. Genetic deletion of the NF cytoskeleton does not rescue the *BPAG1* phenotype. What else is BPAG1 doing for the nerve cell?

- Explain what determines the binding properties of a protein.
- Research/review the role of axons in nerve cell communication.
- What is a neurite? Describe the relationship between a neurite, an axon, and the cytoskeleton?
- Predict whether the organization of the cytoskeleton in a fibroblast would be the same as in a neuron. Explain your answer.
- Conduct a search to learn more about bullous pemphigold, an autoimmune disease that affects humans and other mammals.

METHODS

Mice

Single- and double-mutant mouse strains were maintained according to standards for the care of vertebrate animals.

Transfections and immunofluorescence

NIF 3T3 fibroblasts and adrenal carcinoma, IF minus, SW13 cells were maintained in tissue culture media and transfected with a construct of the neuronal BPAG M1 domain tagged with a carboxy terminal FLAG sequence. The FLAG sequence was detected using an anti-FLAG antibody.

Tissue culture or primary culture cells were fixed, permeabilized, and incubated with various combinations of antibodies as indicated. Antibodies used in this work include antitubulin, antiactin, antivimentin, antineuronal BPAG1 head domain, and antiBPAG specific to the coil–coiled tail domain of multiple BPAG isoforms.

Primary culturing of mouse neurons

Dorsal root ganglia were collected from newborn mice and cultured on coverslips in tissue culture media supplemented with 15% fetal calf serum and nerve growth factor for 5 days. Coverslips were rinsed with buffer, and neurite outgrowths were lysed, extracted using 1% Triton X-100 in cytoskeleton-stabilizing buffer, and fixed in 1% glutaraldehyde.

In vitro microtubule-binding assays

The nBPAG1 M1 domain was expressed in bacteria and purified. Purified bovine brain microtubule protein or rabbit muscle actin was polymerized. Purified M1 domain (0.5–10 μM) was added to the polymerized MTs or actin MFs for 20–30 min, followed by centrifugation at 10,000g for 30 min at 23°C, to pellet the cytoskeletal filaments and bound protein. For depolymerization assays, MTs with or without bound M1 domains, were placed on ice or treated with 2.5 μg/mL colchicine for the indicated times. Samples were taken at each time point, centrifuged, and supernatant and pellet fractions were prepared for SDS-PAGE.

- Tissue culture cells such as mouse 3T3 and human SW13 are typically derived from cancerous cells. Does the origin of the cell make a difference in your interpretation of the data?
- How do primary cultured cells differ from tissue-cultured cells?
- Describe the control experiment(s) to demonstrate that M1 domain binding is specific to MTs.

RESULTS

- Which of the images in Figure 6.1.1a/á – provides you with a lower magnification view of the axon?
- Describe the difference you observe in the cytoskeleton of the wild-type and *BPAG1* null axons in Figure 6.1.1a/á and b/b'.
- Compare the axonal cytoskeleton of the *NF-L* null mouse and the *BPAG1/NF-L* double mutant in Figure 6.1.1c and d. How do these images compare with those of growing neurites in Figure 6.1.1e and f?
- What other cellular components are present in the axons in Figure 6.1.1?
- Develop an argument for or against the following statement regarding the results in Figure 6.1.2, "in 3T3 fibroblasts, >90% of the MT network was decorated by M1-anti-FLAG labeling. Contrastingly, anti-FLAG colabeling was not seen with either IF or actin cytoskeletons."
- Why are four cells visible in the right-hand panel of Figure 6.1.2d and only two in the left-hand panel?
- MTs depolymerize in the cold, and SW13 cells lack an IF cytoskeleton. Interpret the experiment shown in Figure 6.1.2d in view of these facts.
- Summarize the conclusion supported by the data from the assays shown in Figure 6.1.3a and b.
- What evidence in Figure 6.1.3c supports the assumption that cold and colchicine treatments cause MTs to depolymerize?

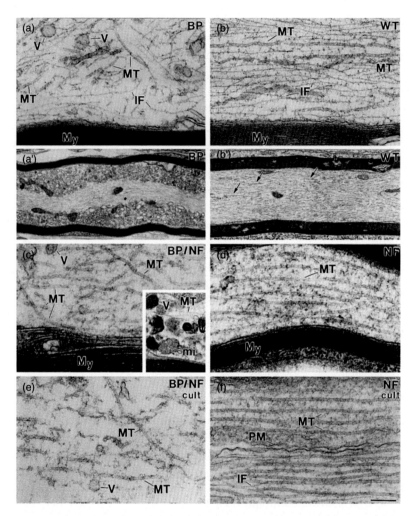

FIGURE 6.1.1 Disruption of the axonal MT cytoskeleton is not rescued by removal of the neurofilament cytoskeleton.

Myelinated axons (a–d) or neurites grown from cultured neurons (e and f) from wild-type *BPAG1* (WT), *BPAG1* null mutants (BP), *NF-L* NF null mutants (NF), or *BPAG1/NF-L* double-null mutants (BP/NF). Abbreviations: mi, mitochondria; V, vesicle; PM, plasma membrane; My, myelin sheath; IF, intermediate filament. Arrows in (b′) indicate microtubules. Scale bar in (f) represents 200 nm in images (a–f), 1 μm in images (a′) and (b′), and 500 nm in the inset image in (c).

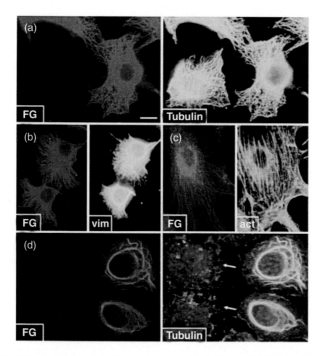

FIGURE 6.1.2 The head domain of the neuron-specific isoform of BPAG1 has the capacity to bind to MTs.
Mouse 3T3 tissue culture cells (a–c) transfected with the M1 domain from the head region of neuron-specific BPAG1 (nBPAG1). The M1 fragment was tagged with FLAG and visualized using an anti-FLAG fluorescent antibody. Cells were double-labeled using antibodies specific to the indicated cytoskeletal filament (vim represents vimentin, an IF protein). d. transfected, IF-minus SW13 cells treated at −20°C for 10 min to initiate MT depolymerization, then fixed and double-stained for M1 and MTs. Scale bar represents 15 μm in (a), 25 μm in (b) and (c), and 12 μm in (d).

- Develop an argument to support the statement, "binding the M1 domain of nBPAG1 protects MTs from depolymerization."
- What is the genotype of the cell in Figure 6.1.4a? How did the genotype of this cell affect its ability to grow neurites?
- What length of incubation time is required to depolymerize MTs using colchicine (Figure 6.1.4c–h)?
- Explain the significance of Figure 6.1.4i.
- What effect does the *BPAG1/NF-l* double mutation have on the stability of MTs exposed to cold or colchicine?
- Which of the panels in Figure 6.1.4 support the conclusion that MTs are required for binding the neuronal nBPAG1 isoform? Explain your answer.
- Develop a hypothesis to account for the pattern of fluorescence in Figure 6.1.4d. How could you test your hypothesis?

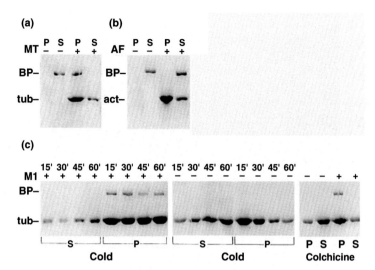

FIGURE 6.1.3 The M1 domain of nBPAG1 binds directly to MTs *in vitro.*
a. MT-binding assays. Assays were carried out using purified M1 domain of nBPAG1 and MTs. Supernatant (S) and pellet (P) fractions were analyzed by SDS-PAGE Coomassie Blue–stained gels. Each assay was performed with (+) or without (−) polymerized MTs. b. F actin–binding assay. Actin was polymerized (AF) and incubated with the M1 domain of nBPAG1 as described previously. c. Binding of the M1 domain increases resistance of MTs to depolymerization by cold or colchicine. MTs ± M1 were placed on ice or treated with 2.5 μg/mL colchicine at 37°C for the times indicated. Following incubation, samples were centrifuged and supernatant (S) and pellets (P) were analyzed by SDS-PAGE as described previously.

- Discuss how the formation of crosslinks between different cytoskeletal elements might benefit a cell.
- Explain how different isoforms of the BPAG protein could be expressed in different cell types in a single organism.
- Refer back to what you learned about the disease, bullous pemphigold. Connect your understanding of the role played by BPAG1 protein to symptoms associated with the loss of the gene for this protein.

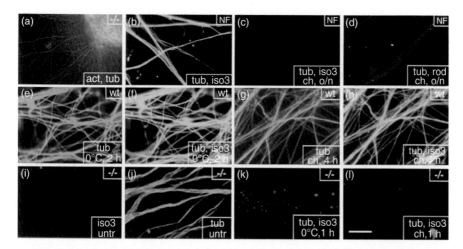

FIGURE 6.1.4 nBPAG1 stability is dependent upon MTs, and MT stability is dependent upon nBPAG1 in cultured neurons.

Primary neuronal cultures were made from WT, *BPAG1* null (BP), *NF-L* null, and *BPAG1/NF-l* double-null (−/−) mutant newborn mice. In each of the assays, *NF-L* null and WT cultures behaved similarly, as did *BP* null and *BP/NF* double-null cultures. The genotype of each culture is indicated in the upper right corner. Where indicated in the lower right corner, some cultures were treated with 2.5 μg/mL colchicine (ch) at 37°C or incubated at 0°C for the times indicated prior to fixation (o/n, overnight). Fixed cells were permeabilized, washed, and then incubated with antibodies specific to β tubulin (tub; red in panel A), α tubulin (tub; green), the actin stain, phalloidin (act; blue), nBPAG1 head domain (iso3; red), or the tail domain of all BPAG isoforms (rod; red). Colocalization of red and green fluorescence is indicated by the color yellow. Scale bar in A represents 50 μm and in all others, it is 10 μm.

The *Moving* Story of a Microtubule Motor Protein [2]

INTRODUCTION

The cytoplasm of a eukaryotic cell is not a static place. Vesicles are formed from one membrane compartment and fuse with another. Membrane tubules of the endoplasmic reticulum elongate and retreat. Mitochondria and chloroplasts travel from one end of the cell to the other. Organelles travel up and down the axon of a nerve cell, a distance that can reach a meter or more! All of these are examples of **intracellular motility**.

Organelles move in association with a cell's cytoskeleton. Organelle transport is an energy-dependent process that requires a special type of mechanoenzyme or **motor protein** to convert the energy stored in a molecule of ATP into the force required to move an organelle. Different cytoskeletal filaments interact with different motors. Classic examples of cytoskeleton–motor protein interactions have been characterized for large-scale cellular movements, such as the role of actin and myosin in muscle contraction, and microtubules (MTs) and dynein in flagellar motility. Do these same motors function in the movement of organelles within the cell? The discovery of a new MT motor protein is the topic of this case study.

- Describe a cellular process that involves vesicle transport.
- Conduct a search for videos showing intracellular motility in action.
- Compare and contrast the cytoskeletal filaments; *actin* and *MTs*.
- Connect the activity of a cytoskeletal motor protein with the properties of an enzyme.

BACKGROUND

Fast axonal transport describes the bidirectional movement of organelles along the length of an axon. The squid *Loligo pealeii* possesses a "giant" (1 mm) axon that contributes to its ability to escape predators by rapidly expelling a jet of water out its funnel, propelling the squid away from danger. The cytoplasmic contents or **axoplasm** of a giant axon can be **extruded** from the surrounding plasma membrane, much like squeezing toothpaste from a toothpaste tube. Under proper buffer conditions, fast axonal transport can be observed to continue in the isolated axoplasm. Research on the properties of cytoskeletal filaments and organelles in axoplasm revealed that

ATP-dependent organelle movement was occurring along single MTs. Axoplasmic intracellular motility became a model system for investigation of the mechanics behind organelle transport.

Squid axoplasm or optic lobes, the equivalent of a squid's brain, could be homogenized allowing for the isolation of MTs, organelles, and soluble factors (proteins). Supernatant from the low-speed centrifugation of homogenized axoplasm (**S1**) contains organelles that undergo movement along MTs in the presence of ATP. Higher speed centrifugation pellets the endogenous organelles and MTs, leaving behind a supernatant (**S2**) consisting of soluble proteins. The S2 supernatant can be used to reconstitute *in vitro* motility when combined with microtubules, organelles, and ATP.

- What contents, other than MTs and organelles, would you expect to find in the axoplasm of the squid giant axon?
- Suggest a reason why brain tissue is considered to be an excellent source for large quantities of MTs and MT proteins.
- Name some of the advantages of an *in vitro* assay system.

METHODS

Preparation of dissociated axoplasm

The giant axon was dissected from the squid *L. pealeii* in running seawater. Axons were stored in liquid nitrogen. To make a dissociated axoplasm preparation, an axon was thawed and the axoplasm extruded onto a glass coverslip. 20 µL of motility buffer containing 2 mM ATP was added and a second glass coverslip was placed on top of the sample and light pressure was applied.

Preparation of vesicles and supernatant from axoplasm

Axoplasm from 6–8 axons was placed in 50 µL of cold motility buffer supplemented with a protease inhibitor cocktail and homogenized by pipetting multiple times, using a 200 µL pipette. The homogenate was centrifuged at 10,000g for 4 min at 4°C, and the supernatant was carefully collected. The pellet was resuspended in 60 µL of the same homogenization buffer and centrifuged as before. Supernatants were combined to form the S1 supernatant.

The S1 supernatant was treated with 20 µM taxol and 1 mM GTP for 25 min at 23°C to induce MT polymerizastion. The taxol-treated S1 supernatant was then layered onto sucrose gradient and centrifuged at 150,000g for 60 min at 4°C using a swinging bucket rotor. The top 50 µL of the gradient, containing the S2 high-speed supernatant, was collected with a 200 µL pipette and kept on ice. Vesicles gathered at the 35% sucrose interface and were collected by puncturing the side of the tube. The MT fraction was discarded.

Preparation of squid optic lobe microtubules

Optic lobes were dissected from 8–10 squid and placed in a buffer containing a protease inhibitor cocktail and homogenized using a Dounce homogenizer. The homogenate was centrifuged at 30,000g for 30 min at 4°C. The supernatant was collected and centrifuged for 140,000g for 90 min at 4°C. The final supernatant was collected and mixed with taxol and GTP to a final concentration of 20 μM and 1 mM, respectively. The mixture was incubated for 20 min at 23°C to promote MT polymerization followed by centrifugation through a 15% sucrose cushion at 27,000g for 1 h. The pellet was gently resuspended in buffer containing taxol and GTP. To remove MT-associated proteins (MAPs) 1 mL of the taxol pellet was incubated with 0.5 mL of a 3 M NaCl solution in buffer containing taxol and GTP for 15 min at 23°C. Salt-extracted MTs were then pelleted by centrifugation at 35,000g for 30 min at 4°C. The pellet was washed with buffer and resuspended as described previously.

In vitro motility assay

Salt-washed squid optic lobe MTs (1 μL from a 500 μg/mL stock) were added to a glass coverslip followed by the addition of 3 μL of the S2 supernatant. If MT motility was being assayed, then a second coverslip was placed on top of the sample and motility was analyzed with video microscopy.

For assaying organelle motility, 4 μL of the organelle fraction was added to 1 μL of salt-washed MTs in the presence or absence of the S2 supernatant. Organelle motility was measured using video microscopy. The influence of various inhibitors was tested by adding 2 μL of a concentrated stock solution to the organelle mix.

Movement of carboxylated latex beads was assayed using coverslips that were treated overnight with poly-D-lysine, to adhere the MTs to the coverslip surface. Beads were diluted in motility buffer and mixed with S2 supernatant in a ratio of one part beads to four parts S2 and incubated for 5 min on ice. MTs and S2-treated beads were added to the poly-D-lysine–coated coverslips and observed using video microscopy.

Electron microscopy

Negative stain images of MTs and vesicles were made by applying one drop of MT, supernatant, or organelle fractions onto a formvar-coated copper grid for 30 s. The grid was washed with several drops of motility buffer and placed on a drop of 1% uranyl acetate for 30 s, then dried with filter paper. Specimens were examined using a transmission electron microscope.

- Speculate on the reason for including a mixture (cocktail) of protease inhibitors in the motility buffer.
- Create a flow chart that outlines the steps required to obtain the S2 supernatant used in these experiments.

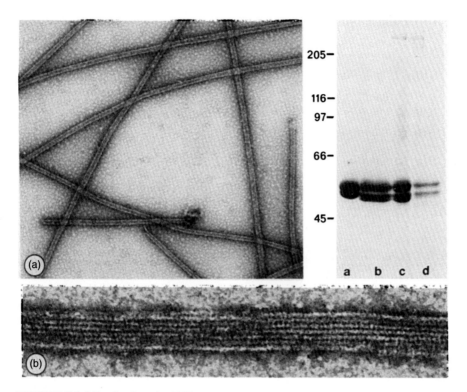

FIGURE 6.2.1 Taxol-polymerized MTs.
a. MTs were extracted with 1 M NaCl to remove any MT-associated proteins (left-hand panel). Magnification: 110,000×. MTs were polymerized using the compound taxol and analyzed by SDS-PAGE. MT proteins from purified bovine brain tubulin (lane a), squid optic lobe MTs after high salt extraction (lane b), squid optic lobe MTs before high salt extraction (lane c), and proteins released from MTs following high salt extraction (lane d). Molecular weight standards are indicated. b. High-magnification view of MTs as prepared in (1). Magnification: 700,000×.

- What was the final concentration of NaCl used to remove MT-associated proteins (MAPs) from the surface of squid optic lobe MTs?
- Characterize the nature of the MT–MT-associated protein interactions given that they can be disrupted by high salt concentration.
- What concentration of MT protein is used in the *in vitro* motility assay?
- Discuss how negative stain EM images differ from more traditional EM images.

RESULTS

- Estimate the size of bovine brain α and β tubulin proteins in Figure 6.2.1a.

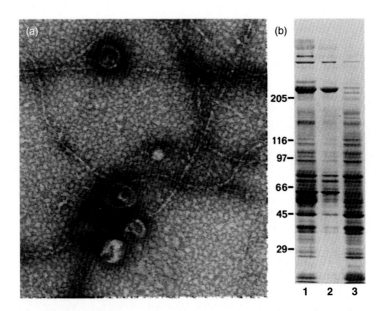

FIGURE 6.2.2 Organelle and high-speed supernatant from squid axoplasm.
The low-speed (S1) fraction of squid axoplasm contains MTs and membrane-bound organelles (a).
b. High-speed centrifugation of squid axoplasm (lane 1) through a sucrose gradient produced an
organelle-rich fraction (lane 2) and a high-speed supernatant free of organelles and MTs (lane 3).
Magnification: 50,000×.

- Describe the appearance of the salt-extracted MTs in Figure 6.2.1a.
- Relate the image of the MT in Figure 6.2.1b to the protein profile revealed by SDS-PAGE in Figure 6.2.1a.
- What level of protein folding is represented in Figure 6.2.1b?
- Identify the bands corresponding to α and β tubulin on the gel in Figure 6.2.2.
- Explain how the gel in Figure 6.2.2 supports the researchers' claim that the S2 supernatant was free from MTs and organelles.
- Estimate the relative speed of the organelle versus the latex bead in Figure 6.2.3.
- Predict what would happen to the movement of the organelle and latex bead in Figure 6.2.3 under the following conditions:
  - Motility buffer contained no ATP
  - Motility buffer contained 0.5 mM ATP
  - Motility contained no S2 supernatant
  - Assay used MTs assembled from bovine brain tubulin
- Propose an explanation for why a latex bead could move like an organelle along an MT.

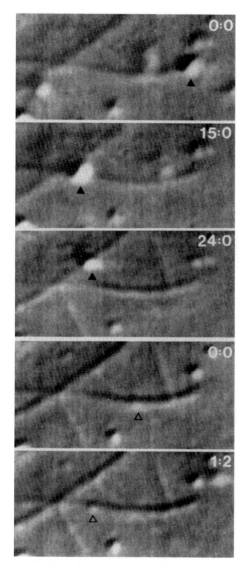

FIGURE 6.2.3 Movement of isolated organelles and carboxylated beads on a single MT.
Still images from a video of an *in vitro* motility assay using MAP-free squid optic lobe MTs, S2
supernatant, and 2 mM ATP. An organelle (open arrow) moves along an MT. A carboxylated latex bead
(closed arrow) moves along one MT and then switches to a different one. Elapsed time is indicated. MTs
are immobilized by treating the coverslip with poly-D-lysine. Magnification: 10,000×.

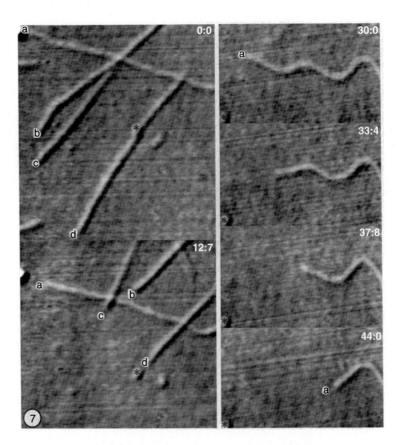

FIGURE 6.2.4 Movement of purified MTs on glass in the presence of S2 supernatant.
Still images from a video showing the movement of four MTs in an *in vitro* motility assay conducted in the presence of S2 supernatant and 2 mM ATP. Elapse time is indicated. Three MTs (b–d indicates their trailing ends) move diagonally from the lower left to the upper right of the field while a fourth MT (a indicates its trailing end) moves horizontally from left to right. An organelle stuck to the cover glass (asterisk) is used as a reference point. Magnification: 12,000×.

- Use the distance from the stationary organelle and MT (a) in the left-hand panels of Figure 6.2.4 to estimate the rate of movement of MTs (d).
- What do you observe about the pattern of movement of MTs (a) in the right-hand panels of Figure 6.2.4? Use your observation to develop a model to explain the movement of MTs on glass.
- Trypsin is a protease, an enzyme that degrades proteins. What can you conclude from the data for the trypsinization experiments in Table 6.2.1?

Table 6.2.1 Comparison of MT-based motility *in vitro* and in dissociated squid axoplasm

Treatment	MTs on glass	Organelles on MTs	Beads on MTs	Dissociated squid axoplasm
None	+ + +	+ + +	+ + +	+ + +
AMP-PNP*				
(5 mM)	–	–	–	+/–
Vanadate				
(20 μM)	+ + +	+ + +	+ + +	+ + +
(100 μM)	+	+	+/–	+
Trypsinization**				
Supernatant	–	–	–	ND
Organelles	+ + +	–	ND	ND
Heat (70°C)†				
Supernatant	–	–	–	ND
Organelles	+ + +	–	ND	ND
Poly-D-lysine				
Coverslip‡	+/–	+ + +	+ + +	ND
Beads§	ND	ND	–	ND

ND, not determined.

*ATP and GTP present in samples at a final concentration of 1 mM.

**S2 supernatant or organelles treated with trypsin (0.5 mg/mL) for 20 min at 23°C. Soybean trypsin inhibitor (7.5 mg/mL final concentration) was applied for 15 min at 4°C to stop the proteolytic digestion.

†Organelles or S2 supernatant heated at 70°C for 5 min, then cooled at 4°C for 5 min. Precipitates pelleted at 10,000g for 5 min at 4°C.

‡Coverslips (22 × 22 or 40 mm) incubated in poly-D-lysine (100 μg/mL or 1 mg/mL) overnight at 4°C, washed 4 times with 50 mL water and air dried.

§Carboxylated beads (100-fold dilution) incubated with 200 μg/mL poly-D-lysine for 1 h at 23°C, pelleted at 10,000g (10 min) and washed 5 times with 1.5 mL water. Beads resuspended in their original volume of motility buffer, and 2 μL of beads incubated with 10 μL of S2 supernatant for 10 min at 4°C.

- Explain why exposure to 70°C heat would result in the effects reported in Table 6.2.1.
- AMP-PNP is a nonhydrolyzable form of ATP that can act as a competitive inhibitor of ATPases. Explain the results in Table 6.2.1 based on this information.
- Research/review current models of MT-based organelle transport. Discuss how the research presented in this case study contributed to those models?

The WASP and the Barbed End [3]

INTRODUCTION

Actin microfilaments (MFs) are involved in many aspects of cell motility, ranging from whole cell migration to the intracellular motility of organelles. Myosin motor proteins are responsible for force production found in muscle contraction, cytokinesis, and vesicle transport. However, MFs are capable of generating force by themselves. Assembly of G actin monomers into F actin polymers generates sufficient force to deform the plasma membrane of the cell, as occurs at the leading edge of lamellipodia. Polymerization of actin is also responsible for the rocket-like propulsion of pathogens, such as the bacteria *Listeria*, which contributes to the spread of the pathogen between host cells.

The rate-limiting step of F actin assembly, is the initial nucleation of G actin monomers into a short filament. Once established, elongation of a filament can occur more rapidly through the addition of G actin monomers. Elongation of the MF will occur as long as the concentration of G actin monomers remains high. At steady state, the addition and loss of subunits is equivalent, resulting in the "treadmilling" or flow of subunits from one end to the other. However, if the monomer concentration is reduced, treadmilling will lead to disassembly of the MF. Elongation of the actin MF is unequal, with more monomers added to one end of the growing filament than at the other. The "fast-growing" end of an MF is defined as the **plus end** with the opposite end of the filament being designated the **minus end**. The difference in polymerization reflects a real difference in conformation of the assembled actin subunits. The difference can be visualized using a method known as "S1 decoration," in which the head domain of myosin proteins (S1) is allowed to bind to an MF. The orientation of the myosin heads creates an arrowhead-like structure where the point of the arrowhead is the minus end of the MF and the barb is the plus end.

The dynamics of the actin cytoskeleton is regulated by a host of **actin-binding proteins** within the cytoplasm and on the surface of cell membranes. Actin polymerization is promoted by the activity of **nucleating proteins** such as **Arp2/3** and polymerizing promoting proteins like **profilin**. Working against these are proteins such as **cofilin** or **gelsolin**, which promote disassembly of MFs. **Capping proteins** bind to one or the other end of an MF, blocking the addition or loss of subunits.

- Describe an example of actin–myosin based motility.
- Conduct a search for a video of *Listeria* rocketing inside a cell.
- Propose an explanation for how actin assembly can generate force.
- Predict what would happen if an MF was capped at its plus end.

BACKGROUND

Polymerization of actin is responsible for propelling *Listeria* inside the cell. Proteins located at one end of the bacterium, the **attachment zone**, bind and activate the host cell's Arp2/3 proteins, promoting actin assembly. As the bacterium moves it leaves behind a **comet tail** composed of MFs. **WASP** is a eukaryotic version of the *Listeria* protein involved in activating Arp2/3 and triggering actin assembly. WASP protein is made up of several distinct protein domains. At the carboxy terminal end, the **central acidic (CA)** domain is thought to be involved in binding to Arp2/3. The **WH2** domain is associated with G actin binding to the MF nucleated through the activity of Arp2/3. The result is elongation of the microfilament and movement of the bacterium.

The rocketing of *Listeria* has been replicated *in vitro*. In the presence of the actin-binding proteins WASP and Arp2/3, liposomes, and beads have moved, driven by the force generated by the polymerization of actin. The development of an *in vitro* assay makes it possible to investigate details of the mechanism of actin rocketing. The function of specific domains of the neuronal form of the WASP protein, **N-WASP**, in Arp2/3 binding and actin polymerization are investigated in this case study, using a modified *in vitro* model system.

- Why would it be important that actin-polymerizing proteins be located only at one end of the *Listeria* bacterium?
- Create a diagram of the *in vitro* actin-rocketing assay that illustrates the relative location of a bead, WASP and Arp2/3 protein, and a growing actin MF. Indicate where in your drawing new G actin subunits are being added to the growing MF. Label the plus (barbed) and minus (point) ends of the MF.

METHODS

Protein expression, purification, and fluorescent labeling

N-WASP fragments were amplified by PCR and expressed as His_6-tagged fusion proteins in *Escherichia coli*. After Ni-NTA purification, the His_6 tag was removed. Actin was purified from rabbit skeletal muscle and labeled with Alexa488 or rhodamine. The Arp2/3 complex was purified from bovine brain. Capping protein, intersectin, and Cdc42/RhoGDI were expressed in *E. coli* and purified. Fluorescent labeling of N-WASP fragments was achieved by incubating with excess maleimide dyes for 2 h prior to overnight incubation at 4°C.

Attachment zone localization assay

Vesicle motility reactions used *Xenopus* egg extract. Eggs were lysed and centrifuged at 300,000*g* for 1 h to obtain a fraction consisting of cytosol and a crude membrane fraction. Cytosol was centrifuged at 541,000*g* for 15 min. Cytosol and membrane fractions were stored frozen.

Crude membranes were mixed with 35 μL of cytosol, 0.3 μL rhodamine-labeled actin, and buffer at 4°C. Samples were warmed to room temperature. After 30–40 min, 2 μL aliquots were viewed and images were acquired. Fluorescent and phase contrast images were acquired from ~50 rocketing vesicles.

Reconstitution of *in vitro* membrane motility

Liposomes composed of 5:75:25 PIP2:PC:PS, were fused to the surface of glass beads (2.3 μm diameter) to create supported lipid bilayers. Lipid-coated beads were incubated with 1.5 μM Cdc42/RhoGDI complex, 2 μM DH-PH, and 100 μM GTPγS for 10 min at room temperature, followed by the addition of 1.5 μM N-WASP for 10 min. Motility reactions were initiated by diluting the suspension of N-WASP/Cdc42/lipid-coated glass beads tenfold into motility buffer containing 6 μM G-actin, 0.075 μM Arp2/3, 0.05 μM capping protein, and 2 μM profilin. Then, 5 min time lapse videos were captured for each bead.

Actin filament capture assay

Beads (2.0 μm) were coated with 1 μM N-WASP in buffer. Actin (8 μM) was polymerized for 2 h and stabilized with 6 μM Alexa488 phalloidin to form "Alexa488 filaments." N-WASP beads (5.73×10^8 beads/mL) were rotated with Alexa488 filaments (4 μM) in the presence of varying concentrations of rhodamine-labeled actin monomers stabilized with latrunculin. After 10 min, the suspension was diluted into buffer, and randomly chosen beads (15–20) were imaged by phase contrast and fluorescence microscopy.

- What is a His_6 tag? Describe how it is used in protein research.
- How does the composition of lipid-coated glass beads compare with the lipid composition of the plasma membrane?
- Compare the protocol for the *in vitro* assay using vesicles derived from egg extract and lipid-coated beads. How are the assays similar? How are they different?

RESULTS

- Conduct a search to determine the function of VASP-enabled homology domains and polybasic regions.

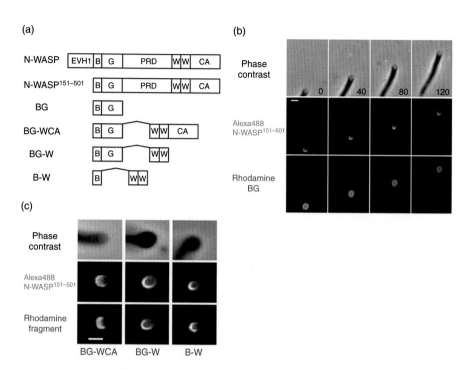

FIGURE 6.3.1 The WASP homology 2 (WH2) domain of N-WASP is the minimum fragment required for localization to the attachment zone of a moving vesicle.
a. N-WASP consists of multiple protein domains. N-WASP fragments were created by deletion of various domains as indicated. Internally deleted domains were replaced with a (Gly-Ser)$_4$ linker. Fragments were labeled with the fluorescent markers Alex488 or rhodamine. Abbreviations: EVH1, enabled-VASP homology domain 1; B, polybasic region; G, GTPase-binding domain; PRD, proline-rich domain; W, WASP homology domain 2; CA, central and acidic domain. b. Alexa488 N-WASP$^{151-501}$ and rhodamine BG fragments were added to the *in vitro* assay mixture and rocketing vesicles were imaged every 40s by phase contrast and fluorescence microscopy. c. Alexa488 N-WASP$^{151-501}$ and the indicated rhodamine-labeled fragments were imaged as described in (b). Images are representative of >50 rocketing vesicles for each rhodamine fragment. Scale bars represent 2 μm.

- Describe the logic behind the choice of the fragments illustrated in Figure 6.3.1a. How do the different constructs contribute to understanding which part of the protein is required for its function?
- Explain why the three rows in Figure 6.3.1b look different.
- What is the long dark streak visible in the phase contrast images of Figure 6.3.1b made of?
- Functional N-WASP is localized to the attachment zone at the point on the vesicle, where actin polymerization is occurring. Which of the N-WASP fragments in Figure 6.3.1b is functioning properly?

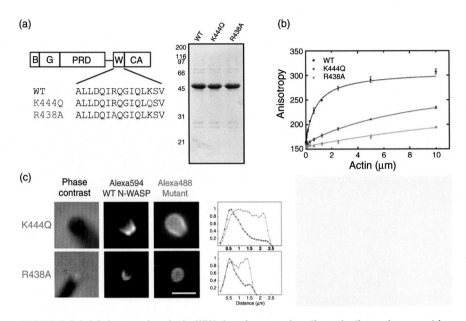

FIGURE 6.3.2 Point mutations in the WH2 domain uncouple actin nucleation and asymmetric localization of N-WASP.

a. One of the WH2 domains was deleted from N-WASP creating the construct WT N-WASP. Point mutations were made within the WH2 domain as indicated (K444Q and R438A). All three constructs were expressed, isolated, and analyzed using SDS-PAGE and Coomassie Blue stain. b. Actin binding by the three N-WASP constructs (WT, black; K44Q, red; R438A, green) was measured by fluorescence anisotropy as a function of increasing actin concentration. c. Alexa594 WT N-WASP and either Alexa488 K44Q or R438A N-WASP were added to the *in vitro* assay mix and rocketing vesicles were imaged as described in Figure 6.3.1. Line scan analysis compares localization of the WT N-WASP (red lines) and the mutant (green lines) constructs. Scale bar represents 2 μm.

- Estimate the rate of vesicle movement from the phase contrast images in Figure 6.3.1b.
- Defend the conclusion that the WH2 domain is the minimum fragment required for proper localization of N-WASP using the data from Figure 6.3.1c.
- Discuss how the phase contrast images contribute to the interpretation of the results in Figures 6.3.1b and c.
- What do the letters and numbers correspond to in the designation "K444Q" and "R438A" in Figure 6.3.2a?
- Predict the effect of each of the point mutations on the chemistry within the protein domain.
- Discuss how the inclusion of the SDS-PAGE gel contributes to Figure 6.3.2a.
- Summarize the actin-binding ability of the three N-WASP constructs.

- Develop an argument for or against the statement, "the deletion of the second WH2 domain had no effect on the actin-binding activity of N-WASP."
- Connect the change in protein chemistry associated with each of the point mutations with the change in actin binding.
- Construct a hypothesis to address the potential relationship between actin binding and amino acid structure of the WH2 domain. Design an experiment that would test your hypothesis.
- Explain the relationship between the images and graphs in Figure 6.3.2c.
- Summarize the effect of the point mutations on the ability of N-WASP to localize to the attachment zone.
- How does the *in vitro* motility assay in Figure 6.3.3 differ from that in Figures 6.3.1 and 6.3.2?
- Research/review the role of each of the components listed in Figure 6.3.3a in supporting *in vitro* motility through actin rocketing.
- Estimate the rate of rocketing for wild-type (WT) N-WASP and K44Q-coated beads in Figure 6.3.3b. Do the rates of movement differ?
- Characterize how the K444Q and R438A constructs differ from WT N-WASP in the *in vitro* motility assay.
- Use the data from Figures 6.3.2 and 6.3.3 to develop a model that explains the behavior of point mutation constructs. What does your model imply about the function of WT N-WASP?
- What impact did dilution with buffer alone have on the attachment of WT N-WASP-coated beads in Figure 6.3.4a (use Figure 6.3.3 as a reference)?
- Suggest a reason why bead attachment is rescued by the addition of 1 μM actin.
- How do capping proteins interact with actin filaments? Apply this information to interpret Figure 6.3.4a.
- Conduct a search to learn more about the compounds *phalloidin* and *latrunculin*.
- Look again at your model for the interaction between WT N-WASP and actin. Where in your model would you find a G actin monomer? Where would you find an F actin filament?
- Figure 6.3.4b and c reflect the ability of WT N-WASP-coated beads to "capture" actin filaments. How do the results of these experiments support the conclusion that N-WASP binds to the barbed end of an actin filament?
- Finalize your model of the function of N-WASP. Use your model to explain the ability of an organelle carrying N-WASP to move by the force of actin polymerization.
- Imagine a model in which N-WASP binds to the pointed end of the actin filament. Predict whether this arrangement could support "actin rocket propulsion."

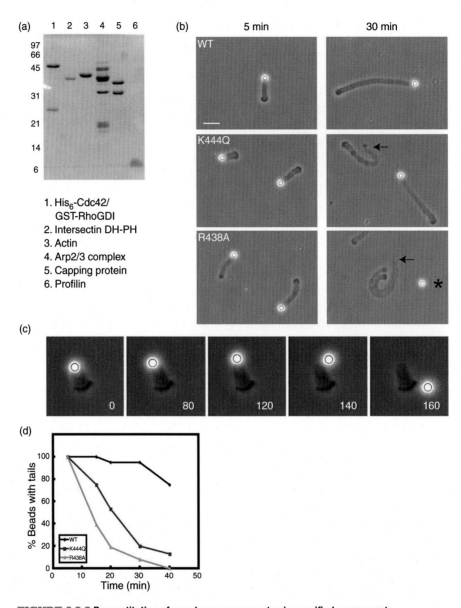

FIGURE 6.3.3 Reconstitution of membrane movement using purified components.

a. Coomassie Blue–stained SDS-PAGE of purified proteins used to reconstitute an *in vitro* rocketing assay using lipid bilayer–coated glass beads. b. Phase contrast images of motility reactions acquired 5 and 30 min after initiation. Arrows indicate detached actin comet tails; asterisk indicates detached lipid-coated bead. Scale bar represents 10 μm. c. Time lapse sequence showing the detachment of an R438A N-WASP, lipid-coated bead from the comet tail. d. Time course of comet tail detachment. At the indicated time points, motility reactions were fixed with 1.5% glutaraldehyde, and the percentage of beads with attached tails was determined (40 beads per time point).

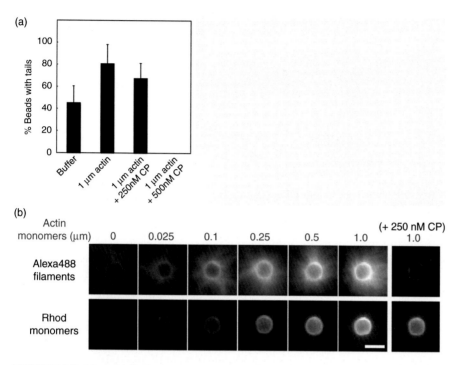

FIGURE 6.3.4 Comet tail attachment requires interactions between WH2 domains and actin filament barbed ends.

a. Motility reactions containing WT N-WASP were allowed to proceed for 10 min and were then diluted fourfold into buffer alone or buffer containing 1 μM actin plus the indicated concentrations of capping protein (CP). After 15 min, reactions were fixed in 1.5% glutaraldehyde and the percentage of bead with attached tails was determined (40 beads counted for each condition; results expressed as mean ± SD percentages from four independent experiments). b. Beads were coated with N-WASP (WT, K444Q, and R438A) and incubated with Alexa488 phalloidin-stabilized actin filaments (± capping protein) and increasing concentrations of latrunculin-stabilized actin monomers. Latrunculin-stabilized monomers are unable to polymerize and phalloidin-stabilized filaments are unable to depolymerize. Images are of WT N-WASP beads with bound Alexa488 actin filaments (upper panels) and rhodamine-labeled actin monomers (lower panels). Scale bar represents 2 μm.

Cilia Grow Where Vesicles Go [4]

INTRODUCTION

Many eukaryotic cells use cilia or flagella in order to move. However, many more cells have a single and non-motile **primary cilium** that extends from their cell surface. Evidence points to a role for primary cilia in development and sensory input. Loss of primary cilia is associated with a range of diseases. The best-known example of a disease that is linked to the function of primary cilia is **polycystic kidney disease (PKD)**. Primary cilia on the epithelial cells of the kidney, bend in response to the movement of fluid, opening calcium channels, and altering gene expression. Defects or loss of these cilia prevents signaling, resulting in changes in the cell cycle, leading to the formation of cysts, and ultimately kidney disease. Primary cilia can also sense light, temperature, gravity, osmolarity, and chemicals and they play a role in establishing bilateral symmetry in development.

At the core of all of these structures is a microtubule (MT) cytoskeleton arranged into a cylindrical **axoneme**. The classic axonemal structure of a motile flagella is a ring of nine outer doublet MTs surrounding a central pair of MTs (9 + 2) stabilized by proteins that serve as "spokes" and "links," and powered by the MT motor protein, dynein. The axoneme of primary cilia lack central pair microtubules (9 + 0) and dynein. Axonemes "grow" from **basal bodies**, centriole-like structures that are also composed of MTs and associated proteins.

The axoneme of a primary cilium is surrounded by a plasma membrane. Given its unique role in sensory transduction, it should not be surprising to learn that the plasma membrane of the primary cilium contains a specialized population of receptors and other signaling proteins. Movement of the proteins and lipids required to build and maintain the primary cilium, occurs through the process of **intraflagellar transport (IFT)**. Particles formed by a collection of IFT proteins carry cargo proteins in the space between the axoneme and the plasma membrane using the MTs of the axoneme and the MT motor proteins, cytoplasmic dynein, and kinesin. Many of the diseases associated with primary cilia are the result of defects in the genes that code for IFT proteins.

- Name some examples of cells that use cilia or flagella to either propel themselves or move material in the extracellular space.

- Axonemes lacking central pair MTs or dynein are paralyzed. Explain how each of these losses would affect the ability of the axoneme to move.
- How is cytoplasmic dynein similar to or different from axonemal dynein?
- Create a diagram that illustrates the likely physical relationship between the axoneme, an intraflagellar particle, a motor protein, and the plasma membrane.

BACKGROUND

The primary cilium represents a functionally distinct region of the cell made up of a cytoplasmic compartment and a specialized plasma membrane. Proteins such as **α** and **β tubulin**, the radial spoke proteins **RSP1-3**, and the IFT protein **IFT46**, are either part of the axoneme or located within the cytoplasmic space of the cilia. However, the proteins **FMG-1** and **PKD2**, are flagellar transmembrane proteins. IFT of protein particles has been shown to target the movement of cytoplasmic ciliary proteins into the cilium, but how do the primary cilia-specific membrane proteins become segregated from the cilia plasma membrane? The experiments described in this case study test the hypothesis that cytoplasmic vesicles play a role in the transport of both membrane and soluble proteins to the primary cilium.

- Where are cytoplasmic and plasma membrane proteins synthesized in the cell?
- Conduct a search for an image of PKD2. Describe its structure.
- How does a newly synthesized transmembrane protein become part of a cell's plasma membrane?

METHODS

Cell culturing and extraction of cytoplasm

Chlamydomonas reinhardtii strains used in this study, included cell wall-less strain cw15 and paralyzed flagella strain pf14. For extraction of cytoplasm, cw15 cells were cultured in a light:dark cycle of 12:12 h at 18°C with constant aeration. Cells were harvested by low-speed centrifugation and deflagellated by pH shock. During the first few minutes of synchronous flagellar regeneration, the cells were harvested by centrifugation and their cytoplasm was extracted by mechanical disruption using a blender. A clarified cytoplasmic extract was obtained by centrifugation at 16,000*g* for 15 min to remove unlysed cells and cellular debris.

Isolation of cytoplasmic vesicles

Separation of vesicle membranes from cytoplasmic extract was achieved by flotation centrifugation through a seven-step OptiPrep gradient. A mixture of

2 mL cytoplasmic extract and 2 mL of 60% OptiPrep was loaded onto the bottom of an ultracentrifuge tube. Then, 1 mL each of 25, 20, 15, 12, 10, and 5% OptiPrep dissolved in buffer were layered over the sample to create a discontinuous gradient. Gradients were centrifuged for 3 h at 200,000*g*. Following flotation centrifugation, membrane bands were visible at each density interface and were extracted manually by pipet.

Immunoelectron microscopy

For immunogold labeling of isolated cytoplasmic vesicles, whole vesicles were resuspended in buffer and transferred to formvar-coated EM grids. Grids were floated on a droplet of blocking buffer and transferred to a droplet of primary antibody for 1 h at room temperature. Grids were then washed and incubated with gold-conjugated secondary antibody and then negatively stained with 2% aqueous uranyl acetate.

For *in situ* immunogold labeling, 10 mL of cells in the process of regenerating flagella were harvested by centrifugation at 500*g*. Cell pellets were resuspended in buffer and fixed using 4% formaldehyde and 0.5% glutaraldehyde for 30 min at room temperature and then postfixed with 0.5% osmium tetroxide. Cells were washed and prepared for electron microscopy by dehydration and infiltration with LR gold resin. Ultrathin sections were cut and incubated with primary antibody followed by incubation with gold-conjugated secondary antibody. Samples were imaged using a transmission electron microscope.

Immunofluorescence microscopy

Chlamydomonas were fixed for 1 h at room temperature in media containing 1% paraformaldehyde. Cells were washed with buffer and transferred to a coverslip. Cells were permeabilized by immersion in −20°C methanol for 5 min, rehydrated in buffer, and then incubated with blocking buffer for 1 h at room temperature prior to overnight incubation with primary antibody at 4°C. Samples were washed, incubated with fluorescent-conjugated secondary antibody, and imaged using a laser scanning confocal microscope.

- Discuss the advantages of the two *Chlamydomonas* strains used in these experiments.
- Outline the protocol for the isolation of cytoplasmic vesicles from *Chlamydomonas*.
- What was the final density of the cytoplasmic extract in the ultracentrifuge tube?
- Compare the protocols for immunofluorescence and immunogold labeling of cells. Discuss the advantages and drawbacks of the two methods.

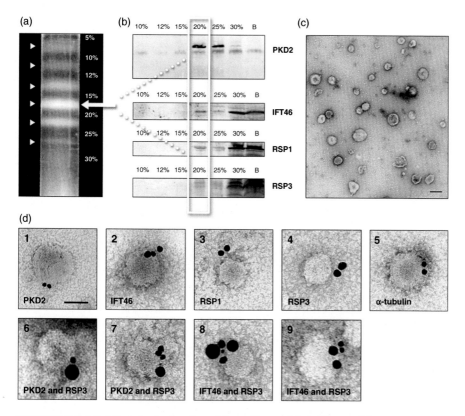

FIGURE 6.4.1 Isolation of cytoplasmic vesicles with associated ciliary proteins.

a. Cytoplasmic extract from the mechanical disruption of *Chlamydomonas* cells was placed at the bottom of a seven-step discontinuous OptiPrep density gradient. During high-speed centrifugation, various populations of membrane vesicles float up through the gradient and collect at the different interfaces according to their buoyant density. Arrowheads indicate bands formed by accumulation of the different membrane populations. b. Isolated gradient fractions were subjected to SDS-PAGE and analyzed by immunoblotting with antibodies specific to ciliary membrane and axonemal proteins. Material from the bottom of the gradient b was also examined. c. Fraction from the 20% gradient interface examined by negative staining and transmission electron microscopy (TEM). Scale bar represents 100 nm. d. Immunogold labeling of intact, negatively stained vesicles from the 20% gradient interface fractions. Antibodies used in single-labeling experiments (images 1–5) or double-labeling experiments (images 6–9) are indicated. PKD2 and IFT46 are labeled with a small gold particle. RSP3 is labeled with a large gold particle. Scale bar represents 50 nm.

RESULTS

- Label the image in Figure 6.4.1a to indicate the direction of movement of the membranes during centrifugation.
- Predict how material at the bottom of the centrifuge tube differs from material near the top of the tube.

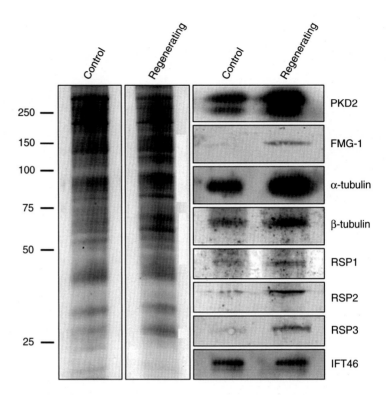

FIGURE 6.4.2 Ciliary membrane and axonemal proteins associate with cytoplasmic vesicles in increased amounts during flagellar regeneration.

The 20% gradient interface vesicle fraction was isolated from control (full length flagella) cells and cells undergoing flagellar growth (regeneration) and analyzed by SDS-PAGE and immunoblotting. The two left-hand panels are silver-stained SDS-PAGE resulting from equal protein loads of control and regenerating vesicle fractions. Panels on the right-hand side are immunoblots of the same samples reacted with antibodies specific to the indicated ciliary proteins.

- Explain the logic behind identification of the membrane fraction from the 20% gradient interface as being specific to the cilia.
- What can you conclude about the IFT46 and RSP proteins based on their migration during density gradient centrifugation?
- Estimate the average size of the membranes in the 20% gradient interface fraction (Figure 6.4.1c).
- Summarize the results from Figure 6.4.1d.
- Predict the results of the gradient density centrifugation experiments shown in Figure 6.4.1. Would they differ if the following modifications were made to the protocol:
  - Treatment of the cytoplasmic extract with high salt prior to centrifugation
  - Treatment of the cytoplasmic extract with detergent prior to centrifugation

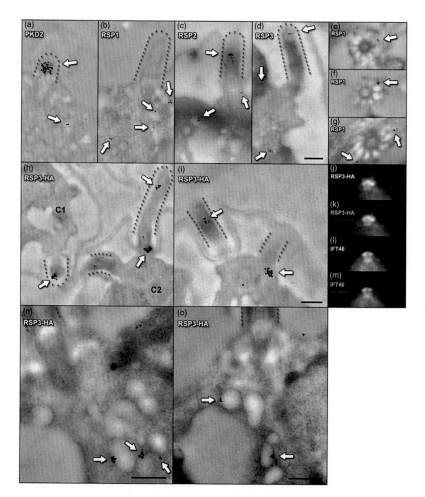

FIGURE 6.4.3 *In situ* immunogold labeling of ciliary membrane and axonemal proteins during flagellar regeneration.

a. White arrows indicate PKD2-specific gold particles clustering on the membrane of a newly forming *Chlamydomonas* flagellum (emphasized by dotted outline) and on membrane vesicles nearby in the cytoplasm. (b–d) In addition to labeling the growing axoneme (c and d), gold particles specific to radial spoke proteins are observed on the surface of membrane vesicles in the cytoplasm beneath the forming flagella (RSP1, RSP2, and RSP3; white arrows). e–g. Cross-sections proximal to the base of newly forming flagella show transitional fibers radiating from the basal body. RSP-specific gold particles are found clustered at the cell membrane where transitional fibers terminate (indicated by white arrows). h. and i. An *rsp3* null mutant cell line was rescued with a human influenza hemagglutinin (HA)-tagged version of RSP3 and immunogold labeled with HA-specific antibodies (indicated by white arrows; C1 and C2 = cell 1 and cell 2). j. and k. Immunofluorescence localization with antibodies specific to RSP3–HA (green), α tubulin (red), and DNA (blue). l. and m. Immunofluorescence localization of IFT46 (green), the centriololar protein, centrin (red), and DNA (blue). Colocalization is indicated by the color yellow. n. and o. RSP3–HA-specific gold particles on membrane vesicles in the cytoplasm beneath growing flagella (indicated by white arrows). Scale bars represent 200 nm.

- Predict the origin of the membrane vesicles isolated from the 20% gradient interface. Justify your answer.
- Compare the protein profile of control and regenerating flagella. Identify any differences in proteins between the two samples. Estimate the molecular weight of three of the proteins you have identified.
- How would you expect the immunoblot of the control sample in Figure 6.4.2 to compare with the blot in Figure 6.4.1b? Explain your answer.
- What do the results from Figure 6.4.2 indicate about flagellar regeneration? What other research questions could you ask about the cell biology of regeneration?
- How are the immunogold-labeling images in Figure 6.4.3a–d different from those in Figure 6.4.1?
- Develop an argument for or against the statement, "vesicle membrane association of PKD2 and RSP is an artifact."
- What is different about the immunogold labeling in Figure 6.4.3h–o compared with Figure 6.4.3b–d?
- Explain how the immunofluorescence images in Figure 6.4.3j–m contribute to interpretation of these experiments.
- Discuss the strengths and weaknesses of the data presented in Figure 6.4.3.
- Propose a model to explain how the tubulin and RSP proteins become associated on the cytoplasmic side of ciliary vesicles.
- Describe how vesicles are transported from their point of origin, through the cytoplasm, and to their target. Can you reconcile the models of intracellular vesicle transport with the model for ciliary vesicle transport?
- The flagella of *Chlamydomonas* have been used as a model system for study of the primary cilium. Debate whether there are limits to the usefulness of this model system.

References

[1] Yang Y, Bauer C, Strasser G, Wollman R, Julien J-P, Fuchs E. Integrators of the cytoskeleton that stabilize microtubules. Cell 1999;98:229–38.

[2] Vale RD, Schnapp BJ, Reese TS, Sheetz MP. Organelle, bead and microtubule translocations promoted by soluble factors from the squid giant axon. Cell 1985;40:559–69.

[3] Co C, Wong DT, Gierke S, Chang V, Taunton J. Mechanism of actin network attachment to moving membranes: Barbed end capture by N-WASP WH2 domains. Cell 2007;128:901–13.

[4] Wood CR, Rosenbaum JL. Proteins of the ciliary axoneme are found on cytoplasmic membrane vesicles during growth of cilia. Curr Biol 2014;24:1114–20.

CHAPTER 7

Organelles

Putting the "Retic" in the Endoplasmic Reticulum [1]

INTRODUCTION

The presence of membrane-bound organelles is a defining characteristic of eukaryotic cells. Phospholipid membranes separate the internal environment of an organelle from the cytoplasm of the cell. The chemistry of the phospholipids, in combination with transport proteins, determines which molecules can cross the membrane of an organelle. Segregation of the inside of an organelle from the rest of the cytoplasm allows for specialization of organelle function.

Purified phospholipid molecules will spontaneously adopt a bilayer structure when placed in an aqueous solution, forming spherical "bubbles" or liposomes. However, organelle membranes, like other biological membranes, are more than just phospholipids. Biological membranes are characterized as fluid mosaics, consisting of a phospholipid bilayer with **integral** and **peripheral membrane proteins**. While some cellular organelles can be described as spheres, others are decidedly not. The endoplasmic reticulum (**ER**) is an example of an organelle that forms a complicated, tubular network that can fill the cytoplasm of a cell. If phospholipid bilayers form spheres, then how does the ER form tubules?

■ Create a list of the membrane-bound organelles you would expect to find in a eukaryotic cell.
■ Describe how the unique function of each of the organelles on your list is helped by the presence of a distinct compartment separated from the rest of the cell.

157

- Explain why phospholipids form bilayers and liposomes.
- Illustrate the concept of the fluid mosaic model of a biological membrane. Label your drawing.

BACKGROUND

The ER is an extensive tubular network of membranes, emerging from the outer membrane of the nuclear envelope, creating a large surface area and enclosing an equally extensive lumenal compartment. There are two forms of ER: the **rough ER (RER)** and the **smooth ER (SER)**. Although the two types of ER have different functions, they are in fact continuous, with lipids and proteins diffusing from one to the other. Ribosomes give the RER its "rough" appearance. Protein synthesis and glycosylation are the primary functions of the RER. In contrast, the SER functions in the synthesis of lipids and steroid hormones, removal of toxins, and the storage of intracellular calcium (Ca^{2+}). Ca^{2+} is a powerful signaling molecule inside cells. Regulation of the release of Ca^{2+} from the ER is controlled by **IP3 receptor** (IP3R), an ER-specific protein. Other ER-specific proteins include p97, Ca^{2+} pumps and channels, translocon associated protein subunits (TRAP α), and a family of proteins known as **reticulons**.

In vitro model systems reproduce cellular processes outside a living cell. Because these systems consist of a simplified collection of components, they are powerful tools for investigating the contributions of the individual components to the cellular process. An *in vitro* model for the formation of ER tubules was developed using membranes isolated from *Xenopus laevis* frog eggs. A population of small (50–100 nm) spherical vesicles was isolated from lysed eggs by centrifugation. The vesicles were washed with a high concentration of salt to remove peripheral membrane proteins and any cytosolic contamination. When incubated at 25°C in the presence of 1 mM ATP and 0.5 mM GTP, the vesicles fuse and form a network.

Identification of the ER as the original membrane source of the small vesicles was established using several methods. The protein composition of the vesicles was examined using immunoblotting. Transmission electron microscopy also helped confirm that these small vesicles were made up of membrane derived from the ER. It was also found that the vesicles contained Ca^{2+}.

- Why does treatment with high salt remove peripheral membrane proteins?
- Research/review the types of proteins that are synthesized on the RER.
- Design an experiment that would confirm that the small vesicle membranes are derived from the ER and not the Golgi complex or nuclear envelope.

METHODS

Vesicle isolation

Eggs were collected from 10 *X. laevis* frogs and washed with 0.2 M NaCl to remove the external jelly coat. The eggs were transferred to a buffer solution and crushed by centrifugation at 10,000 rpm for 10 min at 4°C. The resulting egg extract was collected and centrifuged a second time at 40,000 rpm for 1 h at 2°C to pellet the larger membranous organelles. The supernatant was collected and centrifuged at 100,000 rpm for 1.5 h at 2°C to pellet smaller membranous organelles, leaving soluble cytosolic proteins in the supernatant. The small vesicle membrane fraction was washed twice with buffer plus 200 mM KCl before use in network formation assays.

In vitro network formation

Salt-washed small vesicle membranes were incubated in buffer containing 200 mM KCl, 1 mM ATP, and 0.5 mM GTP at 25°C for 1 h, unless otherwise noted. For low salt incubations the 200 mM KCl was replaced with 50 mM potassium acetate. The membranes were visualized using the lipophilic fluorescent stain, octadecyl-rhodamine.

Aequorin Ca^{2+} assay

Aequorin is a luminescent protein that fluoresces in the presence of Ca^{2+}. The intensity of the fluorescence signal is proportional to the concentration of Ca^{2+} in solution. The amount of Ca^{2+} released during network formation assays was measured in the presence of 80 nM aequorin using a luminometer. Measurements were taken every 3 min.

Sulfhydryl modification and protein isolation

Sulfhydryl reagents act by modifying exposed sulfhydryl (SH) groups in proteins. AuCl$_4$ and maleimide-type reagents bind to SH groups. Maleimide PEG (MP), *N*-ethylmaleimide (NEM), and maleimide neutravidin (MN) are large, membrane-impermeable compounds. Maleimide biotin (MB) is capable of crossing the membrane. The reagent dithiothreitol (DTT) acts to stabilize SH groups and prevent the binding of maleimide reagents.

Salt-washed membranes were incubated in the presence or absence of 5 mM MB for 20 min followed by the addition of 1 mM DTT. For some reactions, MB was pretreated with 1 mM DTT or was reacted in the presence of MP, a competing sulfhydryl reagent. The vesicles were then solubilized with the detergent, Triton X-100. Solubilized vesicle extract was incubated with avidin-coated beads. Biotinylated proteins bound to the avidin beads and were pelleted. Beads were incubated in 2 mM biotin to elute proteins off the beads. Eluted proteins were examined using SDS-PAGE.

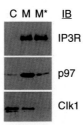

C M M*   IB

IP3R

p97

Clk1

FIGURE 7.1.1 Treatment of *Xenopus* egg membranes with high salt removes cytosolic and peripheral membrane proteins.
Xenopus egg cytosol (C), small vesicle membranes (M), and salt-washed small vesicle membranes (M*) were analyzed by SDS-PAGE and immunoblotting (IB) with antibodies to the integral membrane ER protein IP3-receptor (IP3R), the peripheral membrane ER protein p97, or the cytosolic protein cdc2-like kinase 1 (Clk1).

- Create a flow chart of the steps involved in the isolation of small vesicles from *Xenopus* eggs.
- Why did the researchers include 200 mM KCl in the buffer used to wash the small vesicles?
- Why does it make sense to incubate the small vesicle samples at 25°C instead of 37°C?
- Define the term "lipophilic."
- Predict whether MB will add biotin to a protein in the presence of DTT or MP? Explain your prediction.

RESULTS

- Explain why treatment with high salt has no effect on IP3R (Figure 7.1.1).
- Discuss how effective the high salt treatment is in removing peripheral membrane proteins and cytosolic proteins from the small membrane vesicle sample.
- What conditions are required for network formation to occur (Figure 7.1.2a)?
- How does the release of Ca^{2+} from small vesicles differ under conditions of high and low salt (Figure 7.1.2b)?
- Approximately how much Ca^{2+} was stored in the small vesicles?
- What effect did the treatment with sulfhydryl reagents have on network formation (Figure 7.1.2c)?
- Sulfhydryl reagents are membrane-impermeable compounds that act to modify exposed SH groups on proteins. What can you conclude about the protein(s) involved in network formation based on this statement?

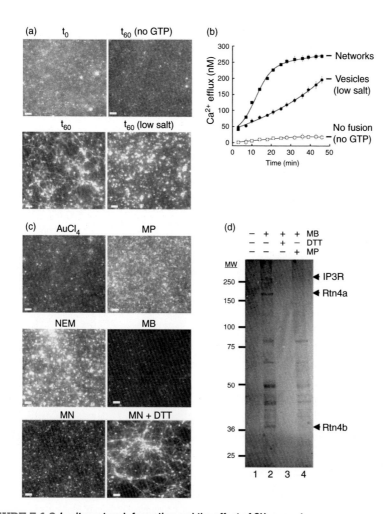

FIGURE 7.1.2 *In vitro* network formation and the effect of SH reagents.

a. Salt-washed small vesicles from *Xenopus* eggs were incubated with 200 mM KCl and 0.5 mM GTP at 25°C and examined immediately (t_0) or after 1 h (t_{60}). Similar membrane samples were incubated in the absence of GTP (t_{60} no GTP) or in the presence of 50 mM potassium acetate (t_{60} low salt). All samples were stained with octadecyl rhodamine and visualized by fluorescence microscopy. b. Ca^{2+} efflux from the small vesicles was measured using aequorin. The reaction was performed in the presence of 200 mM KCl and GTP (networks), low salt and GTP (vesicles) or without GTP (no fusion). Error bars indicated the standard deviation of three experiments. c. Salt-washed membranes were preincubated for 20 min at 25°C with the SH reagents: $AuCl_4$, maleimide PEG–5 kDa (MP), NEM, MB, MN or MN pretreated with 1 mM DTT (MN + DTT). SH reactions were stopped by the addition of 1 mM DTT prior to performing the network formation assay in the presence of 200 mM KCl and 0.5 mM GTP. d. Identification of the protein(s) modified by SH reagents was achieved using MB. The activity of MB could be inhibited by the addition of 1 mM DTT or excess MP. Biotinylated proteins from extracts of vesicles treated with the indicated combination of reagents were analyzed using SDS-PAGE and Coomassie Blue staining. Proteins modified only by the absence of MP were IP3 receptor and the reticulon proteins, Rtn4a/NogoA and Rtn4b/NogoB. Scale bars = 5 mm.

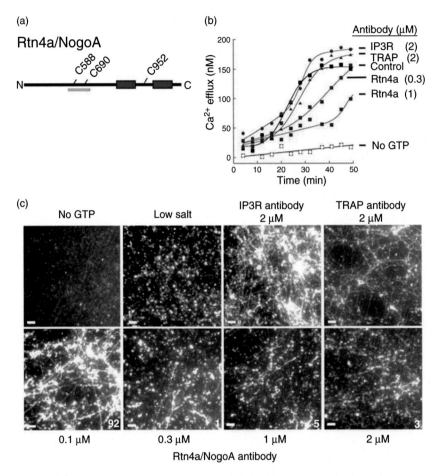

FIGURE 7.1.3 Antibodies to Rtn4A inhibit network formation.
a. Polyclonal antibodies were raised to the cytoplasmic amino terminal domain of Rtn4a (shown in green). Hydrophobic regions of the protein are shown in blue. b. Ca^{2+} release during network formation in the presence of the indicated antibodies was measured. Controls included incubation in the presence and absence of GTP and in the presence of antibodies to IP3R and TRAP α. c. Same as (b) but visualized by fluorescent lipid staining. The number of three-way junctions in the networks is shown in the lower right-hand corner of the images in the lower row. Scale bars = 5 mm.

- Which lane or lanes represent the control for the experiment shown in Figure 7.1.2d?
- Explain the significance of the result shown in lane 4 of Figure 7.1.2d.
- Describe the Rtn4a/NogoA protein based on the schematic diagram shown in Figure 7.1.3a. Do you see anything unusual about this protein?
- How do the data in Figure 7.1.3b relate to the data in Figure 7.1.3c?

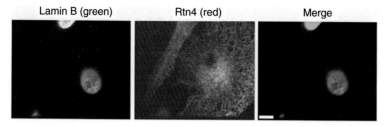

FIGURE 7.1.4 Localization of reticulons.
Xenopus tissue culture cells were stained with antibodies specific to Rtn4 (red) and nuclear lamin B (green). Scale bar = 10 μm.

- Develop an argument to contradict the statement, "the appearance of inhibition by incubation with antibodies is an artifact due to nonspecific, steric hindrance" based on the data in Figure 7.1.3.
- Discuss why increasing amounts of Rtn4A/NogoA antibody reduces, but does not eliminate the formation of membrane networks in Figure 7.1.3c.
- Characterize the staining pattern of Rtn4 and lamin B in Figure 7.1.4. Are reticulon proteins present in the membrane of the nuclear envelope?
- The authors conclude, "reticulons are restricted to the tubular, peripheral ER, consistent with a role in shaping this organelle." Describe which experiment you feel best supports this conclusion.
- List some predictions you can make based on the hypothesis that reticulons are responsible for the formation of the ER membrane network
- Design an experiment to test one of your predictions.

How the Golgi Stacks Up [2]

INTRODUCTION

The Golgi complex is an organelle that consists of multiple, independent membrane compartments called cisternae. Individual cisternae are organized into a stack with each compartment in close physical proximity to the others. The classic appearance of the Golgi complex as a set of "flattened sacs" is conserved across a wide range of eukaryotic cells. Stacking of Golgi membrane compartments creates an orientation where the cisterna closest to the ER is designated the *cis* Golgi while the cisterna farthest away is the *trans* Golgi. The term *medial* Golgi is applied to the cisternal compartments between *cis* and *trans*.

The Golgi complex contributes to the modification and sorting of proteins and lipids as part of a cell's endomembrane system. Each cisterna is associated with a specific modification based on the enzymes that are located within the lumen of that compartment. For example, the enzyme mannosidase I is found in the *cis* Golgi, mannosidase II is present in the *medial* Golgi, and sialyltransferase is located in the *trans* Golgi. A consequence of the packaging of specific enzymes within specific cisternae is the sequential modification of proteins and lipids as they move from the *cis* to *trans* Golgi. Physical stacking of the cisternae is thought to contribute to maintaining the sequential nature of Golgi function.

- Conduct a search to find images of the Golgi complex. Compare the images you find. How are they similar? How are they different?
- What is the significance of the observation that the shape of the Golgi complex is highly conserved?
- Research/review the function of the enzymes mannosidase I, mannosidase II, and sialyltransferase in the Golgi complex.

BACKGROUND

During mitotic cell division organelles of the parent cell must be divided equally into the two, new daughter cells. In order to accomplish this goal, the Golgi complex breaks up into multiple small vesicles or **mitotic Golgi fragments** (MGFs). Once the daughter cells enter interphase, the Golgi complex is reassembled from the membranes of the MGFs. This cycle of disassembly/reassembly of the Golgi complex can be replicated *in vitro* using Golgi membranes isolated from rat liver and cytoplasm isolated from interphase or mitotic cells. Soluble factors in the cytoplasm determine whether the Golgi membranes form cisternae or vesicles while factors on the membranes themselves control the stacking of the Golgi.

How do MGF membranes "know" to reassemble into stacked Golgi compartments? It is likely that proteins play a role in the process given their importance in every other aspect of cellular life. This case study makes use of a cell-free model system for the disassembly and reassembly of the Golgi to identify the proteins responsible for organizing the cisternae into stacks.

One way to identify a protein is by labeling it. The reagent **NEM** reacts with proteins by binding to any exposed SH groups. **BMCC** is another sulhydryl reagent with the same reactivity as NEM; however, it carries a biotin group that allows for detection of any proteins that bind BMCC. Pretreatment of intact Golgi membranes with NEM blocks all exposed SH groups preventing BMCC binding. However, when NEM-treated membranes are incubated with mitotic cytosol, the Golgi fragments revealing a new population of proteins that can react with BMCC. Only proteins that are exposed as a result of Golgi fragmentation will become biotinylated.

Another technique that is utilized in this work is **immunoprecipitation** (IP). Just like the techniques of immunoblotting or immunofluorescence, IP relies on the ability of a primary antibody to bind to a specific protein of interest. In this case, however, the secondary antibody that binds to the primary antibody is chemically linked to a bead. Beads are easy to pellet due to their size, and as they pellet they carry with them all of the beads that have attached to them. If the protein of interest binds to another protein, then that second protein will also be "pulled down" and found in the pellet, a result called coimmunoprecipitation (co-IP). The validity of a co-IP result can be confirmed by reciprocal IP reactions using antibodies to both proteins. If IP with either antibody yields the same result, then it can be concluded that the interaction between the two proteins is not an artifact.

- Predict the outcome of a mitotic cell division where the Golgi complex remained intact.
- Relate the ability of the Golgi to fragment into vesicles with the properties of a biological membrane.
- How might the cytoplasm of a mitotic cell be different from the cytoplasm of an interphase cell?
- Create a list of cellular functions. Name one or more proteins that are involved in each of the functions you listed.
- Which amino acids contain SH groups?
- Create a flow chart or diagram to illustrate the NEM/BMCC experiment.
- What assumption can you make about proteins that are only exposed after Golgi membranes are incubated with mitotic cytosol?

METHODS

Membrane and cytosol isolation

Rat livers were homogenized in buffer. The homogenate was subjected to sucrose density gradient centrifugation. Intact Golgi complexes accumulated at the 0.5/0.8 M sucrose interface and were collected using a Pasteur pipet. Rat liver cytosol remained in the supernatant. Interphase and mitotic cytosols were collected from HeLa cells, a human tissue culture cell line. HeLa cells were harvested and then grown for 24 h in the presence or absence of the microtubule inhibitor, nocodazole. Cells treated with 100 ng/mL of nocodazole were blocked at mitosis. Cells were homogenized and the homogenate was centrifuged at 400,000g for 30 min. Supernatants were collected and stored frozen until use.

Cell-free disassembly/reassembly assays

Purified rat liver Golgi membranes were disassembled by incubation with mitotic cytosol in the presence of an ATP-regenerating system for 30 min at 37°C. Reassembly of the Golgi was accomplished by incubating the disassembled Golgi membrane fragments with interphase cytosol, also in the presence of an ATP-regenerating system. Membranes were recovered by centrifugation through a 0.5 M sucrose cushion.

To examine the role of proteins in Golgi stacking, membranes were treated with 2.5 mM NEM or 0.34 mM BMCC for 15 min followed by a 10 min incubation with 5 mM DTT to quench the reaction. NEM and BMCC bind to SH groups on proteins while DTT protects SH groups. Because BMCC is a biotinylated reagent any proteins that it binds to will become biotinylated and can be detected using streptavidin conjugated to horseradish peroxidase (HRP).

^{32}P-labeled Golgi membranes were prepared by incubating Golgi membranes with interphase or mitotic cytosol in the presence of 100 μCi of [γ-^{32}P]ATP. Membranes were recovered as described earlier. Following SDS-PAGE, radiolabeled proteins were detected by exposing the gel to X-ray film.

Protease and extraction treatments

The orientation of Golgi membrane proteins was examined by digestion with proteases at concentrations that would not cross the phospholipid bilayer. Membranes were incubated with 10 μg/mL trypsin, 10 μg/mL chymotrypsin, or 1 μg/mL proteinase K for 15 min on ice. Golgi membranes were incubated with buffer, a calcium chelator (10 mM EDTA), high salt (1 M KCl), or alkaline pH (100 mM sodium carbonate) to examine the nature of the interaction between the proteins and the membranes. Following either the protease or extraction treatments the samples were centrifuged to pellet the membranes, leaving soluble proteins in the supernatant. The detergent Triton X-114 (TX-114) was used to confirm the location of proteins in the Golgi membrane. TX-114 creates a phase separation between aqueous and detergent-soluble proteins.

Immunoblots and immunoprecipitation

Samples were analyzed using SDS-PAGE and immunoblotting. Antibodies were raised against the complete p65 protein or amino acids 220–234 of the p65 protein sequence. Antibodies to the enzyme mannosidase II and the peripheral Golgi membrane protein GM130 were purchased. Antibody binding to immunoblots was detected using HRP-labeled secondary antibodies.

IP was performed by incubating solubilized membrane samples with primary antibodies and protein-A Sepharose beads. After 60 min incubation at 4°C the beads were washed several times with buffer and the bound proteins were eluted for analysis by SDS-PAGE and immunoblotting.

Electron microscopy

Golgi membrane pellets were fixed in 1% glutaraldehyde, stained with tannic acid, dehydrated, embedded in plastic, and sectioned. Cisternae were defined as membranes with a length that was more than four times their width. A membrane stack was defined as consisting of two or more cisternae in parallel alignment for more than half their length and separated by a gap of no more than 15 nm. The relative proportion of cisternae stacks was determined by counting the number of intersections of each membrane structure with a 4 mm line grid. From these values the relative number of stacked membranes per n cisternae was calculated.

- Explain why inhibiting microtubule polymerization would stop a cell in mitosis?
- Conduct a search for information about the history and ethical questions that surround the story of the HeLa cell line.
- Create a diagram of a vesicle that contains (1) a lumenal protein, (2) a transmembrane protein, and (3) a cytosolic peripheral membrane protein. Predict whether you would expect to find each of the proteins in the supernatant or pellet following protease or extraction treatments. Explain your predictions.
- What type of protein would you expect to find in the detergent phase following TX-114 solubilization of Golgi membranes?

RESULTS

- Which of the treatments listed in Figure 7.2.1 will cause the Golgi to disassemble?

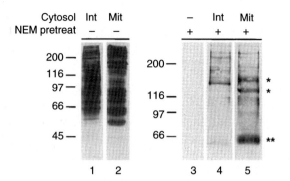

FIGURE 7.2.1 Identification of a protein potentially involved in the formation of Golgi stacks. Golgi disassembly reactions were performed with control (NEM; lanes 1 and 2) or NEM pretreated (lanes 3–5) rat liver Golgi membranes in the presence of interphase cytosol (Int), mitotic cytosol (Mit), or buffer (—). Membranes were incubated with BMCC, and biotinylated proteins were detected using a streptavidin overlay technique. Multiple proteins were labeled by BMCC in the absence of NEM pretreatment, but only three proteins were detected in NEM pretreated samples: a major biotinylated protein at 65 kDa, a protein at 130 kDa, and one at 160 kDa.

- Propose an explanation for the differences in the pattern of labeled proteins in lanes 1 and 2 of Figure 7.2.1.
- Explain why only a few proteins were biotinylated in lanes 4 and 5 of Figure 7.2.1.
- Suggest a reason for including lane 3 in Figure 7.2.1.
- What affect would preincubation of the antibody with peptide have on the immunoblot? Why does this serve as a control for the immunoblot experiments in Figure 7.2.2a?
- Why is p65 present in the homogenate sample, but not in the cytosol sample in Figure 7.2.2a?
- Illustrate the relative locations of p65, GM130, and mannosidase II on the Golgi membrane based on the evidence from the protease experiment shown in Figure 7.2.2b.
- Modify your illustration to take into account the results from the extraction and TX-114 solubilization experiments in Figure 7.2.2b.
- Explain the significance of the results shown in Figure 7.2.3a and b.
- What is happening in the experiment shown in Fig. 7.2.3c that is causing Golgi membrane proteins to become radioactive?
- Use your knowledge of the cell cycle to propose an explanation for the differences in radioactive labeling of Golgi membrane proteins shown in Figure 7.2.3c.
- Construct an argument based on the data in Figure 7.2.3 against the following statement, "GM130 only interacts with p65 during mitosis."

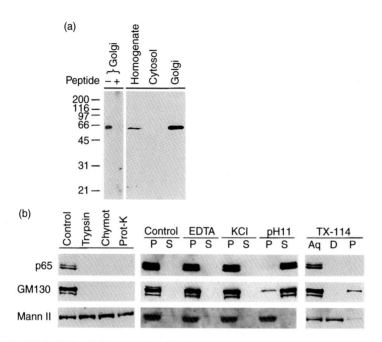

FIGURE 7.2.2 p65 is a tightly associated Golgi peripheral membrane protein.
a. Samples from the isolation of rat liver Golgi membranes were analyzed by immunoblotting using an antibody specific to amino acids 220–234 of the p65 sequence. The antibody was used either with (+) or without (−) preincubation with the 220–234 peptide. b. The location of p65 on the Golgi membrane was examined using protease digestion. Golgi membranes were incubated with buffer (control), trypsin, chymotrypsin (Chymot), or proteinase K (Prot-K). Association of p65 with the Golgi was tested by incubating membranes in the presence of buffer (control), 10 mM EDTA, 1 M KCl, or 100 mM sodium carbonate (pH11). Membranes were recovered by centrifugation following either the protease or extraction treatments and examined by immunoblotting with antibodies to p65, GM130, and mannosidase II (Mann II). Golgi membranes were also extracted using the detergent Triton X-114 (TX-114) to separate peripheral and integral membrane proteins and analyzed by immunoblotting as described earlier.

- Under control conditions what percent of the reassembled Golgi membranes form a stack of three or more cisternae (Figure 7.2.4a)?
- Locate an example of a two-cisternae stack and a three-cisternae stack in the image in Figure 7.2.4a.
- What effect does the loss of p65 have on the formation of cisternal compartments from Golgi membrane fragments?
- Summarize the conclusion that is supported by the data shown in Figure 7.2.4b and c.
- Propose an explanation for the effect of the addition of excess recombinant p65 shown in Figure 7.2.4d.

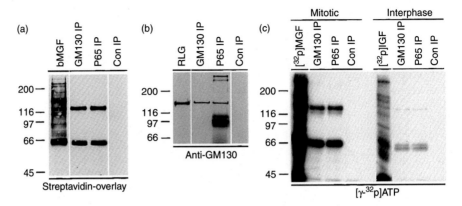

FIGURE 7.2.3 p65 binds with GM130 and is modified in a cell cycle–dependent manner.
a. Biotinylated mitotic Golgi fragments (bMGF) were solubilized and immunoprecipitated with polyclonal antibodies to either GM130 (GM130 IP) or p65 (P65 IP) or the corresponding preimmune serum (Con IP). Bound proteins were analyzed by streptavidin–HRP. b. Rat liver Golgi membranes were solubilized and immunoprecipitated as described earlier; however, the bound proteins were analyzed by immunoblotting with an antibody to GM130. Additional bands in the P65 IP lane are an artifact of the secondary antibody. c. Rat liver Golgi membranes were incubated with [γ-^{32}P]ATP under mitotic or interphase conditions. Membranes were solubilized and immunoprecipitated as described earlier. Bound proteins were analyzed by SDS-PAGE and autoradiography.

- Discuss the evolutionary implication of the observation that rat liver Golgi membranes will reassemble when incubated with interphase cytosol from a human tissue culture cell line.
- Design an experiment to investigate the role of GM130 in *in vitro* Golgi reassembly.
- The Golgi complex in the cells that make up the silk-producing glands in a spider are not organized into the classic stack of cisternae that you would expect to find in other eukaryotic cells. Construct a hypothesis regarding the organization of the Golgi in a spider silk gland. Design an experiment to test your hypothesis.
- The Golgi compartment within the cell is organized into *cis, medial,* and *trans* cisternae. Design an experiment to investigate whether *in vitro* reassembly of Golgi membranes recreates this organization.

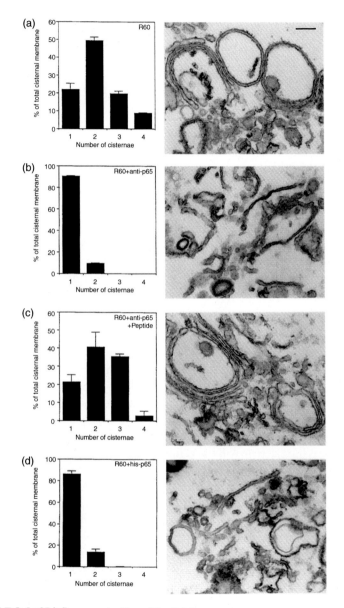

FIGURE 7.2.4 p65 influences stacking of the Golgi.
Rat liver Golgi membranes were incubated with mitotic cytosol for 30 min at 37°C. Membrane fragments were recovered and treated for 15 min on ice with either buffer (a), antibody specific to p65 peptide sequence 220–235 (anti-p65) (b), p65 antibody preincubated with peptide sequence (c), or excess recombinant p65 (his-p65) (d). Treated membranes were switched to interphase cytosol for 60 min at 37°C after which samples were processed for electron microscopy. The amount of cisternal membrane in structures containing one, two, three, or four cisternae was quantitated. Values were normalized and the means plotted. Error bars indicate the standard error of the mean. Scale bar = 0.2 μm.

Case of the Coated Vesicle [3]

INTRODUCTION

Vesicles are membrane-bound organelles that function to transport material throughout the cell. A typical vesicle consists of a phospholipid bilayer surrounding a lumen or interior space. The cargo a vesicle carries is packaged inside the lumen or, in the case of integral membrane proteins and lipids, is part of the vesicle membrane. The nature of the cargo found in a vesicle will differ depending on the vesicle's origin. Vesicles that arise from the RER typically carry proteins and lipids while vesicles that form from the plasma membrane may carry large macromolecules or even other cells.

The life of a vesicle consists of three phases. Phase one involves the formation of the vesicle by budding from a donor membrane. In this process a region of the bilayer "bends" away from the plane of the donor membrane, forming a pocket or bud. **Coat** and **adaptor proteins** located on the cytosolic face of the membrane help to stabilize the shape of the bud and can function to select its cargo. Once all the components have been assembled, the coated vesicle will separate from the donor membrane. The separation step typically requires energy. During phase two of its life a vesicle will travel from the donor membrane to an **acceptor membrane**. Vesicles can move by diffusion or in association with motor proteins that catalyze their transport along elements of the cytoskeleton. Before this step can occur, however, the vesicle must lose its protein coat. Removal of the coat is necessary both to facilitate fusion and to expose other membrane proteins that are involved in targeting the vesicle. Vesicles are targeted to the correct acceptor membrane through the activity of a class of proteins called Rabs. Each membrane compartment in the cell is characterized by a unique set of Rab proteins. In the final phase, the vesicle docks and fuses with the acceptor membrane. Docking occurs when a *v-SNARE* protein located on the vesicles binds to a *t-SNARE* protein located on the acceptor membrane. The interaction between the SNARE proteins holds the vesicle in close proximity to its target. Fusion occurs when the two phospholipid bilayers make contact with one another and merge.

- Explain why the phospholipids that form a biological membrane are organized into a bilayer structure.
- We know that the cytosol can be considered an aqueous environment. Predict what type of environment would exist inside the lumen of a vesicle.

- Could a vesicle "bud" from a donor membrane in the absence of coat proteins? Why or why not?
- Botox® is a toxic protein produced by *Clostridium botulinum* bacteria. *Botulinum* toxin works as a protease that specifically degrades SNARE proteins. Predict how the destruction of SNAREs results in the "reduction of lines and wrinkles" on a person's face.

BACKGROUND

One of the best-characterized examples of vesicle formation occurs during receptor-mediated endocytosis. Clathrin-coated pits on the cytoplasmic side of the plasma membrane develop into clathrin-coated vesicles that carry material from the plasma membrane to endosomes. A clathrin-coated vesicle is formed when multiple clathrin *triskelion* proteins assemble along the surface of a membrane to form a structural scaffold. The adaptor protein AP1 controls the assembly of clathrin on the membrane surface. AP1 serves both to link the clathrin to the membrane and to help select and retain specific cargo within the developing vesicle by interacting with the cytoplasmic tails of receptors or other transmembrane proteins. The final step in vesicle formation is mediated through the activity of the enzyme dynamin, which uses GTP energy to "pinch" the vesicle off the donor membrane. Once the coated vesicle is free in the cytoplasm it goes through an "uncoating" step that removes clathrin leaving the vesicle capable of targeting and fusing with its acceptor membrane.

Vesicle transport within the endomembrane system involves a different class of coated vesicle. Just like clathrin-coated vesicles, these vesicles are formed when adaptor proteins facilitate the assembly of a structural protein complex or scaffold on the surface of a donor membrane. **COP I**–coated vesicles were the first class of "nonclathrin coated vesicles" to be identified. COP I–coated vesicles are generated from Golgi membrane and are believed to be involved in retrograde transport or return of materials that have "leaked" from their proper compartment; for example, the return of RER-specific proteins from the *cis* Golgi using the KDEL receptor.

The proteins involved in the formation of COP-coated vesicles have been identified in yeast and in humans. COP I assembly is mediated by the adaptor protein *ARF1*. ARF1 facilitates the assembly of a **coatomer complex** analogous to the clathrin triskelion. The coatomer is made up of seven different protein subunits. Both ARF1 and the coatomer proteins are soluble cytosolic proteins that require GTP in order to bind to the donor membrane surface. Hydrolysis of GTP to GDP by ARF1 has been linked to the uncoating of COP I vesicles.

A second class of nonclathrin-coated vesicles are involved in endomembrane transport. **COP II**–coated vesicles use the adaptor protein **Sar1** to recruit coat protein complexes to the surface of the ER membrane. **Sec 23** and **Sec 13** protein complexes assemble to form the COP II protein coat and cause the donor membrane to curve forming a vesicle bud. This case examines the role of the adaptor and coat proteins in the formation and transport of COP II vesicles.

- Speculate on why different coat proteins are required for different vesicle transport pathways.
- In addition to their role in endocytosis, clathrin-coated vesicles are also involved in the transport of lysosomal enzymes from the *trans* Golgi. What similarities exist in these two pathways that might explain why clathrin can be used for both?
- Conduct a search for diagrams of the structure of COP I and COP II–coated vesicles. Identify the following proteins in the diagrams:
  - ARF1
  - β COP I
  - γ OP I (Sec 21)
  - Sar 1
  - Sec 12
  - Sec 23
  - Sec 13

METHODS

Cell-free vesicle-budding assay

Transport from the ER to the Golgi was reconstituted using gently lysed yeast cells as a membrane source and a [^{35}S]-methionine–labeled yeast secretory protein, prepro-alpha factor (^{35}SppαF). Washed and extracted yeast membranes were incubated at 10°C for 15 min in the presence of purified Sar1p, Sec13p complex, Sec24p complex, and ^{35}SppαF. Energy was provided in some experiments by the inclusion of an ATP-regenerating system. The effect of the 0.1 mM guanine nucleotides such as GDP or GTP, or the nonhydrolyzable nucleotides GTPγS (**GγS**) or GMP-PNP (**PNP**) were also examined. ^{35}SppαF translocates into the lumen of the ER where core glycosylation occurs, converting the protein to glycosylated proalpha factor (^{35}SgpαF).

Transport between the ER and Golgi was stopped by incubating the membranes at 0°C followed by a medium-speed centrifugation step. Large membranes pellet leaving smaller membranes, such as vesicles, in the medium-speed supernatant (**MSS**). Density gradient centrifugation was used to separate these smaller vesicles from other components in the MSS. The MSS fraction was mixed with an equal volume of 70% buffered Nycodenz and placed on the bottom of a centrifuge tube. 0.5 mL of three solutions of decreasing density (30, 25, and 15%) were layered on top of the supernatant. The sample was centrifuged at 55,000 rpm for 2 h. A purified vesicle membrane fraction containing ^{35}SgpαF was isolated from the 25–15% interface.

Cell-free vesicle fusion to Golgi membranes

Vesicles isolated as described earlier were incubated with yeast membranes, cytosol, 0.2 mM GTP, and an ATP-regenerating system at 25°C for 1 h. Core-glycosylated ^{35}SgpαF proteins that entered the Golgi were modified by the addition of α-1,6-mannose to the protein. Modified proteins were quantified by IP with an antibody that recognizes α-1,6-mannose.

Gel filtration chromatography

Gel filtration chromatography was used to separate soluble proteins from vesicle-associated proteins based on the time required for these two populations to move through a gel filtration column. Supernatant from vesicle formation reactions was loaded onto a 14 mL Sephacryl-S1000 column equilibrated in buffer. The pore size of these Sephacryl beads accommodates spherical particles up to 400 nm allowing for the separation of vesicles from soluble proteins. The column was eluted using the buffer, and 0.5 mL fractions were collected. The presence of ^{35}SgpαF protein in the column fractions was detected and quantified by scintillation counting. Column fractions were also examined by immunoblotting with antibodies to Sec23p, Sec13p, and Sar1p.

Electron microscopy and immunocytochemistry

Purified ER-derived vesicles were fixed in1% glutaraldehyde in buffer for 30 min at 4°C and then centrifuged to form a pellet. The pellet was stained with tannic acid to provide contrast, dehydrated, embedded in plastic, and sectioned. For immunocytochemistry, the membrane pellet was infiltrated with 2.3 M sucrose and frozen. Cryosections of the frozen pellet were cut, placed onto grids, and incubated with antibodies specific to the indicated proteins. Antibody binding was detected using protein A-labeled gold particles.

- Why was it necessary to add Sar1p, Sec13p complex, and Sec24p complex to the vesicle-budding assays?
- Describe what occurs when a protein enters the ER that would explain why the abbreviation for the yeast marker protein used in these experiments changes from ^{35}SppαF to ^{35}SgpαF.
- Explain why incubation at 0°C could be used to stop the formation of vesicles from the ER and their fusion with the Golgi.
- What is the final percent density of Nycodenz in the MSS sample? Which direction do the transport vesicles move during the process of density gradient centrifugation?
- Conduct a search online to find videos or animations that help describe the process of gel filtration. Predict the pattern of elution of a vesicle versus a soluble protein.

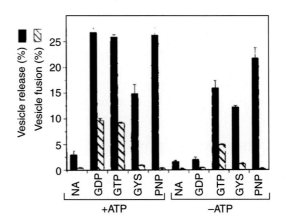

FIGURE 7.3.1 Cell-free vesicle budding requires energy.
Percent of ^{35}SgpαF-containing vesicles (solid bars) released from the ER in the presence or absence (NA) of guanine nucleotides and saturating amounts of Sar1p, Sec23p complex, and Sec13p complex. Vesicle release was assayed in the presence (+) and absence (−) of 1 mM ATP. Vesicle fusion with the Golgi (hatched bars) was determined based on the percent of ^{35}SgpαF immunoprecipitated with an antibody to α-1,6-mannose.

RESULTS

- Which nucleotide, ATP or GTP, is more important for vesicle release from the RER and fusion with the Golgi (Figure 7.3.1)? Explain your answer.
- The compounds GTPγS (GγS) and GMP-PNP (PNP) are competitive inhibitors of GTPases. Describe how a competitive inhibitor affects a protein's function.
- Summarize your observations of the gel in Figure 7.3.2a.
- Explain why the ER membrane would react to all of the antibodies tested in Figure 7.3.2b.
- Discuss the significance of the observation that Sec21p (a COP I protein) and Sec12p (an integral membrane protein in the ER) are not found in any of the vesicle samples.
- Construct an argument from the data presented in Figure 7.3.1a and Figure 7.3.2b to support the conclusion that GTP hydrolysis is required for uncoating vesicles.
- Which fraction(s) contain the ER-derived transport vesicles (Figure 7.3.3d)?
- Is all of the ^{35}SgpαF protein packaged into vesicles? Justify your answer.
- Explain why the majority of Sec23p and Sec13p are found in fractions 18–20 in Figure 7.3.3c.

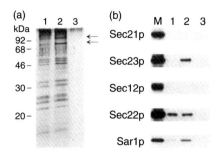

FIGURE 7.3.2 Protein composition of gradient-purified ER-derived vesicles.
a. Vesicles were generated using the cell-free system in the presence of 0.1 mM GTP (lane 1), GMP-PNP (lane 2), or no guanine nucleotide (lane 3) and were purified using a Nycodenz gradient centrifugation. Vesicle samples were run on a 12.5% SDS-PAGE and silver-stained to reveal the protein composition.
b. Immunoblots of yeast ER membrane (lane M) and gradient-purified vesicles produced in the presence of GTP (lane 1), GMP-PNP (lane 2), or no guanine nucleotide (lane 3) reacted with antibodies against the indicated proteins.

- Generate a hypothesis based on the observation that Sar1p staining is detected in Figure 7.3.3b, but not in Figure 7.3.3a.
- Identify the vesicle's phospholipid bilayer, protein coat, and vesicle lumen in Figure 7.3.4a.
- The arrows and arrowheads in Figure 7.3.4e indicate uncoated or partially uncoated vesicles. How does this observation relate to the data presented in Figures 7.3.1–7.3.3 in this case study?
- What new information is provided by the data presented in Figure 7.3.4?
- Does the difference in gold labeling between Figure 7.3.4b and f support or refute the hypothesis you developed regarding the behavior of Sar1p? Explain your answer.
- One of the conclusions from this work is that COP II–coated vesicles are responsible for anterograde transport in the endomembrane system. Explain how the data support this statement.
- Design an experiment to determine whether COP I proteins are excluded from COP II–coated vesicles.

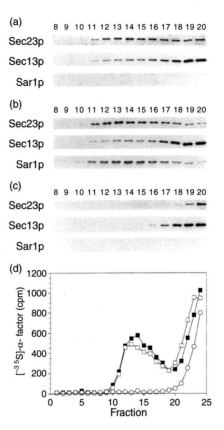

FIGURE 7.3.3 Isolation of ER-derived vesicles by gel filtration chromatography.
Vesicles from yeast ER donor membranes were eluted through a Sephacryl S-1000 column. Column
fractions were immunoblotted with antibodies specific to the indicated proteins for vesicles formed in
the presence of GTP (a), GMP-PNP (b), or no guanine nucleotide (c). d. Scintillation counting was used to
quantify the amount of ^{35}SgpαF present in each fraction: GTP (open squares), GMP-PNP (filled squares),
and no guanine nucleotide (open circles).

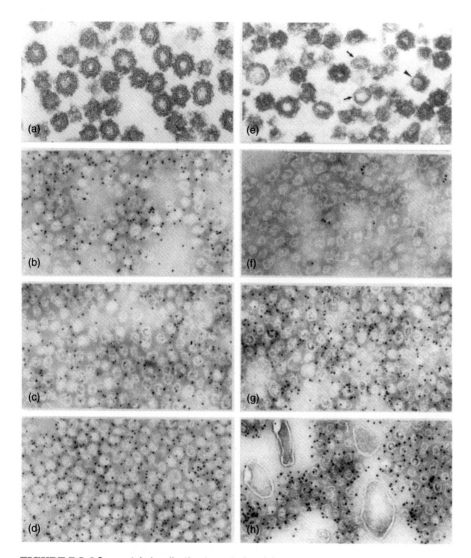

FIGURE 7.3.4 Sec protein localization to coated vesicles.
Purified vesicles formed in the presence of GMP-PNP (a–d) or GTP (e–h) were fixed, embedded, and sectioned for TEM (a and e) or cryosectioned for immunocytochemistry and protein A-gold labeling following incubation with antibodies to Sar1P (b and f), Sec23p (c and g), or Sec 13p (d and h). Images (a) and (e) magnified 103,700×. Images (b–d) and (f–h) magnified 68,400×.

How to Build a Peroxisome [4]

INTRODUCTION

Peroxisomes are membrane-bound organelles that function to break down a variety of toxic compounds that can damage a cell. Peroxisomes contain a collection of enzymes that function to degrade compounds such as alcohols, toxins, and fatty acids. Enzymes found in the plant equivalent of a peroxisome, a glyoxysome, catalyze the formation of acetyl CoA from fatty acids stored within germinating seeds. The peroxisomal enzyme luciferase is responsible for the production of light in fireflies and the enzyme catalase converts highly reactive hydrogen peroxide (H_2O_2) into water. Peroxisomes are also involved in the synthesis of phospholipids that are important for the formation of the myelin sheaths that surround the axons of nerve cells.

Peroxisomes are assembled within the cell. The single membrane that surrounds a peroxisome is characterized by the presence of multiple peroxisomal membrane proteins (**PMPs**). This membrane originates from the **ER**. The enzymes that are critical for peroxisomal function, however, are transported into the lumen of the peroxisome from the cytoplasm of the cell. Several of the PMPs assemble to form a translocon spanning the peroxisomal membrane. As its name implies, a translocon is a type of channel that facilitates the movement (translocation) of proteins across the membrane. Peroxisomal enzymes carry a short peroxisomal targeting sequence (**PTS**) that imports the enzymes through the translocon and into the organelle.

The research question addressed in this case study asks how peroxisomes assemble in a yeast cell. To answer this question, the researchers made novel use of fluorescently labeled proteins. In the **split GFP assay**, proteins that are known to interact are each labeled with half of a functional fluorescent protein. On their own, these half-labeled proteins do not fluoresce. However, if the proteins interact, then the two halves of the fluorescent protein are brought together and will fluorescence. Another name for this protocol is fluorescence complementation. **Fluorescence pulse chase** is a technique that was also used in this work. In this technique the timing of expression of fluorescently labeled proteins is controlled through the use of the **Gal1 inducible promoter**. When yeast cells are grown in a medium that contains the sugar galactose, the Gal1 promoter will be activated and any genes located downstream of the promoter will be transcribed and translated. Switching the yeast to a growth medium that contains glucose instead of galactose shuts down the expression of Gal1-induced proteins.

- Why would a germinating seed need a supply of acetyl CoA?
- Connect the cellular activities associated with the ER to its ability to provide membrane and membrane proteins to form a peroxisome.
- Why does a peroxisomal enzyme need a translocon to move across the membrane of the peroxisome?
- Zellweger syndrome is a genetic disease in which the import of enzymes into peroxisomes is blocked. Predict what effect this disease would have on an individual.

BACKGROUND

Yeast cells are an important model system for the study of eukaryotic cell biology. A wide variety of mutations exist which allow researchers to "dissect" cellular processes. As a result, much is known about the genes that are responsible for various aspects of yeast cell function. Yeast can also be transfected with plasmids carrying genes for modified or novel proteins, making it possible to investigate cellular mechanisms by manipulating or tracking individual proteins. Another way to manipulate the genetics of yeast cells is by mating. Haploid yeast cells fuse forming a new diploid cell and merging the nuclear and cytoplasmic contents of both cells.

Peroxisomal assembly and function requires the expression of a large set of proteins. Functional peroxisomal translocons are formed by the association of two separate protein subunits: a **docking complex** and a **RING finger complex**. Each of these subunits is made up of peroxisomal proteins or Pex. **Pex13p** and **Pex14p** assemble to form the docking complex and **Pex2p**, **Pex10p**, and Pex12p assemble to form the RING finger complex. **Pex3p** and **Pex19p** are PMPs that facilitate the formation of **preperoxisomal vesicles** carrying PMPs from the ER. **Pex1p** and **Pex6p** are involved in fusing preperoxisomal vesicles to form a peroxisome. The import of enzymes carrying the PTS is required to form a **mature peroxisome**.

- What properties define yeast as being eukaryotic cells?
- Diploid yeast cells have 16 chromosomes. How many chromosomes would be present in a haploid cell?
- What does the abbreviation "RING" stand for? What can you learn about RING finger proteins?
- Create a diagram of the steps involved in formation of a peroxisome starting with PMPs in the ER and ending with a mature peroxisome.

METHODS

Yeast strains and DNA

Various haploid mutant strains of *Saccharomyces cerevisiae* were used in this study. Plasmids carrying gene deletions or gene fusions were transfected into the cells using techniques that caused the genes to become integrated into the yeast's genome. Yeast were grown in YP media at 30°C.

Galactose induction for fluorescence pulse chase

Yeast cells were transfected with genes for yellow fluorescent protein (**YFP**) or cyan fluorescent protein (**CFP**)–tagged PMPs or CFP fused to the PTS (**CFP-PTS**) under control of a GAL1 promoter. Transfected cells were grown in YP media containing 2% galactose for 2 h at 30°C. Cells were pelleted by centrifugation, resuspended in buffered saline, then pelleted again, and resuspended in YP media containing 2% glucose. Cells were grown for another 2 h at 30°C. YFP-tagged proteins fluoresce green while CFP-tagged proteins fluoresce red. Overlap of the fluorescent signals from YFP and CFP proteins that are colocalized within the cell will appear yellow.

Mating assays

Yeast cell cultures were grown to early log phase and collected by centrifugation at 3000g at room temperature. Pelleted cells were resuspended in fresh YP media. Ten microliters of each haploid strain were spotted on top of each other on a YP plate and incubated at 30°C.

Microscopy

Cells were examined using fluorescence microscopy. For the split GFP experiments, mated yeast were examined at 24 h intervals for up to 72 h. For the fluorescence pulse chase experiments, cells were mated and then observed after 5, 10, and 24 h. To observe the yeast, a small amount of the cells was removed from the YP plate and resuspended in 1.5 µL of buffered saline. The sample was then applied to a microscope slide and viewed immediately. Separate YFP and CFP digital images were collected and analyzed using image analysis software.

- Conduct a search to learn more about *S. cerevisiae*. How do people use these cells other than in research?
- What would happen to the expression of *Gal1-Pex13-CFP* when cells are grown in the presence of glucose?
- Research how a traditional pulse chase experiment is conducted. How does switching cells from galactose to glucose equate to a fluorescence pulse chase?
- Suggest a reason colocalized proteins appear yellow?

RESULTS

- What color is produced when PEXx-VN assembles with PEXy-VC?
- Explain why the "peroxisomes" in the haploid cell in Figure 7.4.1a are red, but in the diploid cell they are yellow.
- Summarize the results shown in Figure 7.4.1b in terms of the assembly of a mature peroxisome.
- Design an experiment that would work as a control for the specificity of the split GFP assay.
- Predict where you would find CFP-PTS1 labeling in a Δpex3 mutant cell.
- The mutants Δpex3 and Δpex19 cause PMPs to accumulate in the membrane of the ER. Which of the schematic diagrams shown in Figure 7.4.1c reflects the results of mating Δpex3 × Δpex3 or Δpex19 × Δpex19?
- Predict what would happen if the docking complexes and RING finger complexes assembled while in the membrane of the ER.
- The mutants Δpex1 and Δpex 6 can form preperoxisomal vesicles, but not mature peroxisomes. What do the results from the Δpex1 × Δpex1 and Δpex6 × Δpex6 mating experiments tell you about the preperoxisomal vesicles?
- Explain why fluorescence complementation was seen in all of the cells resulting from the Δpex1 × Δpex6 mating in Figure 7.4.1c.
- Where in the cell would the Δpex6 Gal-Pex11-YFP and Δpex1 Gal-Pex13-CFP proteins be located after 2 h incubation in the presence of glucose (Figure 7.4.2a)?
- Explain how the results presented in Figure 7.4.2b support the conclusion that two distinct populations of preperoxisomal vesicles exit the ER and eventually fuse.
- Use your diagram of the steps of peroxisome formation to explain the result for the mating the wild-type Pex11-YFP and Pex13-CFP cells (Figure 7.4.2b).
- Describe what you observe in the images for Figure 7.4.2c. What conclusion is supported by these data?
- Propose an explanation for the lack of colocalization of the fluorescent signal in Figure 7.4.3a.
- Where would you predict the GAL1-CFP-PTS proteins would be found in a wild-type cell (Figure 7.4.3b)?
- Can preperoxisomal vesicles fuse with mature peroxisomes? Use the data in Figure 7.4.3 to defend your answer.
- Modify your diagram of how peroxisomes are formed to reflect the data presented in this case study.

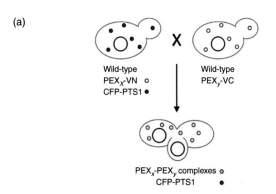

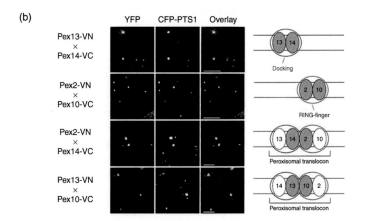

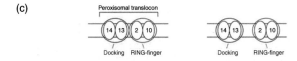

PMPs	wt × wt	Δpex3 × Δpex3 Δpex19 × Δpex19	Δpex1 × Δpex1 Δpex6 × Δpex6	Δpex1 × Δpex6
Pex2p-Pex10p	+	+	+	+
Pex13p-Pex14p	+	+	+	+
Pex2p-Pex14p	+	−	−	+
Pex10p-Pex13p	+	−	−	+

FIGURE 7.4.1 Stages of peroxisome formation can be tracked using split GFP assay.
a. Schematic representation of the experimental protocol. Haploid yeasts, each expressing a different translocon protein (Pex) linked to one half of a GFP (VN or VC), were mated. One of these haploid cells also expressed a red fluorescent protein (RFP) linked to a PTS. b. Wild-type haploid cells expressing either Pex10-VC or Pex14-VC were mated in various combinations with wild-type cells coexpressing Pex13-VN or Pex2-VN and the marker protein CFP-PTS1. The pattern of green fluorescence (YFP) is compared with that of red fluorescence (CFP-PTS1) by merging the two images into an overlay. c. Summary of split GFP experiments using various yeast peroxisome mutants. Haploid cells lacking Pex3 (Δpex3), Pex19 (Δpex19), Pex1 (Δpex1), or Pex 6 (Δpex6) were mated. (−) indicates no YFP fluorescence; (+) indicates YFP fluorescence.

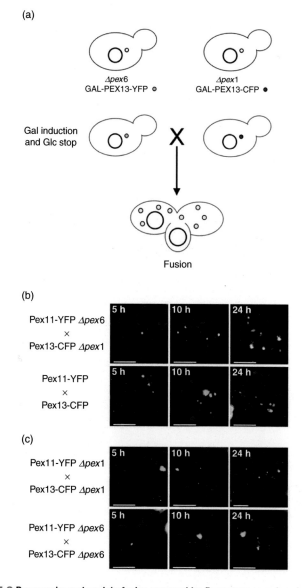

FIGURE 7.4.2 Preperoxisomal vesicle fusion assayed by fluorescence pulse chase.

a. Schematic diagram of the experimental protocol. Plex13 (a docking complex protein) and Plex11 (a RING finger complex protein) were tagged with CFP and YFP, respectively, and their expression was controlled with the GAL1 promoter. Haploid cells mutant for Δpex1 or Δpex6 were grown in the presence of galactose then switched to glucose prior to mating. b. Haploid Δpex1 or wild-type cells that had expressed GAL-PEX13-CFP were mated to haploid Δpex6 or wild-type cells that had expressed GAL-PEX11-YFP and were examined after 5, 10, and 24 h for colocalization of the fluorescent proteins. c. Haploid Δpex1 cells expressing either GAL-PEX11-CFP or GAL-PEX13-YFP were mated. A parallel experiment was conducted using haploid Δpex6 cells expressing GAL-PEX11-CFP or GAL-PEX13-YFP.

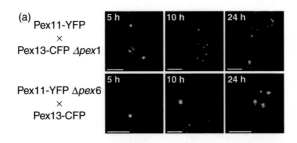

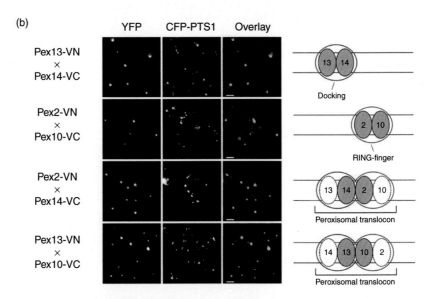

FIGURE 7.4.3 Preperoxisomal vesicles turn into mature new peroxisomes.
a. Haploid wild-type cells that had expressed GAL-PEX11-YFP or GAL-PEX13-CFP were mated with haploid Δpex1 or Δpex6 cells that had expressed GAL-PEX13-CFP or GAL-PEX11-YFP, respectively, and observed 5, 10, or 24 h after mating. b. Split GFP assay for translocon assembly in wild-type cells coexpressing CFP-PTS under the control of the Gal1 promoter. Expression of CFP-PTS was induced in the presence of galactose prior to mating of the haploid cells.

Putting the Squeeze on Mitochondria [5]

INTRODUCTION

Mitochondria are well known as the "powerhouse of the cell" since they provide the majority of ATP energy used by eukaryotes. Given the importance of mitochondria to the life of cells and multicellular organisms, obvious questions are, "How do cells maintain a sufficient population of mitochondria to support their energy needs?" and "How does a dividing cell insure that there are enough mitochondria present in its cytoplasm to be distributed to the two new cells?"

Mitochondria are dynamic organelles that have the ability to grow and divide. Like their prokaryotic ancestors, mitochondria divide through a process known as binary fission. During the process of binary fission a dynamin-like protein *Drp1* moves from the cytoplasm of the cell and accumulates as a spiral "belt" of protein that wraps around the mitochondrium at a specific point along the outer membrane. Drp1 is a GTPase, meaning that it uses the energy stored in a molecule of GTP to trigger the constriction or "pinching in" of the mitochondrial membranes, finally leading to separation of the organelle into two parts.

In order to study how mitochondria behave in eukaryotic cells it is necessary to have easy access to a population of cells. The technique of **tissue culture** allows researchers to grow large numbers of specific types of cells under controlled conditions in the laboratory. Cells can be grown in dishes or on surfaces, such as coverglasses, which facilitate viewing the cells under a microscope. In addition, cultured cells can be manipulated with the addition of reagents, drugs, dyes, or plasmids that will alter the genetic makeup of the cell.

Genetic engineering of cultured cells allows researchers to investigate cellular mechanisms *in vivo* by disrupting or altering the activity of the individual proteins involved in a process. **Transfection** is the addition of genetic information to a cell. Genes that are introduced into cells through transfection can be modified in a number of ways that will result in additions, deletions, or changes to the amino acid sequence of the protein encoded by that gene. Another way to influence the activity of proteins in a cell is through **RNA interference** or RNAi. In this method, researchers are able to suppress expression of a specific protein by triggering the cell to selectively destroy the mRNAs for that protein.

The presence of double-stranded RNA molecules (**siRNAs** or small interfering RNAs) in the cytoplasm triggers a cellular response that targets any RNA molecules that carry the same sequence as that of the siRNA.

Light microscopy is another important method for the examination of cells and cell behavior. There are many different types of light microscopy, but one thing they all have in common is the use of images to provide both qualitative and quantitative data. Fluorescence microscopy is a specialized form of light microscopy that relies on the ability of special molecules to absorb and emit light. Fluorescent dyes can be attached to a variety of molecules or proteins. Alternatively, the sequence for **green fluorescent protein** (GFP), a small protein responsible for bioluminescence in marine invertebrates, can be genetically engineered so that it becomes part of the sequence for a protein of interest. When that modified gene is expressed in a cell, the resulting hybrid protein is fluorescent.

- Research/review the role of mitochondria in cellular respiration.
- Predict what would happen if the number of mitochondria within a dividing eukaryotic cell remained constant.
- Research/review the role of dynamin in eukaryotic cells.
- Compare and contrast the molecules ATP and GTP.
- What effect might the hydrolysis of GTP have on the Drp1 protein?
- Design the siRNA you would need to target an mRNA with the sequence AUUGCUCCAGC.
- GFP has many applications. Look up some of the other ways that GFP is being used today.

BACKGROUND

Mitochondrial fission requires force. Part of that force comes from the activity of the dynamin-like GTPase protein Drp1; however, Drp1 may not be the only protein involved in the process. One current model of mitochondrial fission proposes that constriction is initiated by the polymerization of actin microfilaments at the surface of the mitochondria. The growth of these actin filaments, in turn, is stimulated by *INF2*, a protein that is located on the cytoplasmic surface of the cell's ER. Polymerization of actin filaments can generate sufficient force to alter the shape of a membrane as seen during lamellipodia formation and cell migration. Actin-based force generation is also associated with the activity of the motor protein **myosin**. The sliding filament model of muscle contraction describes the interactions between actin thin filaments and myosin thick filaments. A similar arrangement of myosin pulling on actin is responsible for the constriction of one cell into two cells during the process of cytokinesis. Is there a role for myosin in mitochondrial fission?

- Search for and observe video images of cell migration. Connect the changes you see at the leading edge of the cell with the activity of actin polymerization.
- Explain how a protein like myosin can generate force.
- Review/research the sliding filament model of muscle contraction.
- Relate the sliding filament model of muscle contraction to the process of cytokinesis.

METHODS

Cell lines and labeling

The human bone cancer cell line U2OS was used in this work. Some of the cells were transfected with *GFP-INF2-A149D*, a gene for INF2 (the actin polymerization-promoting protein) which had been genetically modified to produce a protein that would be constantly active and that would carry its own built-in fluorescent tag through the addition of the sequence for GFP. Other cells were transfected with a combination of a red fluorescently tagged mitochondrial protein (*mito-dsRed*) and a green fluorescently tagged myosin protein (*GFP-MIIA*).

Inhibition of Myosin II

The role of myosin in mitochondrial fission was tested in several ways. Cells were treated with 50 μM **blebbistatin**, a drug that is known to inhibit the activity of myosin for varying amounts of time. In order for this drug to enter the cytoplasm it is necessary to dissolve it in the solvent DMSO. In a different experiment, cells were transfected with siRNAs specific to the two forms of myosin associated with muscle contraction and cytokinesis: *Myosin IIA* and *Myosin IIB*. An siRNA that did not target any mRNA was used as a control. Finally, the behavior of myosin and mitochondria was observed directly through the use of the fluorescently tagged proteins mito-dsRed and GFP-MIIA.

Microscopy

Confocal fluorescence microscopy was used to obtain images of the cells used in these experiments. Confocal microscopy provides very high–resolution images of fluorescently labeled cells. A confocal microscope creates optical sections of a sample, filtering out light coming from above or below that section or plane of focus. Removing the extra out-of-focus light improves the quality and enhances the resolution of the image. The fluorescent dye Mitotracker® was used to stain and observe the mitochondria. Confocal images of cells transfected with the mito-dsRed and GFP-MIIA proteins were captured every 4 s for a period of 10 min, creating a time lapse video.

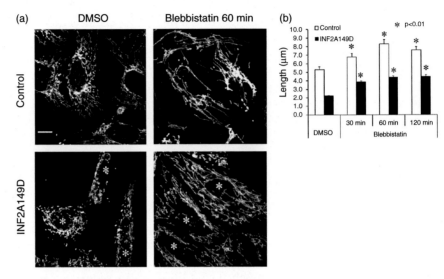

FIGURE 7.5.1 Blebbistatin treatment increases mitochondria length.
a. Control U2OS cells (top) or U2OS cells expressing GFP-INF2-A149D (bottom) were treated with DMSO alone (left) or 50 μM blebbistatin (right) for 60 min and then stained with Mitotracker. Asterisks indicate the cells expressing INF2-A149D. Scale bar = 10 μm. b. Quantification of mitochondrial length in control or GFP-INF2-A149D expressing cells treated with DMSO or 50 μM blebbistatin for the indicated times. $N = 103–344$ mitochondria. $*p < 0.01$ by Student's t test.

- List some of the concerns scientists must consider when using a genetically modified protein such as GFP-INF2-A149D
- How would the addition of siRNA for myosin II affect the cell?
- Predict what effect inhibition of mitochondrial fission would have on the length of mitochondria.
- Research/review the excitation:emission spectra for the Mitotracker mitochondrial stain.

RESULTS

- Describe what you are seeing in the images in Figure 7.5.1a.
- Why did the authors include the images labeled "DMSO"?
- How did the authors know which cells in Figure 7.5.1a were expressing the activated form of INF2?
- What is the average length of the mitochondria in a U2OS cell?
- How does the expression of INF2-A149D affect mitochondrial length in an otherwise normal cell (Figure 7.5.1b)?
- How did exposure to blebbistatin affect mitochondrial length?

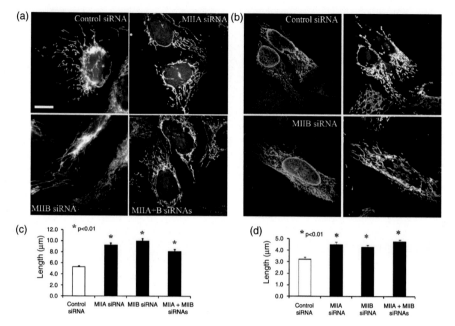

FIGURE 7.5.2 **Depletion of myosin II increases mitochondrial length.**
a. Mitotracker staining of U2OS cells expressing the indicated siRNAs. Scale bar = 10 μm. b. Control or MIIB siRNA-expressing cells were transfected with GFP-INF2-A149D (green) and stained with Mitotracker (red). Mitotracker staining alone is shown in the right-hand panels. c and d. Quantification of mitochondrial lengths for the experiments shown in (a) and (b), respectively. Data are from three experiments; n = 261–368 mitochondria for (c) and 329–407 mitochondria for (d). Error bars represent SEM.

- Explain why the images in Figure 7.5.2b are in color, but the images in Figure 7.5.2a are not.
- Create a diagram or other illustration that represents the variables being tested in the experiment shown in Figure 7.5.2b.
- What is the average length of the mitochondria in cells expressing myosin II siRNA (Figure 7.5.2c)?
- Look closely at the graphs shown in Figure 7.5.2c and d. What do you observe? Summarize the conclusion supported by these data.
- Propose an explanation for the difference in the results of blebbistatin versus siRNA treatment.
- The abbreviation SEM stands for standard error of the mean. Look up how to calculate this value.
- Approximately how much time is required for mitochondrial fission to occur (Figure 7.5.3)?
- Which of the results presented here do you find most convincing for the role of myosin II in mitochondrial fission? Explain your answer.

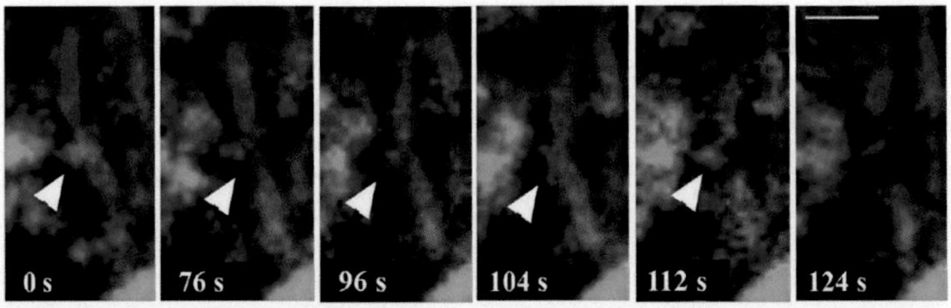

FIGURE 7.5.3 Myosin II accumulates at mitochondrial restriction sites.
Still images from a movie of a living U2OS cell transfected with mito-dsRed and GFP-MIIA imaged using confocal microscopy. The white arrowhead points to myosin II accumulation (green) at a fission point on the mitochondria (red). Total elapsed time is indicated in each frame. Scale bar = 2 μm.

References

[1] Voeltz GK, Prinz WA, Shibata Y, Rist JM, Rapoport TA. A class of membrane proteins shaping the tubular endoplasmic reticulum. Cell 2006;124:573–86.

[2] Barr FA, Puype M, Vandekerckhove J, Warren G. GRASP65, a protein involved in the stacking of Golgi cisternae. Cell 1997;91:253–62.

[3] Barlow C, Orici L, Yeung T, et al. COPII: a membrane coat formed by SEC proteins that drive vesicle budding from the endoplasmic reticulum. Cell 1994;77:895–907.

[4] van der Zand A, Gent J, Braakman I, Tabak HF. Biochemically distinct vesicles from the endoplasmic reticulum fuse to form peroxisomes. Cell 2012;149:397–409.

[5] Korobova F, Gauvin TJ, Higgs HN. A role for myosin II in mammalian mitochondrial fission. Curr Biol 2014;24:409–14.

CHAPTER 8

Exocytosis

Coat Proteins and Vesicle Transport [1]

INTRODUCTION

The pathway of exocytosis starts with the synthesis of a protein on the **rough endoplasmic reticulum** (RER). From there, the protein must travel through each of the compartments of the **Golgi complex**, finally arriving at the **trans Golgi network** (TGN) before being transported to the plasma membrane, lysosome, or vacuole. The collections of membrane-bound organelles that participate in exocytosis are part of the **endomembrane system**. It is generally accepted that proteins move along the exocytic pathway inside small transport vesicles. Transport vesicles are formed through a process of "budding" from the membrane of the RER or Golgi, capturing protein cargo in the process. Fusion of the vesicles with the membrane of the target organelle releases the protein cargo into that compartment. These vesicles carry a collection of peripheral proteins on the cytoplasmic side of their membranes, forming a "coat." COPI and COPII are the two distinct types of coated vesicles associated with transport between the RER and the Golgi complex.

Three models were developed to explain the role of coated vesicles in exocytosis. The first model proposed that COPII vesicles were responsible for the anterograde (forward) movement of materials in the cell. The second model proposed that both COPI and COPII vesicles moved material in parallel, independent of anterograde pathways. The third model proposed a sequential role for coated vesicles. In this model the COPII vesicles carry material from the RER to an intermediate compartment. COPI vesicles are then formed from

193

the intermediate compartment to carry the material to the *cis* Golgi. The work presented in this case study seeks to determine which of these models is correct.

- Research/review the types of proteins that are synthesized on the RER.
- Summarize how a newly synthesized protein is modified from the time it enters the lumen of the RER to the time it reaches the *trans* Golgi network.
- Describe another example of a coated vesicle found in cells.

BACKGROUND

Vesicular stomatitis virus (VSV) is an envelope virus that typically infects animals such as horses and cattle. One of the important viral proteins produced when the virus infects a host cell is a transmembrane glycoprotein called VSV-G. VSV-G has become a model protein for use in studies of exocytosis. A **temperature-sensitive** (ts) version of VSV-G has been developed. **Ts-045-G** protein misfolds when it is exposed to a temperature of 39.5°C. At this **non-permissive temperature,** the misfolded protein accumulates in the RER. Shifting the temperature to 31°C, the **permissive temperature**, allows the ts-045-G protein to fold properly and move to the cell's plasma membrane. The movement of VSV-G through the cell can also be blocked by incubating cells at 15 or 20°C, causing the protein to accumulate in an intermediate compartment between the RER and Golgi or the *trans* Golgi network, respectively.

The gene for ts-045-G was engineered to include the sequence of **green fluorescent protein (GFP)**. GFP is a small, 238 amino acid protein that fluoresces when exposed to blue light. The sequence for GFP was added to the carboxy terminus of ts-045-G, making a chimeric protein (ts-G-GFP$_{ct}$). The relative location of the cargo protein ts-G-GFP$_{ct}$ could then be followed and compared with the location of different coat proteins inside the cell. **Colocalization** or the overlap of fluorescence signal from two or more fluorescent labels is used as an indication that the things being tracked are closely associated in the cell.

- How does a virus gain an "envelope?" What implications does this have on the synthesis and secretion of VSV-G protein?
- Sketch out the pathway that VSV-G would take starting at the RER and ending at the plasma membrane?
- Conduct an online search for images of the structure of VSV-G. Describe the structure of the viral protein.
- Speculate on why an elevated temperature might cause a protein to misfold.

METHODS

Cell culture and gene expression

Vero cells, a tissue culture cell line derived from monkey kidney epithelial cells, were cultured under standard conditions. DNA encoding the chimeric ts-045-G:GFP protein, ts-G-GFP$_{ct}$, was microinjected into the cells' nuclei.

Microinjected cells were incubated at 39.5 °C for 8–12 h prior to visualization and analysis. Injected cells were also incubated at 4, 15, 20 and 31 °C, depending upon the experiment. In some experiments the drug cycloheximide was added to the cell cultures. Cycloheximide reversibly inhibits protein synthesis.

Antibodies and immunofluorescence

Antibodies used in this study include the anti-COPI antibody ß′-COP, the anti-COPII antibody Sec13p, the RER-specific antibody calnexin, the intermediate compartment antibody ERGIC-53, and the *trans* Golgi network-specific antibody galT, which recognizes galactosyltransferase. A tubulin-specific antibody was also used. Fluorescent secondary antibodies were used as appropriate. Fluorescent images were captured using an inverted fluorescence microscope. Images were quantified using image analysis software that made it possible to count the number of ts-G-GFP$_{ct}$ fluorescent structures. *In vivo* analysis of vesicle movement was achieved by capturing images for 0.2 s every 3–5 s. The temperature of the stage of the microscope was controlled.

■ Outline the steps that must occur between injection of the chimeric ts-045-G:GFP DNA and accumulation of ts-G-GFP$_{ct}$ in the RER of the cell.

■ Debate whether results generated from an immortalized monkey kidney cell line can be applied to all eukaryotic cells.

RESULTS

■ Study the images in Figure 8.1.1b, d, and f. Use these data to describe (in writing or drawing) the relationship between the major endomembrane compartments involved in exocytosis.

■ Develop an argument for or against the following statement based on the data in Figure 8.1.1, "ts-G-GFP$_{ct}$ is a valid tool for studying transport of ts-045-G *in vivo*."

■ Study the images of ts-G-GFP$_{ct}$, COPI, and COPII distributions in the cell 6 min after the shift to 31 °C in Figure 8.1.2a–c. Summarize your observations.

■ Estimate the size of the coated vesicles that are carrying ts-G-GFP$_{ct}$.

■ Confirm the labeling of ts-G-GFP$_{ct}$ structures described in the figure legend for Figure 8.1.2. Look for examples of colocalization (or the absence of colocalization) not labeled in the images shown in Figure 8.1.2a′–c′ here. What conclusion is supported by these data?

■ Explain why colocalization between ts-G-GFP$_{ct}$ and the COPS was only analyzed in the "flat peripheral region of the cell." What effect might limiting counting of vesicles to that area have on the data?

■ What was the purpose of including cycloheximide in the experiment shown in Figure 8.1.2? Predict what the consequence of omitting cycloheximide might have been.

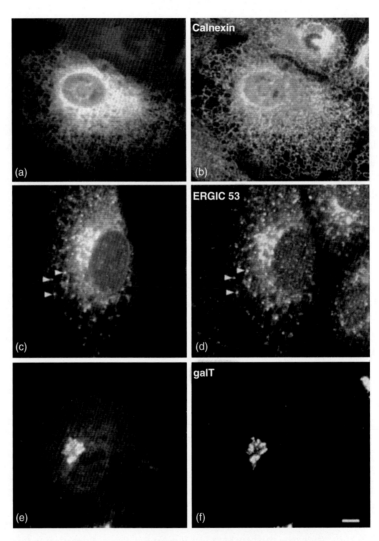

FIGURE 8.1.1 Colocalization of ts-G-GFP$_{ct}$ with exocytic membrane markers.
Tissue culture cells expressing ts-G-GFP$_{ct}$ DNA for 8 h at the nonpermissive temperature were either directly fixed (a and b) or kept at 15°C for a further 3 h in the presence of cycloheximide (c and d) or at 20°C (e and f) before fixation. Cells were immunolabeled with antibodies against calnexin (b), ERGIC-53, (d) and galactosyltransferase (galT; f) to visualize the RER, intermediate compartment, and Golgi complex, respectively. The corresponding GFP fluorescence image is shown in (a), (c), and (e). Arrowheads indicate colocalizing structures. Bar = 10 μm.

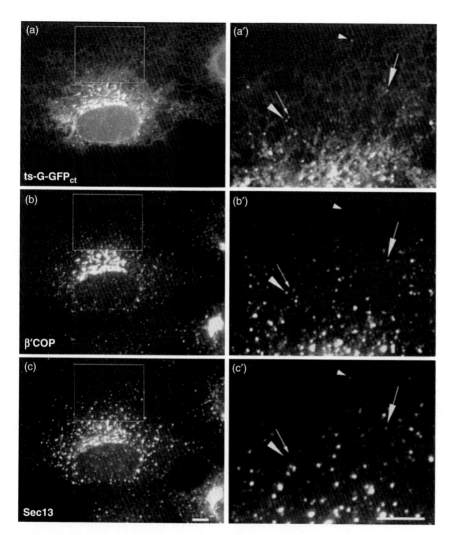

FIGURE 8.1.2 Colocalization of ts-G-GFP with COPI and COPII after release from the RER.
Tissue culture cell expressing ts-G-GFP$_{ct}$ for 8–9 h at 39.5°C was shifted to 31°C for 6 min in the presence of cycloheximide before fixing and processing for immunofluorescence. a. GFP fluorescence of ts-G-GFP$_{ct}$ protein. b. COPI fluorescence based on reaction with a monoclonal antibody specific to ß'COP. c. COPII fluorescence detected by an antibody to Sec13p. (a'–c') 3× magnification of the indicated area in (a–c). ts-G-GFP$_{ct}$-containing structures labeling for COPII only are indicated by small arrowheads, for COPI only by large arrowheads, both coats by large arrows, and no coats by small arrows. Bar = 10 μm.

- Describe the relationship between the data in Figure 8.1.3a and Figure 8.1.2.
- In the area graph in Figure 8.1.3a the number of each type of COP-coated vesicle containing ts-G-GFP$_{ct}$ at any time point can be calculated by subtracting the lower from the upper edge of each region. The total number of ts-G-GFP$_{ct}$ vesicles is the sum of all regions at each time point. Using this information calculate the following:
    - Number of COPI only vesicles at 6 min
    - Number of COPII only vesicles at 10 min
    - Total number of vesicles at 2 and 10 min
    - Number of COP-negative vesicles at 10 min
    - Number of COPI + II vesicles at 4 min
- Propose an explanation for the change in ts-G-GFP$_{ct}$ positive structures over time.
- Discuss the significance of the error bars in Figure 8.1.3a.
- Summarize the conclusion supported by the data shown in Figure 8.1.3b.
- Compare the data in Figure 8.1.3c with that in Figure 8.1.3b. How do the results from the experiment in Figure 8.1.3c influence your conclusion?
- What step(s) in the process of exocytosis is represented in Figure 8.1.4?
- Estimate the speed at which the vesicle complex is traveling in Figure 8.1.4.
- Connect the data in Figure 8.1.4 to the information in Figure 8.1.3.
- Research/review the role of microtubules in intracellular motility.
- Predict which microtubule motor protein is responsible for the vesicle movement shown in Figure 8.1.5.
- Is the data presented in Figure 8.1.5 sufficient to support the conclusion that coated vesicles move from the RER to the Golgi along microtubules? Justify your answer.
- Refer back to the three models described in the "Introduction" to this case study. Which of the models is most supported by the data? Debate whether the work described here can be used to eliminate one or both of the other models.

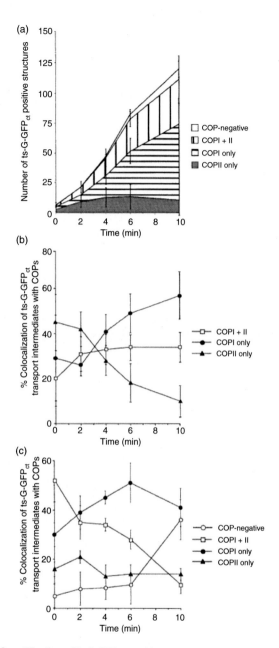

FIGURE 8.1.3 Quantification of ts-G-GFP$_{ct}$–positive structures colocalizing with COPS.
ts-G-GFP$_{ct}$ was expressed in tissue culture cells as described in Figure 8.1.2. After accumulation of ts-G-GFP$_{ct}$ at 39.5°C, cells were shifted to 31°C for various periods of time, fixed, and immunolabeled. Only unambiguously identifiable vesicular structures containing ts-G-GFP$_{ct}$ were scored for their colocalization with COPI or COPII. a. Area graph showing the total number of ts-G-GFP$_{ct}$-containing structures in the flat peripheral region of the cell at different time points. Error bars show the mean standard deviation of 15–19 cells counted for each time point. b. The relative COP coat distributions of the ts-G-GFP$_{ct}$ structures expressed as a percentage of the total number of ts-G-GFP$_{ct}$-positive structures counted in a selected area of the cell for each time point in the experiment in (a). c. Cells were treated and quantified as in (b), except that they were incubated for 3 h at 15°C in the presence of CHX to accumulate ts-G-GFP$_{ct}$ in the intermediate compartment before shifting to the permissive temperature. Error bars represent the mean standard deviation for measurements from 15–19 cells per time point.

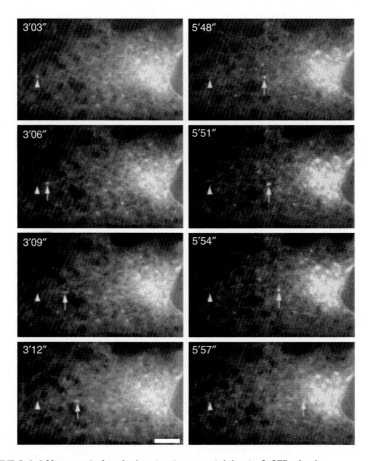

FIGURE 8.1.4 Movement of vesicular structures containing ts-G-GFP$_{ct}$ *in vivo.*
Tissue culture cells accumulating ts-G-GFP$_{ct}$ in the RER at 39.5°C were shifted to 31°C in the presence of CHX, while on the microscope stage and images were recorded every 3 s. The movement of one vesicle complex from its starting point (arrowhead) is shown (arrows) at the indicated time points after the temperature shift. Note the 2.5 min time interval between the left and right-hand panels during which vesicle movement was undirected (hovering). Bar = 10 μm.

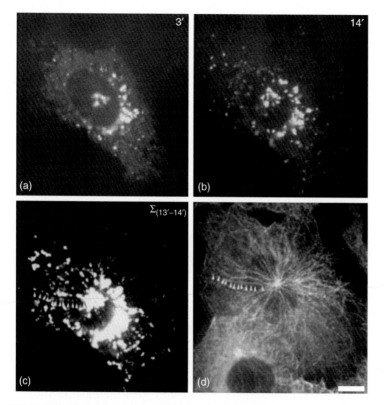

FIGURE 8.1.5 Microtubules facilitate early exocytic membrane transport.
Cells expressing ts-G-GFP$_{ct}$ that had been accumulated in the intermediate compartment by incubation at 15°C were incubated for 1 h at 4°C to depolymerize their microtubules without causing a change in the Golgi complex. Cells were shifted to 31°C while on the microscope stage and images were recorded every 3 s. Within 6 min, vesicular complexes began to move in a directed salutatory manner toward a point near the nucleus from which newly polymerized microtubules emanated (as visualized later by immunofluorescent labeling). d. The first (a) and last frames (b) of the sequence at the indicated times after the shift to 31°C are shown. c. The last 20 frames (a period of 1 min) were superimposed to reveal the path taken by the vesicles (arrowheads) toward the microtubule-organizing center. The microtubule path does not exactly match that taken by the vesicles, possibly because the microtubules also moved during that time. Bar = 10 μm.

Endomembrane Transport in the Absence of a Cell [2]

INTRODUCTION

Protein function is just as much about *where* a protein is located as it is about the shape of that protein. Proteins that are synthesized on the surface of the **RER** travel through a series of membrane compartments known as the **endomembrane system**, which functions to modify and sort proteins to their correct destinations inside and outside the cell. The endomembrane system is made up of the RER, the **Golgi complex**, vacuoles, endosomes, lysosomes, and **small transport vesicles**.

Protein synthesis starts in the cytoplasm when an initiation complex is formed between a mature mRNA and a small ribosomal subunit. With the addition of the large ribosomal subunit and additional tRNAs, the process of translation begins. Proteins that are destined to enter the endomembrane system carry a signal peptide that acts like an address label to direct these mRNAs to the surface of the RER. As translation continues, the proteins move through a translocon or protein channel embedded in the RER membrane and enter the interior space or lumen of the RER. Soluble proteins are released into the lumen while membrane proteins become anchored in the RER membrane by their stop transfer sequences.

One of the main modifications that occur to proteins that travel through the endomembrane system is **glycosylation**. As proteins enter the lumen of the RER glycosylating enzymes catalyze the formation of a covalent bond between the side chain (R group) of the amino acid asparagine and a large oligosaccharide made up of the sugars glucose, mannose, and **N-acetylglucosamine** in a process known as core glycosylation. As these core-glycosylated proteins move from the RER to the Golgi complex, distinct populations of enzymes located in each of the compartments or cisternae catalyze changes to the original oligosaccharide. Mannosidases, located in the *cis* Golgi, remove some of the mannose from the oligosaccharide while **GlcNAc transferases**, located in the *medial* Golgi, add N-acetylglucosamine to it. Enzymes in the final compartment of the Golgi, the *trans* Golgi, complete the glycosylation of a protein with the addition of galactose, fucose, and sialic acid.

- Sketch the layout of a generic eukaryotic cell illustrating the relative positions of the nucleus, RER, Golgi, endosomes, lysosomes, a vacuole, and the plasma membrane. Label your drawing.
- Research/review the steps involved in translation of an mRNA.

- Research/review the signal hypothesis.
- Explain why newly synthesized proteins move through a translocon and not directly through the membrane of the RER.
- Viruses can hijack a cell's endomembrane system to transport viral proteins to the plasma membrane. Along the way, the viral proteins can be glycosylated just like a cellular protein. Provide an explanation why viral proteins can be glycosylated.

BACKGROUND

The sorting and transport of proteins through the endomembrane system is a complicated process that involves many steps and even more proteins. One way to investigate the variables that control the behavior of proteins is through the development of a **cell-free system** that models cellular behavior, but allows researchers the ability to easily manipulate components of the system and observe the effects. Another method that can be used to study a cellular process involves the use of "genetic dissection." Mutations of proteins involved in a process or pathway can be used to examine how the loss of that protein impacts the cell. This case study combined the use of mutant cell type with the use of a cell-free system to investigate the transport of a protein from one compartment of the Golgi to the next.

The cells that were used in this study were **CHO 15B** cells. *CHO* is an abbreviation for Chinese hamster ovary cells and *15B* designates a mutation that results in the loss of the enzyme **GlcNAc transferase 1**. GlcNAc transferase 1 is located in the *medial* cisternae of the Golgi where it catalyzes the addition of GlcNAc (N-acetylglucosamine) to the oligosaccharide chains of glycoproteins. Only proteins that have traveled through the *medial* Golgi will carry GlcNAc. Isolation of proteins that are moving through the endomembrane system can be achieved by breaking open the cells in a process known as homogenization. Homogenization produces a cell lysate that can be further purified by **sucrose gradient density centrifugation**. In this method, a centrifuge tube is filled with layers of a buffer solution with increasing sucrose concentration from 2.0 M to 0.8 M. Centrifugation of the sample in a swinging bucket rotor causes the cell contents to move through the gradient to a location that matches their buoyant density. Differences in the buoyant density of the various organelles will cause them to separate along the gradient in a reproducible manner. After centrifugation is complete, the bottom of the centrifuge tube is pierced and the contents or **fractions** collected as they drip out of the tube. Because the cellular contents have spread out along the gradient, the organelles and cellular material found in each fraction will differ.

To study the movement of proteins through the Golgi it was important to ensure the CHO 15B cells were producing a known protein in high concentration. The CHO 15B cells were infected with **VSV**. When VSV infects a cell it takes over the cell's protein synthetic machinery and forces the cell to make large quantities of a viral protein, **G protein**. VSV-G protein is synthesized on RER, glycosylated as it transports through the Golgi, and eventually ends up in the cell's plasma membrane. G protein is a transmembrane protein with its amino terminal end in the lumen of the RER and its short carboxy terminal end in the cytoplasm.

- Discuss the benefits and drawbacks of the use of cell-free model systems versus genetic dissections in investigating a cellular process.
- Describe some ways that a genetic mutation could result in the "loss" of a protein.
- Predict which organelle in a eukaryotic cell would be the "heaviest." Which fraction would you expect to contain that organelle following sucrose gradient density centrifugation?
- Research/review the viral life cycle for VSV.
- VSV is an example of an envelope virus. Conduct a search to find other examples of envelope viruses.
- Sketch a picture of a VSV-G protein, as it would appear in the *cis* Golgi. Label the carboxyl and amino terminals of the protein. Indicate which region of the G protein would be glycosylated.

METHODS

CHO cells and VSV

CHO cells were grown in tissue culture dishes. Twenty dishes of confluent CHO 15B cells (4–6×10^7 cells per dish) were infected with 5–10 pfu (plaque-forming units) of VSV per cell. The cells were allowed to grow for 4 h in the presence of the virus. Wild-type CHO cells were grown under the same conditions, with the exception that these cells were not infected with VSV. Wild-type or mutant cells were removed from the surface of the dishes by the addition of 0.05% trypsin in buffer. Cells released from the surface of the dish were collected by centrifugation and rinsed to remove any residual trypsin.

Homogenization and gradient purification of subcellular fractions

CHO 15B or wild-type cells were lysed using a dounce homogenizer. The resulting crude homogenate could be used immediately or stored frozen at $-80°C$ for up to 2 months. Purification of Golgi membranes was achieved through the use of sucrose density gradient centrifugation. Crude cell lysates were diluted into a buffer to make a solution with a final concentration of 1.4 M sucrose. The cell lysate solution was layered on top of the 1.6 M sucrose layer in a centrifuge tube. Sucrose solutions of 1.2 and 0.8 M were layered on top of the lysate. Gradients were centrifuged for 2.5 h at 90,000g. Golgi membranes were collected from the interface between 0.8 M and 1.2 M sucrose layers.

Purified Golgi membranes derived from CHO 15B cells were designated to be the **donor fraction** since these Golgi carry the VSV-G protein. Golgi membranes purified from wild-type CHO cells were designated as the **acceptor fraction**.

Cell-free system for Golgi transport

Transport between donor and acceptor Golgi membranes was carried out *in vitro* using a cell-free model. 5 μL each of donor and acceptor membrane fractions were mixed in a test tube containing buffer, cytoplasmic extract isolated from wild-type CHO cells, an ATP regenerating system to ensure a constant supply of energy, and radioactively labeled GlcNAc (^{3}H-GlcNAc). The reaction mixture was incubated at 37°C for 60 min. The reaction was stopped by placing the mixture on ice followed by the addition of 1% Triton X-100, a detergent, to dissolve the membranes. G proteins were immunoprecipitated using a G protein-specific antibody, and the amount of radioactivity was measured. Alternatively, following the incubation step, the Golgi membranes were treated with 5 μL of 1 mg/mL trypsin in the presence or absence of 0.1% Triton X-100. Samples of intact and trypsin-treated membranes were immunoprecipitated with an antibody to G protein and prepared for SDS polyacrylamide gel electrophoresis. The gel was imaged using autoradiography to detect the radioactive G protein.

In vivo inhibition of protein synthesis

VSV-infected CHO 15B cells were released from the surface of the tissue culture dishes and treated with 100 μg/mL cycloheximide (CHX) for increasing amounts of time. CHX inhibits protein synthesis. Cells were sampled at 0, 7.5, 20, and 60 min in CHX and a cell lysate was prepared to assay donor activity.

- What is the definition of "wild-type?"
- Assume that you have a 2.5×10^{10} pfu/mL stock solution of VSV. What volume of stock virus would you add to a confluent dish of 5×10^7 cells to achieve a final infection level of 8 pfu per cell?
- What is trypsin? Conduct a search to learn more about this protein.
- Look for an image of a Dounce homogenizer. Explain how this instrument can break open cells.
- Create a diagram to explain how G proteins located in the CHO 15B donor Golgi become radioactive G proteins in the presence of wild-type CHO acceptor Golgi.

RESULTS

- Predict where you would expect to find Golgi membranes using the information provided in the top graph in Figure 8.2.1.
- Approximately which fraction is most likely to contain the highest concentration of G protein? Justify your answer.

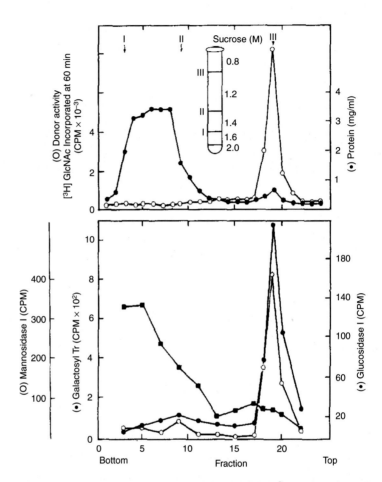

FIGURE 8.2.1 The CHO cell-free system supports the addition of ³HGlcNAc to G proteins.
CHO 15B cells infected with VSV were homogenized and their cellular contents separated and purified using sucrose gradient density centrifugation. The total amount of protein present in each fraction was measured (filled circles; top graph). The ability of G proteins from the mutant CHO 15B cells to incorporate ³HGlcNAc (donor activity) was measured after incubating each fraction in the cell-free system (open circles; top graph). Golgi-specific proteins mannosidase I (open circles; lower graph) and galactosyl transferase (filled circles; lower graph) accumulate at the 0.8 and 1.2 M sucrose interface following centrifugation. The endoplasmic reticulum-specific protein glucosidase (filled squares; lower graph) was detected in lower fractions.

- Based on the information provided in Fig. 8.2.1, where is the greatest concentration of proteins located in CHO cells? Discuss why this result may make sense.
- Is it possible that G proteins could be present in fractions #5–10? Why or why not?
- Explain why the apparent size of G protein has decreased in lane 2 of Figure 8.2.2.

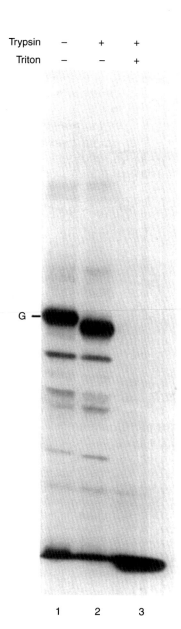

Trypsin   −   +   +
Triton   −   −   +

G —

1   2   3

FIGURE 8.2.2 G protein is transported in sealed Golgi vesicles as a transmembrane protein with the proper orientation.

CHO 15B donor Golgi membranes were incubated with wild-type CHO acceptor Golgi membranes in the cell-free system in the presence of ^{3}HGlcNAc. The assay was stopped after 60 min by the addition of 100 mM Na$_2$EDTA. The Golgi membranes were examined using SDS-PAGE and autoradiography. ^{3}HGlcNAc-labeled G protein appears as a major band (lane 1) in the Golgi membrane sample immediately following incubation. When Golgi membranes were treated with trypsin (lane 2) or trypsin and 0.1% Triton X-100 (lane 3), the size of the G protein band shifted.

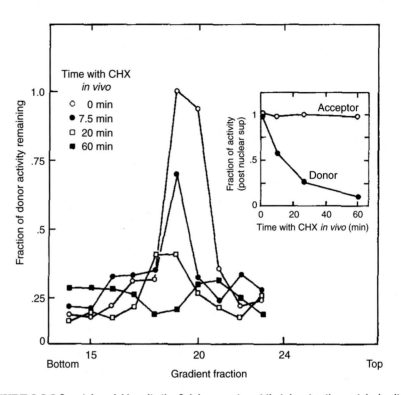

FIGURE 8.2.3 G protein quickly exits the Golgi compartment that donates the protein *in vitro* when protein synthesis is inhibited *in vivo*.

Golgi membranes were isolated from VSV-infected CHO 15B after incubation with the protein synthesis inhibitor, CHX for 0, 7.5, 20, or 60 min. Only the top portion of the sucrose density gradient (fractions 14–23) is shown. The incorporation of ^{3}HGlcNAc is normalized to the value obtained for fraction 19 at 0 min in CHX. The effect of CHX treatment on total donor activity in crude cell lysate was examined for both VSV-infected CHO 15B cells (inset; filled circles) and wild-type CHO cells (inset; open circles).

- What accounts for the dramatic change in size of the G protein in lane 3 of Figure 8.2.2?
- Defend the following statement using data from Figure 8.2.2, "transported G protein is retained in sealed Golgi vesicles with its carboxy-terminal domain on the outside and its oligosaccharide chains on the inside."
- How long does it take for approximately half the G protein to no longer be available for transport?
- Propose an explanation why treatment of wild-type CHO cells with CHX did not affect the ability of acceptor membranes to function.
- Revisit your diagram of the potential interaction between the Golgi membranes from CHO 15B (donor) and wild-type CHO (acceptor). Use your diagram to explain the data shown in Figure 8.2.3.

Extra Large Export: A Case for Cisternal Maturation [3]

INTRODUCTION

Collagen is an abundant protein found in the extracellular matrix of animals. You probably have heard about collagen in the context of tendons and skin. Collagen forms a stiff fiber that is made up of three long polypeptide chains that assemble into a twisted helix, stabilized in part by hydrogen bond formation between the side chains (R groups) of proline. The enzyme **prolyl hydroxylase** reacts with the proline in collagen to help to promote proper folding of the triple helix inside the RER. Activity of prolyl hydroxylase requires **ascorbic acid**, better known as vitamin C, as a cofactor.

Collagen is synthesized and secreted from fibroblast and osteoblast cells. Because collagen is destined to exit the cell, translation of **preprocollagen** polypeptide mRNAs is targeted at the RER. After entering the RER the signal sequence is removed and initial folding of individual polypeptide strands occur. Proper alignment of the three polypeptides is made possible by the presence of globular propeptides located at the amino and carboxyl-terminal ends of each polypeptide. The final product is a 300 nm long microfibril of **procollagen (PC)**. Procollagen is then transported from the RER to the Golgi complex and eventually out of the cell. Misfolded microfibrils are unable to travel to the Golgi and will accumulate inside the RER. Only after PC has left the cell do PC peptidases cleave off the globular propeptides, allowing the microfibrils to assemble into fibrils that finally assemble into collagen fibers forming an important part of the extracellular matrix.

- What is the difference between a "polypeptide" and a "protein?"
- Conduct a search for images or diagrams of the molecular structure of collagen microfibrils and fibers. Can you find micrographs of collagen as part of the extracellular matrix.
- Examine the chemical structure of the amino acid proline. Explain why proline is capable of participating in the formation of a hydrogen bond.
- *Scurvy* is a disease caused by lack of vitamin C in a person's diet. Some of the symptoms include bleeding of the gums and tooth loss, inability of wounds to heal, and joint pain. Suggest why lack of vitamin C might cause these symptoms?
- How do the mRNAs for secretory proteins get targeted at the RER?

■ What was the likely function of the "pre" portion of the protein preprocollagen?
■ Predict what would happen if the globular propeptides were removed from PC while it was still inside the cell?

BACKGROUND

The endomembrane system of a eukaryotic cell consists of a series of membrane-bound organelles that can be thought of as "connected" by the movement of proteins and lipids between them. Proteins, like PC, that must be exported out of the cell are synthesized on the RER, travel through the individual cisternae or stacks of the Golgi complex, and finally exit the cell at the plasma membrane. Two models have been proposed to explain the transport of proteins through the Golgi complex. The **vesicular transport model** proposes that Golgi cisternae are stable membrane compartments. Each cisterna contains a characteristic set of enzymes that function to modify proteins as they pass through the compartment. Based on this model, proteins are carried forward from one compartment to the next in small, membrane-bound transport vesicles. In contrast, the **cisternal maturation model** states that Golgi cisternae are dynamic organelles that are constantly being formed from the RER and that change their function or "mature" with time. Proteins remain within a Golgi cisterna and are modified as a result of changes in the population of enzymes. According to this model a *cis* Golgi compartment "matures" into a *medial* Golgi cisterna and a *medial* Golgi compartment "matures" into a *trans* Golgi cisterna due to the backward transport of enzymes between the cisternae.

Cisternal maturation helps to explain how very large protein aggregates can move through the Golgi complex. Proteins like PC are too large (300 nm) to fit into a transport vesicle (50–90 nm) and yet these proteins are glycosylated just like the smaller, soluble proteins that are known to travel in transport vesicles. The cisternal maturation model predicts that protein aggregates would remain in the lumen of a single Golgi cisterna as that membrane compartment matures from *cis* to *trans*. The synthesis and transport of PC by chick embryonic fibroblast cells was examined in this case study with the goal of testing the validity of the cisternal maturation model.

■ List the organelles that are a part of the endomembrane system.
■ Create a diagram that illustrates the two models of protein transport through the Golgi complex.
■ Research/review the nature of transport vesicles in the endomembrane system. How do COPI vesicles differ from COPII vesicles?
■ Describe the relationship between a protein's amino acid sequence and its function.

METHODS

Primary culture of chick embryo fibroblast cells

Primary cultures of fibroblast cells were isolated from a 15-day-old chicken embryo tendons (ET) by incubation in the presence of 3 mg/mL collagenase in tissue culture media. Cells were harvested from the media by filtering through cheesecloth followed by centrifugation. The fibroblast cell pellet was resuspended in tissue culture media and plated into glass chamber slides coated with 30 μg/mL fibronectin. Cells were incubated for 18 h at 37°C before use in any experiments.

Immunofluorescence

Cultured ET fibroblast cells were fixed with 2% paraformaldehyde and permeabilized with 0.1% Triton X-100. The cells were rinsed with buffer and incubated with antibodies that recognize the carboxy-terminal domain of PC (LF68 antibody) and the Golgi-specific integral membrane protein **giantin**. The cells were then incubated with fluorescent-labeled secondary antibodies.

Immunoelectron microscopy

ET fibroblast cells were fixed and permeabilized as described previously. The cells were incubated with a monoclonal antibody specific to the folded form of PC (hCL(1)) followed by incubation with a horseradish peroxidase (HRP)-conjugated secondary antibody. In the presence of 0.001% hydrogen peroxide (H_2O_2), the HRP enzyme catalyze a reaction with 0.075% 3,3'-diaminobenzidine tetrachloride (DAB-HCl) to form an insoluble precipitate that appears as a dark electron-dense region in the electron micrograph.

PC "wave" protocols

Movement of PC from the RER through the Golgi was synchronized using two different protocols. Both protocols relied on the ability of the iron-chelating compound 2,2'-dipyridyl (DPD) to inhibit the proper folding of PC resulting in the retention of unfolded PC in the RER. The "exiting wave" protocol followed the movement of the population of PC proteins already present in the Golgi. ET fibroblast cells were treated with 0.3 mM DPD for 0–60 min, fixed, and reacted with a monoclonal antibody to folded PC. Since DPD inhibits the folding of PC, the only proteins that the antibody can react with are proteins that had already left the RER. The "incoming wave" protocol also used 0.3 mM DPD to block the exit of PC from the RER; however, in this protocol the cells were incubated in the DPD until the PC proteins already present in the lumen of the Golgi had exited the cell. Washing out the DPD and supplementing the media with ascorbic acid (vitamin C) allowed PC to fold properly and exit the RER as a single wave of protein. Localization of the folded PC was achieved using the same fold-specific antibody. *Cis* to *trans* polarity of the Golgi in both protocols relied on the presence of clathrin-coated vesicles as markers for the *trans* Golgi compartment.

Assay of PC release

Cells were grown as described previously and treated with DPD following either the "exiting wave" or "incoming wave" protocols. Tissue culture media was collected at time points after the addition (exiting wave) or removal (incoming wave) of 0.3 mM DPD. Fresh media were added to the cells following each collection. Proteins present in the media were precipitated using 10% trichloroacetic (TCA) acid and were analyzed by SDS-PAGE and immunoblotting using a polyclonal antibody to PC (Sp1.D8). The intensity of the stained bands was quantitated using image analysis software.

- Use your own words to define the terms *primary culture* and *immortalized cell line*.
- What effect did incubation with collagenase have on the tissue? Why was this step included?
- Describe how immunoelectron microscopy is similar to immunofluorescence microscopy. How are these two methods different?
- Why was it important for the researchers to add fresh media to the culture dishes in the PC release assay? Discuss whether the results would have been the same if they used different dishes of cells for each time point.
- Research the difference between a *monoclonal* and a *polyclonal* antibody.

RESULTS

- What effect does treatment with DPD have on PC polypeptides? How does this manifest in the localization of PC in the cell (Figure 8.3.1a–f)?
- Compare the pattern of PC staining in Figure 8.3.1a and c. What do you observe? Propose an explanation for your observation.
- How would you characterize the appearance of the Golgi and PC in the immunoelectron micrographs shown in Figure 8.3.1a–c?
- Connect the effect of DPD treatment with the appearance of images in Figure 8.3.1a–e.
- Why is there no electron-dense labeling in Figure 8.3.1e?
- How do the results from the "incoming wave" protocol compare with the "exiting wave" protocol?
- Determine the average size of the following organelles in Figure 8.3.1a–h:
  - Golgi cisternae (without PC)
  - Distensions of Golgi cisternae containing PC
  - Clathrin-coated vesicles
  - Distended RER
  - Transport vesicles

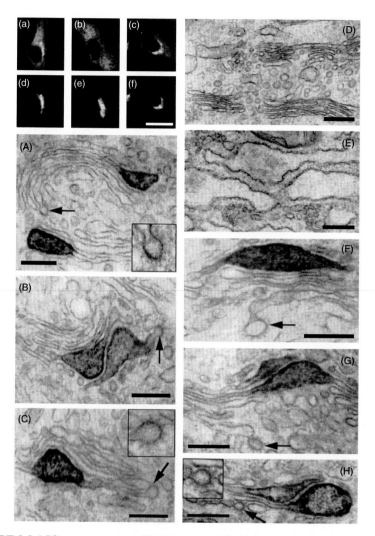

FIGURE 8.3.1 PC aggregates stay within the lumen of the cisternae as they are transported through the Golgi.

Immunofluorescence images of chick embryo fibroblasts stained for PC (a–c) and the Golgi marker protein giantin (d–f). PC colocalizes with the RER and Golgi in control cells (a and d). Incubation in the presence of DPD results in the loss of PC staining of the Golgi (b and e). However, colocalization is recovered after removal of DPD (c and f). Immunoelectron microscopy of fibroblast cells treated with DPD using the "exiting wave" protocol. Visualization of the relative location of PC aggregates in the Golgi complex before (A), or after 10 min (B), 30 min (C), or 60 min (D) treatment with DPD. The *cis* to *trans* orientation of the Golgi complex was determined based on the presence of clathrin-coated vesicles (see arrows and inset images). E. The RER appears distended following the 60 min treatment with DPD. Visualization of PC aggregates following the "incoming wave protocol." Fibroblasts were pretreated for 60 min with DPD, washed to remove the DPD, and fixed at 10 (F), 20, and 40 min posttreatment. Scale bars a–f = 3.5 μm. Scale bars A–H = 200 nm.

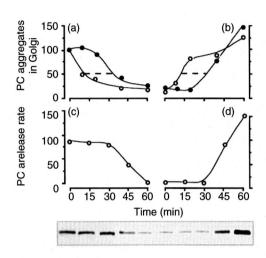

FIGURE 8.3.2 Endomembrane transport results in PC release into the extracellular space.
The number and location of PC aggregates were tracked using either the "exiting wave" protocol (a) or the "incoming wave" protocol (b). Numbers of aggregates in either the *cis* (open circles) or *trans* (filled circles) compartments of the Golgi were counted at the indicated times relative to the addition (a) or removal (b) of 0.3 mM DPD. More than 60 distensions were analyzed per time point in three different experiments. Standard error of the mean (SEM) never exceeded 10%. The amount of PC released from fibroblast cells into the tissue culture media was assayed for both the "exiting wave" protocol (c) and the "incoming wave" protocol (d). Extracellular media were removed at the indicated times and fresh media were added to the cells following the addition (c) or removal (d) of DPD. The amount of PC protein present in the media was determined using SDS-PAGE and immunoblotting (bottom panel).

- Does the size of PC-containing distensions change from one level of the Golgi to the next? What can you conclude from this observation?
- Predict whether the time required for PC aggregates to move through the Golgi complex, the *transit time,* will be different in cells under the "exiting wave" protocol versus cells under the "incoming wave" protocol.
- Use the graph in Figure 8.3.2a to determine the time it takes for the number of PC aggregates to decrease by half in the *cis* and *trans* Golgi cisternae. Estimate the transit time across the Golgi complex as the difference between these two values.
- Use the graph in Figure 8.3.2b to determine the time it takes for the number of PC aggregates to *increase* by half in the *cis* and *trans* Golgi cisternae. Calculate the transit time based on the "incoming wave" protocol.
- Discuss how transit time data relate to your prediction for the two treatment protocols.
- Summarize the results shown in Figure 8.3.2c and d. Include the bottom panel in your summary.

- Formulate a hypothesis for why the number of PC aggregates and the amount of released PC in Figure 8.3.2b and d are greater than in Figure 8.3.2a and c.
- Explain the relationship between the data in Figure 8.3.2a and b and the data in Figure 8.3.2c and d.
- The researchers conclude that the behavior of PC aggregates in ET fibroblast cells is consistent with the cisternal maturation model of Golgi transport. Design an experiment using PC expression in ET fibroblast cells that would further test the cisternal maturation model.

References

[1] Scales SJ, Pepperkok R, Kreis TE. Visualization of ER-to-Golgi transport in living cells reveals a sequential mode of action for COPII and COP I. Cell 1997;90:1137–48.

[2] Balch WE, Dunphy WG, Braell WA, Rothman JE. Reconstitution of the transport of protein between successive compartments on the Golgi measured by the coupled incorporation of N-acetylglucosamine. Cell 1984;39:405–16.

[3] Bonfanti L, Mironov AA Jr, Martinez-Menárguez JA, et al. Procollagen traverses the Golgi stack without leaving the lumen of the cisternae: evidence for cisternal maturation. Cell 1998;95:993–1003.

Endocytosis

Following the Fate of a Phagosome [1]

INTRODUCTION

Phagocytosis describes the process by which a cell captures and ingests foreign particulate material from its extracellular environment. Cell types such as macrophages or neutrophils are considered "professional phagocytic cells" as they function to eliminate foreign material and pathogens as part of an organism's immune response. All cells, however, have the ability to undergo phagocytosis when stimulated to do so.

The process of phagocytosis occurs in stages. Phagocytosis starts when a pathogen binds to a receptor protein on the surface of the phagocytic cell. The interaction between the pathogen and receptor sends signals into the phagocytic cell triggering the assembly of actin microfilaments near the plasma membrane (PM). As the actin microfilaments grow they cause the membrane to push outward, forming structures known as pseudopods that surround the pathogen. Eventually the tips of the pseudopods connect and fuse resulting in the **internalization** of the pathogen in a membrane vesicle or **phagosome** located within the cytoplasm of the cell. Pump proteins (vacuolar-type H-ATPases or V-ATPases) in the membrane of the phagosome transport hydrogen ions from the cytoplasm into the lumen (interior) of the organelle resulting in **acidification** of the phagosome. In the final step, the phagosome fuses with the cell's lysosomes, forming a **phagolysosome** which functions to finally destroy the pathogen.

Acidification of phagosomes is an important part of phagocytosis. The pH of the external environment of a typical cell is neutral (~pH 7.4) while the pH of a lysosome is acidic (~pH 4.6). The acidic environment of the lysosome activates various hydrolytic enzymes that contribute to the breakdown of materials brought into the lysosome. Tracking the acidification of phagosomes was made possible by the observation that the properties of the fluorescent dye fluorescein isothiocyanate (**FITC**) are pH sensitive. Fluorescent dyes work by absorbing light energy at one wavelength, the excitation wavelength, and then releasing some of that energy as light of a longer wavelength, the emission wavelength. The ability of FITC to fluoresce is inhibited when it is exposed to an acidic environment. This change is quantified by measuring the ratio of intensity of FITC fluorescence (emission wavelength = 505 nm) generated by excitation wavelengths of 430 and 470 nm. The intensity of light emitted by FITC at an alkaline pH is greater when the dye is excited by 470 nm light compared with 430 nm light. Conversely, at an acidic pH, the intensity of FITC emission is greater when excited by 430 nm light.

- What is the origin of the phospholipids that make up the phagosome membrane?
- Outline the steps involved in phagocytosis of a bacterium by a macrophage cell.
- Why would a pump protein be required to move hydrogen ions into the phagosome?
- Connect the concepts of protein folding and pH optimums for enzyme function.
- Describe how the properties of a biological membrane contribute to the process of phagocytosis.
- Predict what might occur if a phagosome remained at pH 7.4 prior to fusing with a lysosome (pH 4.6).
- Model how the ratio of fluorescence intensity at 430 and 470 nm would change with pH.

BACKGROUND

Phagocytosis is normally associated with the detection and destruction of foreign material such as bacterial pathogens; however, some bacteria can hijack the process of phagocytosis to gain entry into cells. One example of this is the bacterium *Listeria monocytogenes*, a foodborne pathogen responsible for serious illness and even death. *Listeria* gains access to your body by binding to the epithelial cells that line your gut. The bacterial protein *internalin A* or **InlA** contains an amino acid sequence that allows it to attach to the cell adhesion proteins *E-cadherin* present on the surface of epithelial cells. E-cadherin-InlA binding triggers a series of reactions in the epithelial cell leading to endocytosis of the bacterial cell. Once inside the phagosome, *Listeria* secretes phospholipases and a protein *listeriolysin*, which inserts into the membrane of the phagosome, forming a pore. *Listeria* is now able to escape from the phagosome, replicate within the cytoplasm of the cell, and eventually spread to other cells.

Phagocytosis of *Listeria* was examined using **Caco-2** tissue culture cells derived from human intestinal epithelium. Rather than use intact *Listeria* bacteria, the researchers took advantage of the ability of InlA to induce phagocytosis by attaching the portion the InlA protein required to bind E-cadherin to a bead. In effect, the Caco-2 cells were tricked into engulfing the beads. The addition of fluorescent probes to the surface of the bead allowed the researchers to track bead movement from outside the cell, into the phagosome, and finally into the lysosome.

- Learn more about *Listeria* and food safety by researching the story of the 2011 *Listeria* outbreak associated with cantaloupes in the US.
- Predict what effect listeriolysin pore formation would have on the phagosome.
- List some of the possible advantages of using a protein-coated bead instead of *Listeria* bacteria in an experiment.

METHODS

InlA-coated beads
The extracellular domain of the bacterial protein InlA was expressed in *Escherichia coli*, isolated, and purified. Purified InlA protein was covalently attached to the surface of 2 μm polystyrene beads. The InlA-coated beads were either left unlabeled or further modified by the addition of fluorescent dyes Alexa488 or FITC.

Determination of Alexa488-InlA bead internalization
Alexa488-InlA beads were added to Petri dishes containing Caco-2 cells that had been cooled to 4°C. Beads were allowed to attach to the cells; the cells were then rinsed to remove any unbound beads. The cells were then placed into a 37°C incubator for varying amounts of time. Cells were fixed with 4% formaldehyde at each time point and 2 μg/mL of anti-Alexa488 antibody was added to the Petri dish and incubated for 30 min. Binding of the anti-Alexa488 antibody to Alexa488 molecules has the effect of quenching (suppressing) fluorescence. Fluorescence intensity of the Alexa488-InlA beads was measured using a microscope with a 450–490 nm excitation filter and a 505 nm emission filter. Average fluorescence intensity was determined using digital image analysis software.

Determination of acidification using FITC-InlA beads

FITC-InlA beads were added to Caco-2 cells as described earlier, with the exception that the cells were not fixed. Instead, the cells were removed from the incubator at differing times and immediately examined using a fluorescence microscope with a 430 and 470 nm excitation filter and 505 nm emission filter. The ratio of intensities of fluorescence generated by the beads in response to each excitation wavelength was calculated using digital image analysis software.

Determination of phagosome/lysosome fusion

Unlabeled InlA-coated beads were added to Caco-2 cells that were prelabeled with the fluorescent dye Lyso-Tracker. Lyso-Tracker accumulates in cellular organelles that have an acidic pH such as late endosomes and lysosomes. The labeled cells were rinsed with 4°C buffer prior to the addition of the unlabeled InlA-coated beads. After any unbound beads were removed, the cells were incubated at 37°C for different amounts of time. At each time point the cells were examined using a fluorescence microscope to determine the location of the beads relative to the fluorescently labeled membrane compartments. Colocalization of the InlA-coated beads with a Lyso-Tracker-labeled organelle was used as an indicator of phagosome/lysosome fusion.

Kinetics of phagocytosis

More than 100 beads were analyzed at each time point for each of the experiments. The fraction (percent) of internalized, acidified, or fused beads was calculated based on the criteria described earlier. The fraction of beads in each condition was graphed as a function of time. The graphs were then used to calculate $t_{1/2}$, the time required for half the beads to reach a specific stage of phagocytosis.

- Explain why the extracellular domain of InlA was used for these experiments.
- How does incubating cells at 4°C influence the process of phagocytosis?
- Why is anti-Alexa488 antibody binding a useful assay for the internalization step of phagocytosis?
- Defend the assumption that colocalization of beads with Lyso-Tracker-labeled organelles is evidence that phagosome/lysosome fusion has occurred.

RESULTS

- Interpret and summarize the results shown in Figure 9.1.1.
- Explain why the researchers chose to use 2 µg/mL of the anti-Alexa488 antibody in their subsequent experiments.
- How do the data in Figure 9.1.1b correspond to your model of the effect of pH on FITC fluorescence?

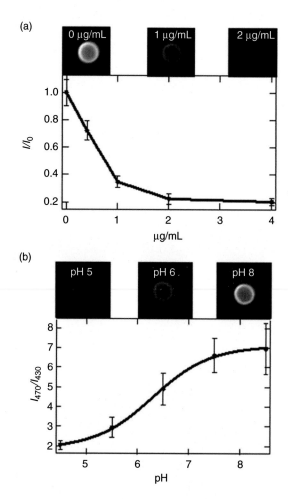

FIGURE 9.1.1 Quantitative changes in fluorescence intensity occur in response to anti-Alexa488 antibody binding or FITC–pH exposure.

a. InlA-Alexa488 beads were incubated in the presence of increasing concentrations of anti-Alexa488 antibody. Fluorescence intensity of individual beads was measured before (I_0) and after (I) exposure to antibody. b. Fluorescence intensity of InlA-FITC beads was measured at pH values from 4.5 to 8.5 using excitation wavelengths of 430 and 470 nm.

- All of the beads in Figure 9.1.2a are in proximity to the cytoplasm of the Caco-2 cell (the yellow line is the cell's PM). Propose an explanation why some beads were quenched and other beads were not.
- Summarize the conclusion supported by the data presented in Figure 9.1.2b.

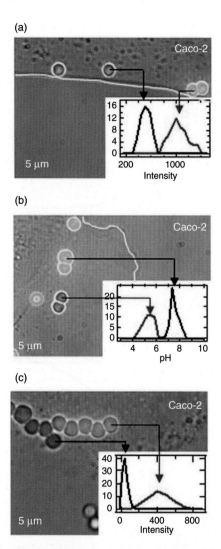

FIGURE 9.1.2 Internalization, acidification, and phagosome/lysosome fusion in intestinal epithelial cells CaCo-2.

a. Caco-2 cells were incubated with Alexa488-InIA beads for 30 min followed by a 30 min incubation with the anti-Alexa488 antibody. Beads that were still present outside the cells reacted with the Alexa488 antibody (black line on graph) while beads that had been internalized did not (red line on graph). b. Caco-2 cells were incubated with FITC-InIA beads for 35 min. The ratio of 470/430 fluorescence intensity was measured for individual beads and the pH that a bead was exposed to was calculated based on the data shown in Figure 9.1.1(b). Beads exposed to an acidic environment appeared less intense (red line on graph) compared with beads in a more neutral environment (black line on graph). c. Caco-2 cells labeled with Lyso-Tracker dye were incubated with unlabeled InIA-coated beads for 135 min. Unlabeled beads show little fluorescence intensity (black line on graph) compared with beads that are colocalized with a labeled organelle (red line on graph). The yellow line indicates the edge of the cell.

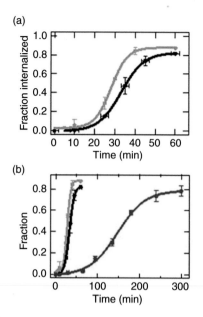

(a)

(b)

FIGURE 9.1.3 Rates of internalization, acidification, and phagosome/lysosome fusion in CaCo-2 cells.

a. Fraction of internalized Alexa488-InlA beads (gray line) and acidified FITC-InlA beads (black line) in Caco-2 cells as a function of time. b. Fraction of unlabeled InlA-coated beads to colocalize with Lyso-Tracker-stained organelles (red line). Internalization and acidification (gray and black lines, respectively) are shown for comparison.

- Using Figure 9.1.1b as a standard curve calculate the pH for the following:

Bead	Intensity at 430 nm	Intensity at 470 nm
1	145	882
2	232	675
3	151	756

- Concanamycin A is a drug that specifically inhibits V-ATPases. Predict the likely outcome if the experiment shown in Figure 9.1.2b was conducted in the presence of concanamycin A.
- List reasons only some of the beads in Figure 9.1.2c are colocalized with the lysosome.
- Use the data in Figure 9.1.3a to calculate the approximate time required for half the beads to be internalized ($t_{1/2}$ internalization).
- What is the approximate time interval between $t_{1/2}$ internalization and $t_{1/2}$ acidification?
- How long does it take for 50% of the acidified phagosomes to fuse with the lysosomes?
- Propose an explanation for the differences in the rates of internalization, acidification, and fusion of phagosomes with lysosomes.

Catching a Receptor by the Tail [2]

INTRODUCTION

Cell membranes contain a host of proteins with diverse functions that support the life of a cell. **Receptors** are a special class of proteins that function by binding a specific **ligand** molecule. When a ligand binds to its receptor, the receptor can change conformation, transmitting a signal into the cell. In some cases the receptors will remain on the surface of the cell and the ligand will eventually diffuse away. In other cases, ligand binding to a receptor triggers a series of events leading to internalization of the receptor:ligand complex in a process known as **receptor-mediated endocytosis**. This type of endocytosis requires that the receptor be "captured" into a **clathrin-coated pit**. Clathrin is a peripheral membrane protein that self-assembles into cage-like structures that cause membranes to bend. The region of membrane surrounded by clathrin will eventually form a vesicle that will separate or "pinch off" from the plasma membrane (PM) and travel into the cell.

Receptor proteins are **transmembrane** proteins. Transmembrane receptor proteins are embedded in the phospholipid bilayer of the PM with a hydrophobic region of the protein spanning the bilayer and hydrophilic regions extending out on both the intracellular (cytoplasmic) and extracellular sides of the membrane. The extracellular domain of a receptor protein is associated with ligand binding. In contrast, the cytoplasmic domain of a receptor may have multiple functions. One of these functions may be to interact with the proteins responsible for forming the clathrin-coated pit.

- Why would the membrane-spanning domain of a receptor protein need to be hydrophobic?
- Research/review the activity of the low-density lipoprotein (LDL) receptor.

BACKGROUND

Epidermal growth factor (**EGF**) is a protein that signals cells to undergo mitotic cell division and differentiation. The cellular effect triggered by this signal is so significant that binding of EGF to its receptor (**EGF-R**) results in the internalization and destruction within lysosomes of both receptor and ligand. By destroying its EGF receptors the cell can downregulate its sensitivity to EGF.

The EGF receptor is a transmembrane protein. Within the cytoplasmic domain of the receptor there is a region, near the membrane, that functions as a kinase. There is also a stretch of 230 amino acids, near the end of the protein, which is associated with regulation of endocytosis. When the EGF receptor binds to the EGF ligand, individual receptor proteins move within the plane of the membrane and interact to form dimers. The kinase domains phosphorylate tyrosine amino acids on the cytoplasmic tail of the adjacent receptor and may be involved in phosphorylation of other, nearby proteins.

Removal of EGF and EGF receptors from the surface of the cell occurs through receptor-mediated endocytosis and the activity of clathrin-coated pits. Although individual clathrin proteins can self-assemble, the ability of clathrin to associate with a membrane is controlled by the activity of other proteins. One of these clathrin-associated proteins is **AP-2**, a clathrin adaptor protein that functions to promote the assembly of clathrin at the surface of the PM. AP2 is composed of multiple polypeptide subunits (α_a, α_c, β_2, μ_2, and σ_2).

- What is the function of a kinase?
- Use the EGF receptor to explain the different levels of protein folding.
- Describe how a transmembrane protein like EGF-R can move in the plane of a membrane.

METHODS

Cell culture

The mouse cell culture line B82L was used in these experiments. Mouse cells were transfected with a gene for intact human EGF-R or a truncated version of the protein that is missing the regulatory cytoplasmic tail domain. Control cells were transfected with a gene unrelated to the EGF or EGF receptor. Culture dishes containing the mouse cells were either chilled at 4°C or maintained at 37°C prior to incubation with human EGF for 60 min. Following the incubation period, the media surrounding the cells were removed and replaced with fresh media that did not contain the EGF. The culture dishes were then kept at 37°C for an additional 60 min. Cells were lysed at specific time points during the 37°C incubation period.

Immunoprecipitation

Cell lysates were incubated with the antibody 13A9 that is specific for the extracellular domain of the EGF-R, allowing the antibody to bind to any EGF receptor proteins present in the solution. The lysate + antibody solution was then mixed with **protein A–Sepharose beads** in the presence of the detergent **Triton X-100**, which dissolves any membranes. Protein A is a bacterial protein that is capable of binding to the Fc or stem domain of mammalian antibody proteins. This interaction does not affect antibody binding since the Fc domain is not involved in antibody specificity. Linking protein A to a Sepharose bead allows the researcher to **immunoprecipitate** or "capture" the EGR-R proteins from the cell lysate solution by centrifuging the mix, pelleting the Sepharose beads and all the proteins attached to them. Using this method it is possible to immunoprecipitate not only the proteins specifically recognized by the antibody (in this case the EGF-R), but also any other proteins that bind to that protein. The identity of the proteins isolated through immunoprecipitation can be revealed through the technique of **Western blotting** in which proteins are separated using polyacrylamide gel electrophoresis, transferred to a nitrocellulose membrane, and then reacted with antibodies that recognize specific proteins.

Surface plasmon resonance detection

Surface plasmon resonance (SPR) is a method that can detect interactions between molecules or proteins occurring in real time. The technique measures small changes in the refractive index at the surface of a sensor chip that occur in response to binding events. In this work, the EGF-R antibody 13A9 was attached to the surface of the sensor chip. Cell lysates from cells transfected with the EGF-R gene or a control gene were then injected into the detection apparatus. Binding between the EGF-R and the 13A9 antibody causes a change in SPR that is measured as a change in resonance units (RU). The EGF-R immobilized on the surface of the sensor chip could then be used to test the ability of purified AP2 to bind. Protein binding to the EGF-R can be measured as the difference in surface resonance before and after the addition of purified AP2 (ΔRU).

- Outline the steps involved in the transcription of the EGF-R in a mouse cell.
- Where in the mouse cell would the EGF-R be translated?
- Illustrate the steps involved in immunoprecipitation.
- Why was it necessary to add Triton X-100 to the solution prior to immunoprecipitation?
- Explain the difference in data obtained using immunoprecipitation and SPR.

RESULTS

- Predict how transmembrane proteins would behave in cells maintained at 4°C.
- Summarize the effect of temperature on the EGF-R (Figure 9.2.1).
- What region of the EGF-R protein is phosphorylated in the presence of the EGF?
- How does the binding of the EGF to EGF-R influence the behavior of AP proteins?
- How does the binding of the EGF to EGF-R influence the formation of a clathrin-coated pit?
- Explain why the graph in Figure 9.2.2b is different from those in Figure 9.2.2 a, c, and d.

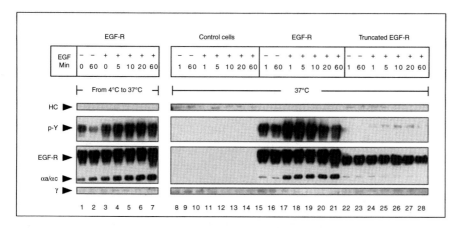

FIGURE 9.2.1 Immunoprecipitation allows for the identification of proteins bound to the EGF-R.
(Lanes 1–7) Cells expressing intact EGF-R were chilled to 4°C for 60 min in the absence of EGF (–) and
then exposed to EGF (+) and warmed to 37°C for another 60 min. (Lanes 8–28) Mouse cells expressing
no EGF-R (8–14), intact EGF-R (15–21), or truncated EGF-R (16–28) were maintained at 37°C in the
absence (–) and presence (+) of EGF for the times indicated. Samples from each time point were
blotted with antibodies specific to clathrin heavy chain (HC), phosphorylated tyrosine (p-Y), human EGF-R
(EGF-R), the α_a, and α_c subunits of AP2, and the γ subunit of a different clathrin AP, AP1 (γ).

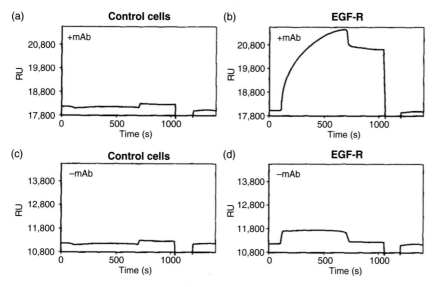

FIGURE 9.2.2 Evidence of EGF-R protein binding using SPR detection.
Cell lysates from control (a and c) and the EGF-R expressing (b and d) B82L mouse cells were injected
through a sensor chip coated with the EGF-R monoclonal antibody (+mAb) 13A9. Binding of the EGF-R
proteins to the surface of the sensor chip causes an increase in SPR units (RU). No change in RU is
detected in the absence of the mAb (−mAb). Addition of an alkaline pH solution at 1000 s removed all
proteins from the surface of the chip.

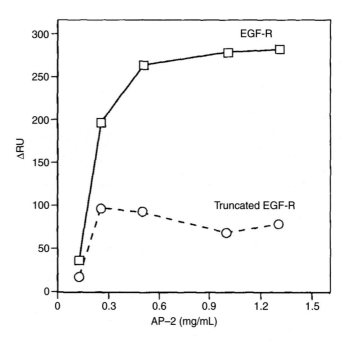

FIGURE 9.2.3 Binding of AP2 to the EGF-R.

SPR detection was used to measure the interaction between the EGF-R and purified AP2. Intact or truncated EGF-R was immobilized on the surface of a sensor chip using the 13A9 antibody. The sensor chip was exposed to different concentrations of purified AP2. Binding was measured as the change in RU (ΔRU) before and after AP2 addition.

- Look closely at the scale of the Y-axis in Figure 9.2.2a versus c. Use your understanding of SPR to explain the difference you observe.
- Why would exposure to alkaline pH alter the binding of proteins to the surface of the sensor chip?
- Describe the nature of the interaction between the EGF-R and AP2 shown in Figure 9.2.3.
- Where does AP2 bind on the EGF-R? What data support your answer?
- Create a diagram that illustrates the following properties of the EGF receptor
    - Binding site for the EGF
    - Transmembrane protein
    - Dimerization
    - Tyrosine kinase domain
    - Autophosphorylation of cytoplasmic domains
    - AP2 binding
    - Recruitment into a clathrin-coated pit

Can Clathrin Bend a Membrane? [3]

INTRODUCTION

Endocytosis involves the capture of material from the surface of a cell and transport of that material into the cytoplasm. Material is brought into the cell inside membrane-bound endocytic vesicles that are formed from the phospholipid bilayer of the cell's plasma membrane (PM). Imagine the PM as the surface of a balloon. Pushing a finger into the balloon's surface will cause the membrane to bend forming an indentation or pit. But a membrane, like a balloon, cannot change shape without force being applied. Peripheral membrane proteins present on the cytoplasmic side of the membrane are thought to produce the force necessary for inward bending of the PM.

Clathrin is one of the proteins involved in the process of endocytosis. A molecule of clathrin is made up of six proteins: three heavy-chain proteins and three light-chain proteins. The quaternary structure of clathrin is described as a *triskelion* with each of the heavy-chain proteins projecting out from a central hub. Multiple clathrin triskelions can self-assemble to form a complex honeycomb-like structure or **coated pit** on the surface of the PM. Clathrin-coated pits are associated with initial bending of the membrane; however, the assembly of clathrin at the surface of the PM requires the function of an additional set of proteins known as **APs**.

AP2 functions in the formation of clathrin-coated pits during endocytosis in a wide range of cell types. AP2 forms a bridge between the inner leaf of the PM and clathrin through its ability to bind to phosphatidylinositol phospholipids. Other proteins associated with clathrin coat formation include **epsin**, *AP-180, synaptojanin, amphiphysin, intersectin, endophilin, auxilin,* and *syndapin*. Each of these proteins is composed of a series of folding domains that regulate protein:protein or protein:lipid binding. The number and variety of proteins involved in the formation of a coated pit make it difficult to know precisely which proteins are necessary to induce a change in the shape of the membrane.

Endocytic vesicles are formed when clathrin-coated pits deepen, becoming structures known as **buds**. The GTPase protein *dynamin* then triggers separation of the buds from the rest of the PM, forming clathrin-coated vesicles. In a final step, the uncoating proteins **auxilin** and **Hsc70** catalyze removal of the clathrin coat using the energy released through hydrolysis of ATP.

- Review/research the different forms of endocytosis. How are they similar? How are they different?
- Explore the structure of a clathrin molecule using online resources (e.g., http://www.rcsb.org/pdb/101/motm.do?momID=88).
- Select one of the APs listed earlier. Conduct a keyword search of a scientific database using the name of the APs you chose. How many citations did you recover?
- Diagram the steps of endocytosis starting with a clathrin-coated pit and ending with an endocytic vesicle.
- Why would it be necessary for the clathrin coat to be removed from the surface of an endocytic vesicle?

BACKGROUND

One way to gain insights into the detailed mechanisms of complex cellular processes is by controlling the number of variables involved. **Cell-free systems** allow researchers to investigate biological reactions *in vitro* or outside the cell. The underlying assumption in these types of studies is that biological molecules will behave in an identical manner whether they are in a cell or in a test tube. A cell-free system gives a researcher control over the individual components present in a reaction. Variables such as time, temperature, and concentration are all defined by the experimental protocol. Genetically engineered or recombinant proteins can be used in place of the normal or wild-type proteins. If a cellular process can be replicated under the artificial conditions of a cell-free system, then it can be concluded that the combination of variables used in that experiment mimic the minimal conditions required for that process to occur inside a living cell.

In this work, the researchers developed a cell-free system that combined liposomes and proteins to model the formation of clathrin-coated buds and vesicles. Liposomes are vesicle-like structures made up of a phospholipid bilayer membrane. The mixture of phospholipids used to construct the liposomes determines the composition of the bilayer. For some of the experiments shown later, the researchers included the chemically modified phospholipid, Ni^{2+}-*NTA-DOGS*. In addition to clathrin, a genetically modified form of the AP *epsin* (H_6-ΔENTH-epsin$^{144-157}$) was included in the assays. This form of epsin is lacking the ENTH phospholipid-binding domain normally found in the protein. In its place, this epsin protein was engineered to include a **His tag**. His tags are a sequence of multiple (typically six to nine) histidine amino acids that are added to a recombinant protein. Histidine can bind to metal ions such as nickel (Ni), cobalt, or copper. The goal of these experiments was to determine the minimal conditions required to trigger clathrin-coated bud and vesicle formation through bending of the liposome membrane.

- Explain why it would be necessary to include an AP in this assay.
- How does the adaptor H_6-ΔENTH-epsin[144-157] bind to the liposome membrane if it is lacking the ENTH phospholipid-binding domain?
- Suggest a reason the researchers felt it was necessary to remove the membrane-binding domain of epsin.

METHODS

Cell-free system for clathrin binding

Phospholipids isolated from pig brain were resuspended by shaking in a buffer solution to form liposomes. For some binding experiments a chemically modified lipid *Ni²⁺-NTA-DOGS* was included in the phospholipid mixture. Clathrin, AP2, and Hsc70 proteins used in the binding assays were purified from pig brain. A genetically modified version of the epsin AP H_6-ΔENTH-epsin[144-157] was expressed and purified from *E. coli*.

Sedimentation assay

Liposomes, with or without Ni^{2+}-NTA-DOGS, were incubated with H_6-ΔENTH-epsin[144-157] or purified clathrin for 30 min at 25 °C. The mixture was centrifuged for 15 min at 90,000g to pellet the liposomes, leaving unbound proteins in the supernatant. For the sequential addition of proteins the liposome pellet was first incubated with H_6-ΔENTH-epsin[144-157], centrifuged to remove unbound protein, then resuspended and incubated with clathrin at either 4 or 37°C followed by a final round of centrifugation.

Electron microscopy

Negatively stained images of liposomes were obtained by placing 5–10 μL of liposome solution onto the surface of a carbon-coated grid. The sample was allowed to sit for 1 min and then was fixed using 3% glutaraldehyde and stained with 2% uranyl acetate. A heavy-metal stain surrounds the sample and provides contrast with the surface of the liposomes. Alternatively, liposome pellets were fixed and processed for plastic embedding and sectioning. Ultrathin sections were stained with uranyl acetate and lead citrate.

- Predict what would happen when a solution of phospholipids is exposed to an aqueous solution.
- Create a flow chart that outlines the steps involved in a two-step sedimentation assay.
- What influence would the difference in incubation temperatures have on the binding of proteins to liposomes?

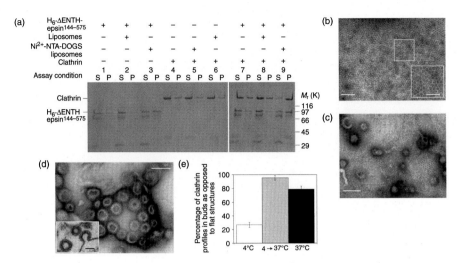

FIGURE 9.3.1 Specificity of clathrin binding to liposomes and the formation of clathrin-coated buds *in vitro*.

a. Sedimentation assay. SDS-PAGE of supernatants (S) and pellets (P) following incubation in the presence (+) or absence (−) of various combinations of liposomes and proteins. (b–d) Negatively stained electron micrographs of clathrin-coated liposomes. Liposomes were incubated with H_6-ΔENTH-epsin$^{144–157}$ followed by incubation with clathrin at 4°C and immediately fixed and stained (b) or allowed to warm to 37°C for 3 min (c) or 20 min (d) before fixation and staining. Inset images show high magnification (b) or ultrathin section (d) of the liposome surface. e. Quantification of the number of clathrin-coated buds formed on the surface of liposomes incubated with H_6-ΔENTH-epsin$^{144–157}$ followed by incubation with clathrin under the temperature conditions indicated in the graph. Scale bars represent 200 nm (main image) or 100 nm (inset image).

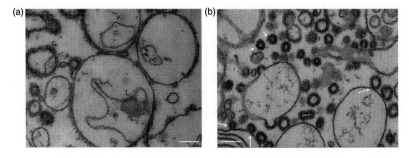

FIGURE 9.3.2 Clathrin-coated bud formation.

Liposomes were incubated with H_6-ΔENTH-epsin$^{144–157}$ followed by incubation with clathrin at 4°C (a) or 37°C (b). Long arrows indicate clathrin-coated buds attached to liposomes by narrow necks. Short arrows indicate clathrin-coated vesicles.

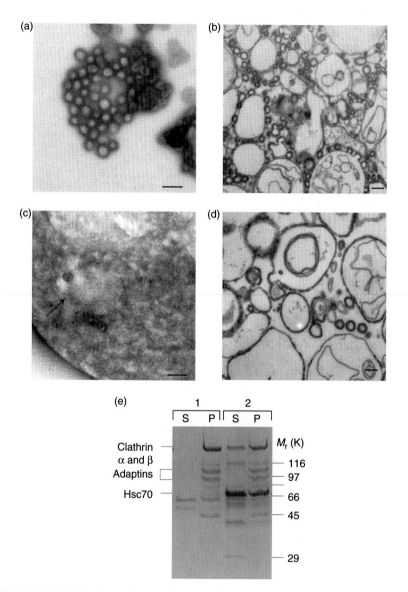

FIGURE 9.3.3 Effect of uncoating proteins on clathrin binding to liposomes and bud formation.
Liposomes incubated at 37°C in the presence of H_6-ΔENTH-epsin[144–157], AP2, and clathrin were
negatively stained (a) or sectioned (b) revealing the presence of clathrin-coated buds. The appearance of
the same type of liposomes changed following incubation with the uncoating proteins, Hsc70, and auxilin
and 2 mM ATP (c and d). e. Sedimentation assay for clathrin binding to liposomes in the absence (1) or
presence (2) of Hsc70.

RESULTS

- What experimental conditions are required before the H_6-ΔENTH-epsin$^{144-157}$ protein will bind to a liposome?
- What conditions favor clathrin binding to liposomes?
- Propose an explanation for the small amount of clathrin seen in the pellets in experiments #4–8.
- Compare and contrast your observations for Figure 9.3.1b, c, and d.
- Summarize the conclusion supported by the data in the graph in Figure 9.3.1e.
- Match the images in Figure 9.3.2 with the equivalent images in Figure 9.3.1. How does the method of sample preparation and visualization influence the data?
- What controls would you propose to support the data presented in Figure 9.3.2?
- Formulate an explanation for the effect of temperature on bud formation in the cell-free system.
- Suggest a reason removal of the clathrin coat might require ATP.
- Where is clathrin located in the sedimentation assay (Figure 9.3.3) in the absence of Hsc70?
- What happens to clathrin and the adaptins when liposomes are incubated with Hsc70 and auxilin?
- Justify why the uncoating experiment shown in Figure 9.3.3 was necessary.
- Defend the authors' conclusion that "clathrin polymerization alone is sufficient to generate buds."

Modeling Membrane Fission [4]

INTRODUCTION

Endocytosis relies on the ability of a cell to bend a region of its membrane into a pit that then separates from the membrane, forming a vesicle that can carry its contents into the cell. A host of proteins are associated with causing the membrane to bend inward. Separation of the newly formed vesicle from the rest of the cell membrane, however, requires a unique set of proteins. Key among these is the protein **dynamin**. The structure of dynamin includes a region associated with lipid binding and a region that functions as a GTPase. Dynamin proteins can self-assemble into rings that bind to the lipids that make up the cell membrane and uses GTP energy to change conformation, causing a constriction or squeezing of the membrane, eventually leading to fission (separation).

Investigations of cellular processes, like endocytosis and membrane fission, use cells as the experimental system. Using cells, researchers can gain insights into a process as it occurs in its normal *in vivo* setting. Cells, however, are very complicated systems that are composed of a host of proteins or other factors that may influence the outcome of an experiment. One way scientists can control for these variables is by moving the experiment out of the cell and into an artificial cell-free or *in vitro* model system. An *in vitro* model allows the researcher complete control over the components, amount, and timing of an experiment. It also opens up a range of experimental methods that can be applied to studying a question.

- What types of endocytosis occur in a eukaryotic cell?
- Describe the similarities and differences between each type of endocytosis.
- Explain how different regions of a protein can have different functions.
- Relate the fluid mosaic model of membrane structure to the behavior of a cell membrane during dynamin-mediated membrane fission.
- Evaluate the pros and cons of the use of an *in vitro* model system for studying a biological process.

BACKGROUND

Research into the role of dynamin in membrane fission and vesicle formation during endocytosis has been limited to the use of inhibitors or gene mutations that alter the activity of the protein *in vivo*. The development of an *in vitro* model system allows for experiments not easily conducted within a cell. The work described here makes use of a small silica bead coated with a phospholipid bilayer as an artificial cell membrane. The researchers developed a method of coating these beads with multiple layers of phospholipid bilayer, creating what they called a "supported bilayer with excess membrane reserve" or SUPER membrane template. The membrane-coated beads could then be exposed to various experimental conditions to test the functional properties of dynamin. If the *in vitro* model works, then dynamin should be able to trigger membrane fission and the production of small membrane-enclosed vesicles using the lipids that present on the SUPER beads.

- What components (other than a source of membrane) would be required for an *in vitro* model of membrane fission?
- One of the advantages of an *in vitro* model is that it allows researchers to control most, if not all, the variables in an experiment. Make a list of the variables you could manipulate using this model system.

METHODS

SUPER templates
Phospholipid bilayers (SUPER templates) were formed on the surface of 5 μm silica beads by mixing the beads with liposomes. The phospholipids used in the construction of the liposomes were matched to the composition of the inner leaf of the cell membrane. Some of these phospholipids were labeled with the red fluorescent molecule RhPE, making it possible to visualize the artificial membrane that coated the silica beads. Changes in the location or intensity of fluorescence were markers for changes in the membrane surrounding the bead.

Dynamin protein
Dynamin protein was isolated from tissue culture cells transfected with the dynamin gene. Some dynamin proteins were labeled with the green fluorescent molecule BODIPY. Various combinations of dynamin and GTP were incubated with SUPER templates to examine dynamin's influence on membrane fission. In some experiments, dynamin and GTP were mixed before their addition to SUPER template beads. In others, the beads were incubated in a buffer containing either 1 mM GTP, GMPPCP, or GDP prior to the addition of dynamin. GMPPCP and GDP are examples of nonhydrolyzable forms of the nucleotide. These molecules can bind to the active site of dynamin, but, unlike GTP, they will not provide energy to the protein.

Dynamin assays
The activity of dynamin in this *in vitro* assay was examined using fluorescence and electron microscopy. A drop containing the SUPER template beads was

placed between two glass coverslips. Dynamin (with or without nucleotide) was added to the edge of the drop and allowed to diffuse. Time lapse images were taken using a fluorescence microscope. The amount of fluorescence within a given region was measured using image analysis software. Electron microscopy was used to obtain high-resolution images of dynamin protein and its association with the SUPER membranes. A method known as **negative staining** was used. In this method a small amount of the sample is placed on the surface of a coated copper grid. Material within the sample will settle onto the surface and stick to the coated grid. Excess sample is rinsed away and a 1% solution of uranyl acetate, a heavy metal solution commonly used in electron microscopy, is added. The stain provides contrast with the sample by accumulating in the spaces surrounding the material, hence "negatively" staining.

The ability of dynamin to produce vesicles from the surface of SUPER template beads was measured using a sedimentation assay. SUPER template beads are relatively large and heavy while the vesicles generated by membrane fission on the surface of the beads are small. SUPER template beads were incubated in buffer for 30 min at 25°C. The beads were centrifuged at a low speed that pelleted the beads, but left the vesicles in the supernatant. As the number of vesicles present in the supernatant increases, so does the intensity of phospholipid-specific fluorescence. The fluorescence intensity of the supernatant was measured. A sample of the supernatant was also added to a copper grid, stained with 1% uranyl acetate, and examined using negative stain electron microscopy.

- Research the various applications of liposomes in medicine.
- Why would it be important to ensure that the phospholipids used to coat the beads matched the composition of the inner leaf of the cell membrane?
- Develop an argument why isolation of dynamin from tissue culture cells is better than isolation from whole tissue such as brain or liver.
- Illustrate how the method of negative-staining works (Hint: imagine preparing a grid using a sample of microscopic Cheerios™).

RESULTS

- Calculate an approximate rate of membrane tubule formation in the presence of dynamin and GTP using the data provided in Figure 9.4.1a.
- Determine the diameter of a membrane tubule.
- What evidence presented in Figure 9.4.1 supports the conclusion that dynamin is responsible for the formation of membrane tubules?
- Does the formation of a membrane tubule require energy?
- How does the presence of GTP in the buffer solution (Figure 9.4.2) alter the behavior of dynamin in this assay?

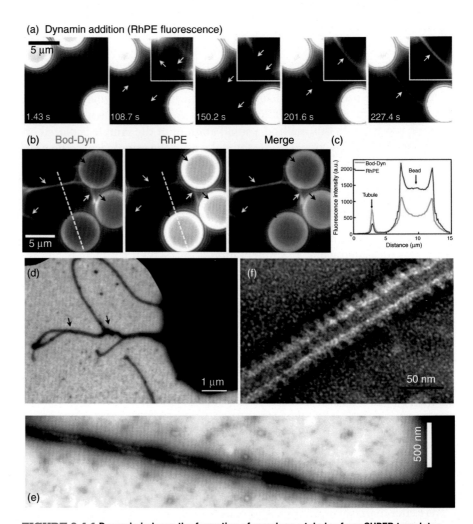

FIGURE 9.4.1 Dynamin induces the formation of membrane tubules from SUPER templates.
a. Time lapse images showing the effect of adding dynamin, in the absence of nucleotide, to the membrane templates. Insets show portions of the frame adjusted for contrast. b. Fluorescence of BODIPY–dynamin (Bod–Dyn) and membrane (RhPE) on membrane tubules and SUPER templates. c. Fluorescence intensity across the dotted line shown in (b). d and e. Negative stain electron microscopy (EM) of a dynamin-coated coiled membrane tubule at low (d) and high (e) magnification. f. Negative stain EM showing the dynamin scaffold on an individual membrane tubule. *Part a: Refer to Movie S3 at http://www.cell.com/supplemental/S0092-8674%2808%2901495-5*

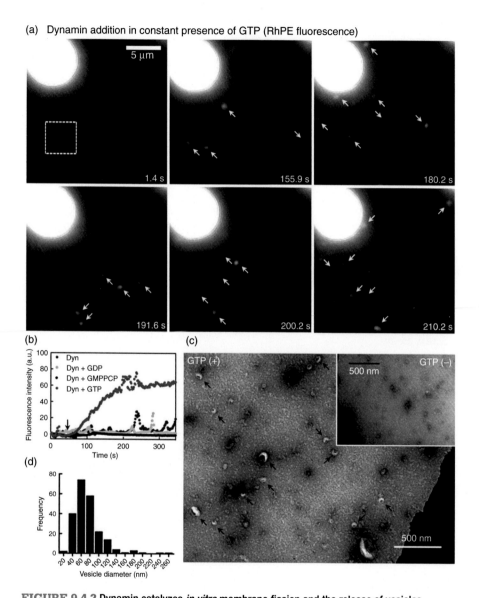

(a) Dynamin addition in constant presence of GTP (RhPE fluorescence)

5 μm

1.4 s

155.9 s

180.2 s

191.6 s

200.2 s

210.2 s

(b)

- Dyn
- Dyn + GDP
- Dyn + GMPPCP
- Dyn + GTP

Fluorescence intensity (a.u.)

Time (s)

(c)

GTP (+)

GTP (−)

500 nm

500 nm

(d)

Frequency

Vesicle diameter (nm)

FIGURE 9.4.2 Dynamin catalyzes *in vitro* membrane fission and the release of vesicles.
a. Time lapse images showing the effect of the addition of dynamin to SUPER templates in the presence
of 1 mM GTP. b. Mean fluorescence intensities in a small area of the solution – white square in (a) –
monitored after dynamin addition to SUPER templates in buffer alone (Dyn) or with GMPPCP, GDP, or GTP.
The black arrow indicates the time of addition. c. Negative stain EM of the solution after the addition of
dynamin and GTP. The inset shows a negative stain EM of the solution after the addition of dynamin in the
absence of GTP. d. Size distribution of vesicles ($n = 221$) measured from EM micrographs. *Part a: Refer to
Movie S4 at http://www.cell.com/supplemental/S0092-8674%2808%2901495-5*

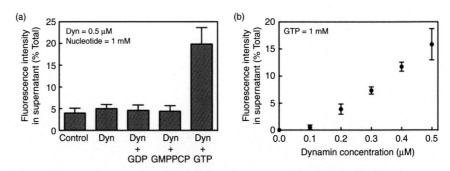

FIGURE 9.4.3 Biochemical analysis of membrane fission by sedimentation assay.
a. Nucleotide dependence for dynamin-catalyzed membrane fission. SUPER templates were incubated in the presence or absence of dynamin (Control) with or without nucleotide. b. Dependence of membrane fission on dynamin concentration. Data are corrected for background fluorescence in the supernatant of control samples.

- Provide an explanation for the data presented in Figure 9.4.2b.
- What are the arrows pointing to in Figure 9.4.2c?
- Estimate the mean and mode for the diameter of the vesicles generated in this assay.
- Create a flow chart describing the sedimentation assay used in Figure 9.4.3.
- Is it significant that fluorescence was detected in the *Control*, *Dyn*, *Dyn + GDP*, and *Dyn + GMPPCP* samples in Figure 9.4.3a?
- Describe the relationship between dynamin concentration and vesicle formation as shown in Figure 9.4.3b.
- The authors conclude, "dynamin alone is sufficient to catalyze membrane fission and vesiculation from a fluid membrane." Discuss whether it is likely that dynamin works alone inside a cell.

References

[1] Blanchette CD, Woo Y-H, Thomas C, Shen N, Sulchek TA, Hiddessen AL. Decoupling internalization, acidification and phagosomal-endosomal/lysosomal fusion during phagocytosis of InlA coated beads in epithelial cells. PLoS One 2009;4:e6056.

[2] Boll W, Gallusser A, Kirchhausen T. Role of the regulatory domain of the EGF-receptor cytoplasmic tail in selective binding of the clathrin-associated complex AP-2. Curr Biol 1995;5:1168–78.

[3] Dannhauser PN, Ungewickell EJ. Reconstitution of clathrin-coated bud and vesicle formation with minimal components. Nat Cell Biol 2012;14:634–9.

[4] Pucadyil TJ, Schmid SL. Real-time visualization of dynamin-catalyzed membrane fission and vesicle release. Cell 2008;135:1263–75.

Cell Walls and Cell Adhesion

Biofilms and Antibiotic Resistance [1]

INTRODUCTION

The dental plaque that develops on your teeth is a **biofilm**. The colorful microbial mats that grow around the geysers and hot springs in Yellowstone National Park are another example of biofilms. A biofilm is created when a population of bacteria secretes a material called extracellular polymeric substance (EPS), which allows the bacteria to attach to surfaces and to each other. EPS is made up of a combination of polysaccharides and proteins. The xanthan gum you see listed as a thickening agent in a wide variety of food products is a type of EPS.

In addition to helping attach bacteria to a surface, biofilms also protect cells from attack by the immune system or antibiotics. This fact makes biofilms a problem when it comes to controlling bacterial infections. Illnesses such as chronic urinary tract infections, chronic inner ear infections, and pneumonia associated with cystic fibrosis are all linked to disease-causing (**pathogenic**) bacteria that are capable of forming biofilms. Biofilm formation is also a serious concern with implanted medical devices.

- Think about the last time you had your teeth cleaned at the dentist's office. What did the dental hygienist have to do in order to remove the plaque biofilm from your teeth?
- How would the formation of a biofilm along the surface of your lung affect your health?
- Research/review how bacteria synthesize and secrete material.

241

BACKGROUND

Antibiotics help control or eliminate bacterial infections by inhibiting growth or killing the bacteria. Different antibiotics use different pathways to impact bacterial cell growth. Antibiotics such as **Gentamicin (Gm)** and **Tobramycin (Tb)** work by inhibiting bacterial protein synthesis. The antibiotic **Ciprofloxacin (Cip)** works by inhibiting the bacterial DNA-specific enzymes required for cell division. The effectiveness of an antibiotic is measured by **minimal bactericidal concentration (MBC)** or the lowest concentration of antibiotic required to kill the cells.

Bacteria can live either as free-swimming (**planktonic**) individual cells or as part of a biofilm community of multiple cells. Cells living in a biofilm are much more resistant to exposure to antibiotics than planktonic cells. The Gram-negative bacteria *Pseudomonas aeruginosa*, is an example of pathogenic bacteria that gains resistance to antibiotics through its ability to form a biofilm. Like other Gram-negative bacteria, *P. aeruginosa* has a thin peptidoglycan cell wall that separates an outer, lipopolysaccharide membrane from the inner, plasma membrane. Between the two membranes is a periplasmic space that is filled with proteins and glucose polymers known as **glucans**, which are associated with the osmolarity of the cell.

In an attempt to understand how the formation of a biofilm acts to protect bacterial cells from the effects of antibiotic treatment, researchers generated a library of over 4000 random mutations in *P. aeruginosa* and then screened each mutant strain to identify cells that formed biofilms, but were missing the enhanced resistance to antibiotics found in the normal, wild-type *P. aeruginosa*. One *P. aeruginosa* mutant, **45E7**, was found to have increased sensitivity to the antibiotic Tb when grown as a biofilm. The experiments presented in this case study address the basis for the difference in antibiotic resistance between wild-type and 45E7 mutant *P. aeruginosa*.

- Explain how inhibition of protein synthesis would lead to the death of a cell.
- Predict which antibiotic, Cip or Tb, would kill bacteria quicker.
- How might the MBC change for a planktonic cell versus a cell in a biofilm?
- The 45E7 mutation was caused by the insertion of a transposon DNA sequence into the middle of the *ndvB* gene. Explain how this addition might affect expression of the *ndvB* gene.

METHODS

Bacterial strains

P. aeruginosa was grown on LB media at 37°C. Transposon mutants were generated using *Escherichia coli* cells carrying the transposon Tn5-B30. Just 1 mL of *P. aeruginosa* was incubated at 42°C for 15 min prior to the addition of 0.25 mL of *E. coli*. The cells were pelleted and then resuspended in 50 μL of LB broth, spotted onto an LB plate, and incubated at 30°C for 24–48 h. Individual transposon mutants were inoculated into sterile 96-well microtiter plates containing minimal arginine medium. Biofilms were allowed to form for 24 h after which the medium was replaced with fresh minimal arginine medium supplemented with 50 μg/mL Tb. After 24 h of exposure to Tb, the biofilms were allowed to recover in fresh media lacking Tb for an additional 24 h. Viability was assessed by plating a small volume of each culture onto an LB plate. Living bacterial cells will grow and divide forming a bacterial colony on the surface of the agar plate.

Antibiotic resistance

The minimal bactericidal concentration for biofilm-grown cells (MBC-B) was determined by exposing 24-h-old biofilms to various concentrations of antibiotic for 24 h, followed by a 24 h recovery period and plating for viability. The minimal bactericidal concentration for planktonic bacteria (MBC-P) was determined by adding antibiotic to bacteria at the time of inoculation, incubating for 24 h, followed by plating for viability. In these experiments, the number of bacteria used in planktonic cultures and microtiter plate-grown biofilms was roughly equivalent: ($\sim 10^7$ cfu per mL or well).

Viability of 24-h-old biofilms was determined using the BacLight Live/Dead stain from Molecular Probes. Living cells with intact membranes can exclude propidium iodide and will stain with the compound syto-9, appearing green. Dead cells or cells with damaged membranes are unable to exclude staining by propidium iodide and will appear red.

- What does the term "wild type" refer to?
- Suggest a reason the cell cultures were allowed to recover in the presence of fresh media after removal of the antibiotic.
- What assumption is being made when scientists use the number of bacterial colonies (cfu) on an agar plate, as a measure of cell density?
- Calculate the number of colonies you would expect to see if you inoculated an LB plate with 3 μL of planktonic cells from a culture with the cell density provided earlier.
- Which part of a cell is being changed when the cell is mutated?
- What is the current price of Molecular Probe's BacLight kit?
- Propidium iodide can only stain a cell with damaged membranes. Why would damage to a bacteria's plasma membrane indicate that the cell is dead?

RESULTS

- Summarize the trend you see in the data for the *wild-type P. aeruginosa* in Table 10.1.1.

- How is this trend similar or different for the mutant 45E7 cells?
- Use Table 10.1.1 to determine the minimal bactericidal concentration of Gm required to kill the 45E7 cells.
- If you had to treat someone with a wild-type *P. aeruginosa* infection, which of the antibiotics shown in Table 10.1.1 would you use?

Table 10.1.1 Antibiotic Sensitivity of Biofilm-Grown and Planktonic *P. aeruginosa* Strains

Strain	Tb		Gm		Cip	
	MBC-P	MBC-B	MBC-P	MBC-B	MBC-P	MBC-B
Wild type	8	400	40	500	4	50
45E7	8	25	40	60	4	6
Fold change	1×	16×	1×	8×	1×	8×

MBC-P, MBC of planktonically grown cells; and MBC-B, MBC of biofilm-grown cells. All antibiotic concentrations are in μg/mL. Values are based on results from at least three separate experiments.

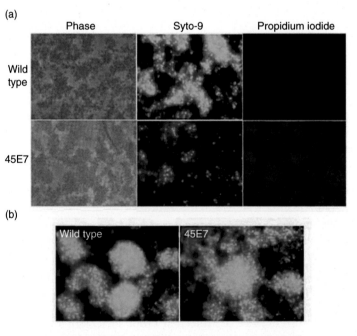

FIGURE 10.1.1 Assay for antibiotic sensitivity of wild-type and 45E7 biofilms.
a. Leftmost panel shows a phase contrast micrograph of the architecture of 24-h-old biofilms formed by wild-type and 45E7 bacteria in the absence of antibiotic. Biofilms were then treated with 20 μg/mL Tb for an additional 24 h followed by staining to detect cell viability. Viable cells with intact membranes stain green in the presence of Syto-9; however, cells with damaged membranes are stained by propidium iodide and appear red. Similar results were observed in three separate experiments. b. Syto-9 staining of wild-type and 45E7 biofilms before treatment with Tb.

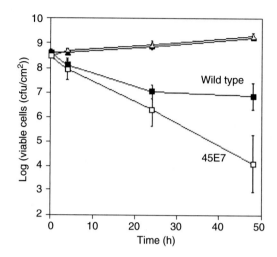

FIGURE 10.1.2 Comparison of antibiotic sensitivity of wild-type and 45E7 48-h-old biofilms.
48-h-old biofilm colonies of wild-type (filled symbols) and 45E7 mutant (open symbols) cells were transferred to solid media without (triangles) or with 10 μg/ml Tb (squares).

- Discuss the significance of the following statement from Table 10.1.1, "Values are based on results from at least three separate experiments."
- Summarize the key information in the figure legend for Figure 10.1.1 using your own words.
- What variable was changed after the first 24 h in this experiment?
- Suggest a reason why researchers chose to use 20 μg/mL Tb in this experiment. (Hint: refer back to Table 10.1.1).
- Why did the researchers include the images labeled "Phase" in part (a) of Figure 10.1.1?
- How do the images in part (a) of Figure 10.1.1 relate to the numbers shown in Table 10.1.1?
- Describe the purpose of the images shown in part (b) of Figure 10.1.1. In your opinion, is the use of only Syto-9 acceptable?
- What was measured to generate the data shown in Figure 10.1.2?
- Using the results in Figure 10.1.2, develop an argument in support of the following statement, "the effect of the 45E7 mutation is limited to antibiotic resistance."
- The 45E7 mutation was found to disrupt the gene *ndvB* that codes for the protein glucosyltransferase, the enzyme responsible for synthesis of glucans. Develop a hypothesis to explain the possible relationship between glucans and antibiotic resistance.

DIY ECM: Cortactin and the Secretion of Fibronectin [2]

INTRODUCTION

The ability of cells to move through their environment is a key property for many normal biological processes. Cell migration is also associated with pathologies such as cancer. Cells move through a series of steps that starts with the extension or **protrusion** of a **lamellipodium**. The lamellipodium attaches to the surrounding substrate using cell adhesion proteins on the cell's surface. In the next step, the cell body contracts, pushing the cytoplasm forward, into the lamellipodium. Finally, the "tail" end of the cell detaches from the substrate allowing the process to repeat.

Cells are surrounded by a substrate or **extracellular matrix (ECM)** that is made up of a collection of fibrous proteins. **Collagen** is one example of an ECM protein, and so is **fibronectin**. Collagen fibers assemble from organized bundles of collagen proteins. Collagen provides strength and elasticity to the surrounding tissue. Fibronectin, on the other hand, functions by binding to other components of the ECM and to cell surface receptors or **integrins**. Fibronectin serves as a point of attachment between cells and ECM, supporting the ability of a cell to migrate through its environment.

Where does the ECM come from? Since the ECM is made up of proteins, it should not be surprising to learn that it is a product of protein synthesis and secretion by the cells themselves. In many cases, cells interact with an ECM that is the product of **autocrine** secretion. In other cases, cells interact with an ECM that was produced by other cells. Migration of cells along an existing ECM pathway is important for the development of multicellular organisms.

- Name some examples of cells that migrate by crawling.
- Outline or draw the stages of cell migration described previously.
- Research/review how the cytoskeleton contributes to lamellipodia protrusion and contractions of the cell body.
- Which proteins control the attachment of a cell to the ECM?
- Describe the relationship between integrins, collagen, and fibronectin.
- Define the term "autocrine" using your own words.
- Outline the steps involved in the synthesis and secretion of an ECM protein such as fibronectin.

BACKGROUND

The human tissue culture cell line HT1080 is derived from cancerous fibroblast cells. These cells develop lamellipodia and migrate while in culture. Previous research has shown that suppression of **cortactin** expression (**cortactinKD**) results in the loss of lamellipodia stability and inhibition of cell migration. Cortactin is a cytoplasmic protein that is associated with cell motility. Cortactin accumulates near the leading edge of lamellipodia and functions to rearrange the actin cytoskeleton. However, cortactin has also been linked with the regulation of membrane trafficking and secretion in cells. What is the connection between the various functions ascribed to cortactin? Does the loss of cortactin prevent a cell from secreting ECM or cell adhesion proteins? This case presents evidence of a surprising answer to that question.

- Where would you find fibroblast cells in your body? What role do they play?
- Discuss the importance of understanding cell migration in the context of cancer research.
- Conduct a search for an image of cortactin. Describe the levels of protein folding you observe.

METHODS

Cells and cell culture

HT1080 cells were grown in tissue culture media supplemented with 10% bovine growth serum (BGS) unless otherwise indicated. Short-hairpin RNAs (shRNAs) were used to knock down expression of cortactin. Control cells were constructed using shRNA made up of scrambled oligonucleotide sequences or cortactin-knockdown cells rescued with an shRNA-insensitive mouse cortactin cDNA. Overexpression of cortactin was achieved by transfection of cells with an extra copy of the cortactin cDNA.

Single-cell motility assay

Live cell imaging was performed using a phase contrast microscope and video camera. Cells were cultured overnight in a six-well plate. Prior to imaging, the tissue culture medium was replaced with fresh media and images were taken every 5 min for a total of 4.5 h.

Lamellipodia persistence assay

The behavior of lamellipodia from the various cell lines was assayed using kymograph analysis. Images of a defined region of the leading edge of a cell were taken with a 40× objective lens every 6 s for a total of 20 min. Change in the leading edge boundary of the cell over time is represented by peaks and valleys in the kymograph. The width of a peak is an approximation of lamellipodial persistence or the length of time the lamellipod remains attached to the substrate.

Immunodectection procedures

Cortactin, integrin, actin, fibronectin, and GAPDH were all examined by immunoblotting with monoclonal or polyclonal antibodies specific to the indicated protein. Actin and GAPDH were included as loading controls to demonstrate that equivalent amounts of protein were present on each blot.

Antibody 12G10, specific to the active form of the integrin β1 subunit was used for flow cytometry analysis of the various cell lines. Antibodies were reacted with intact cells in order to measure the relative amounts of protein on the cell surface. Detection of the presence of IgG was used as a control.

The 12G10 antibody was also used to stain various cortactin cell lines grown in the presence or absence of fibronectin. **Total internal reflection fluorescence (TIRF)** microscopy was used to visualize integrin binding at the interface between the cell and substrate. TIRF microscopy creates high-resolution images without the background staining found in traditional immunofluorescence microscopy by exciting only those fluorophores located within a very narrow plane, typically associated with a cell boundary.

- Conduct a search for a diagram that illustrates how shRNA can be used to silence or knock down expression of a protein.
- Describe how a kymograph works.
- Explain why actin and GAPDH were selected as loading controls.

RESULTS

- Which of the cell lines described in the legend to Figure 10.2.1 would be the control? Justify your answer.
- What is the approximate speed of cell migration for the Sc, KD, and OE cell lines in the absence of exogenous fibronectin (Figure 10.2.1a)?
- Develop an argument in support of the conclusion that the Res cell line represents the "rescued" phenotype using data from Figure 10.2.1a.
- What effect do increasing fibronectin concentrations have on migration of the various cell lines?
- Propose a model to explain the relationship between fibronectin concentration and speed of cell migration shown in Figure 10.2.1a.
- How does increasing the concentration of collagen influence the speed of cell migration (Figure 10.2.1b)?
- Does the rate of migration of the cell lines differ depending on the type of ECM protein present? What does this tell you about the process of cell migration?
- Speculate on the reason the graphs in Figure 10.2.1a and b converge to a single point.

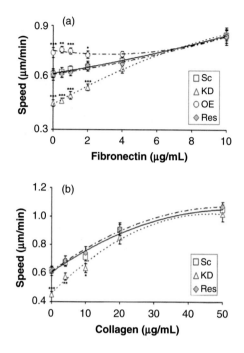

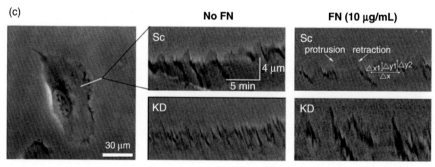

FIGURE 10.2.1 Cortactin-knockdown cell motility and lamellipodial defects are rescued by exogenous extracellular matrix.

If loss of cortactin results in defective ECM secretion, the addition of exogenous ECM should rescue the cell migration phenotype. a. and b. Single-cell migration assays on fibronectin-coated (a) or collagen-coated (b) plates. c. Representative kymographs of lamellipodial behavior over time on uncoated or fibronectin-coated plates. Lamellipodial persistence is measured as the length of time between protrusion and retraction. Scale bars indicate distance and time axes. Abbreviations: Sc, scrambled oligonucleotide; KD, cortactin-knockdown; OE, cortactin overexpression; and Res, cortactin-knockdown cells rescued by expression of an shRNA-insensitive mouse cortactin. $n \geq 45$ cells from three different experiments for each cell line.

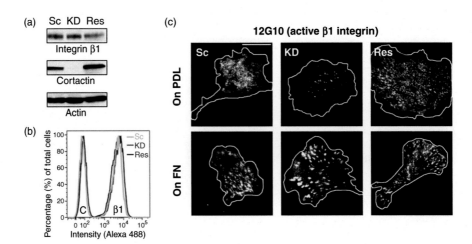

FIGURE 10.2.2 Cortactin expression affects adhesion formation but not total or cell surface expression of β1 integrin.

β1 integrins participate in adhering cells to collagen and fibronectin. Loss of integrin function could account for the phenotype in the cortactin-KD cells. (a) Immunoblot analysis of total cell lysate from Sc, KD, and Res cells. Blots were reacted with antibodies to integrin β1, cortactin, or actin. (b) Flow cytometry analysis of cell surface expression of IgG (c) and integrin β1 (β1) for cortactin-KD and rescued cells. (c) Representative images of the cell–substrate interface using TIRF microscopy. Activated β1 integrin proteins are visualized using the 12G10 monoclonal antibody. Cells were grown in the absence (PDL) or presence (FN) of fibronectin. Cell boundaries are outlined in white.

- Explain how to interpret the images in Figure 10.2.1c.
- Connect the results from the kymographs with the results from the cell migration studies. What can you conclude?
- What is the purpose of including the actin immunoblot in Figure 10.2.2a?
- Summarize the conclusions supported by the results shown in Figure 10.2.2a.
- Explain the significance of the data in Figure 10.2.2b?
- Develop an argument for or against the following statement, "Knockdown of cortactin expression blocks the secretion of integrin proteins to the cell surface."
- Predict whether cells will migrate more quickly on the ECM generated by cells from the same cell line? Explain your prediction.
- What conclusion can you draw from the experiments testing cortactin-KD cell migration on an Sc or Res-generated ECM (Figure 10.2.3)?
- Generate a hypothesis to address the observation that cortactin-KD cell migration is significantly slower on a KD-generated ECM (Figure 10.2.3).
- Is synthesis of fibronectin inhibited in cortactin-KD cells (Figure 10.2.4)? Justify your answer.

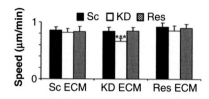

FIGURE 10.2.3 Autocrine-produced extracellular matrix from cortactin-expressing cells rescues motility defects in cortactin-knockdown cells.
The ability of the ECM produced by each of the cell lines to support cell migration was tested using single cell migration assays for all combinations of autocrine-derived ECMs and cell lines. Sc, KD, and Res cells were grown to confluency after which the cells were removed, leaving the ECM intact. ECM-coated plates were then seeded with Sc, KD, or Res cells and the rate of migration was measured. $n > 50$ cells from three independent experiments.

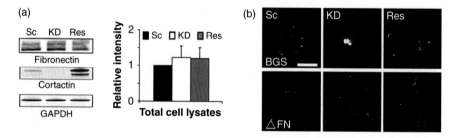

FIGURE 10.2.4 Internalized fibronectin is used for motility.
The ECM generated by autocrine secretion from KD cells could be defective due to the inhibition of expression or the failure to secrete fibronectin. (a) Immunoblot of total cell lysate from Sc, KD, and Res cells. Blots were reacted with antibody to fibronectin, cortactin, and the enzyme GAPDH. Staining intensities for fibronectin were normalized to GAPDH. (B) Immunofluorescence images of fibronectin localization in Sc, KD, and Res cells grown in standard culture media containing BGS, a source of exogenous fibronectin (BGS; top row) or fibronectin-depleted media (ΔFN; bottom row).

- If the ECM generated by cortactin-KD cells is lacking a component, how do you explain the results for the migration of the Sc and Res cells on a KD-generated ECM?
- Fibronectin appears in punctate structures near the nucleus of the cells in Figure 10.2.4b. Discuss the significance of this observation.
- What is the most likely source of fibronectin inside the KD cell in Figure 10.2.4b (top row)? Use the data from Figure 10.2.4 to support your answer.
- Provide an explanation for there being more areas of colocalization (white) in the KD cell in Figure 10.2.5 than in the Sc and Res cells.
- What cellular process is likely to get involved in the transport of fibronectin to the membrane compartments shown in Figure 10.2.5?

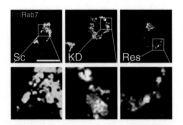

FIGURE 10.2.5 Cortactin promotes fibronectin recycling from the late endosome/lysosome compartment.

Cortactin cell strains were cultured in complete media containing BGS and triple-stained using antibodies to fibronectin (red), the late endosomal marker protein Rab 7 (green), and the lysosomal marker protein LAMP (blue). White indicates triple colocalization. Magnification of the boxed areas is shown further. Scale bar represents 30 μm.

- Describe the evidence supporting the conclusion that cortactin regulates the transport of fibronectin *out* of the late endosome/lysosome (Figure 10.2.5).
- Look back at the outline you made of the synthesis and secretion of fibronectin. Add to your drawing to illustrate how the HT1080 cells are interacting with the fibronectin ECM and how cortactin contributes to the process.
- Look again at Figure 10.2.1b. Design an experiment to test whether collagen is internalized and "recycled" by HT1080 cells.

Bundling the Brush Border [3]

INTRODUCTION

Multicellular organisms exist because populations of cells have the ability to organize, specialize, and stabilize. **Cell adhesion** describes the interactions that connect cells to their extracellular surroundings and to neighboring cells. Whether discussing a tight junction that Ziplocs® two cells together or a hemidesmosome that anchors a cell to the extracellular matrix, cell adhesion involves proteins.

Cell adhesion molecules (CAMs) are proteins that directly participate in the anchoring of one cell to another or to the extracellular matrix. Integrins are cell adhesion proteins that connect a cell to the extracellular matrix. Other CAMS stabilize the connection between cells. **Cadherins** are an example of a class of cell adhesion proteins that mediate calcium-dependent cell–cell attachment. In the presence of calcium, the extracellular domain of a cadherin protein on the surface of one cell associates with the extracellular domain of a cadherin on the surface of an adjacent cell. The cytoplasmic domain of the cadherin protein binds to the cell's cytoskeleton. Cell adhesion is the result of the overlap between the cadherin proteins. Binding can occur between two identical cadherins (**homophilic binding**) or between two different cadherins (**heterophilic binding**).

- Research/review the different forms of cell adhesions.
- Integrins and cadherins are both transmembrane proteins that have a single membrane–spanning domain. Create a drawing to illustrate how an integrin protein might anchor a cell to the extracellular matrix.
- Outline the path an integrin or cadherin protein would take from synthesis to the plasma membrane.
- What determines the type of cadherins present on the surface of a cell?

BACKGROUND

The lining of your intestine consists of a sheet of epithelial cells held together by tight junctions and adherens junctions that form a seal that separates your bloodstream from the contents of your gut. These epithelial cells are **polarized**, meaning that the top (apical) end of the cell that extends into the lumen of your gut appears different from the bottom and sides (basal–lateral) portions of the cell that extends below the cell adhesion junctions and anchors the cell to the extracellular matrix. **Microvilli** cover the apical surface of intestinal epithelial cells forming what is known as the **intestinal brush border (IBB)**. Each microvillus is supported by a bundle of actin microfilaments that cause the plasma membrane to project outward, increasing the surface area of the apical region. The structure of the microvilli is stabilized by proteins that cross-link the actin filaments that form the core of the microvillus and by proteins that connect the actin cytoskeleton to the plasma membrane.

- Suggest a reason it would be important to create a barrier between the lumen of your gut and your blood.
- The apical domain of an intestinal epithelial cell, functions to absorb nutrients. Describe how the structure of the apical domain relates to its function.
- How does a cell "absorb" nutrients such as glucose?
- Conduct a search for images of the structure of the actin cytoskeleton in a microvillus.
- Predict how the loss of microvilli from the surface of the intestinal epithelium would affect a person's health.

METHODS

Cell culture

CACO-2$_{BEE}$ cells were cultured at $37°C$ and 5% CO_2 in tissue culture media supplemented with 20% fetal bovine serum. CACO-2$_{BEE}$ cell lines were created by transfecting the cells with constructs that either targeted PCDH24 mRNA for destruction by RNA interference (PCDH24-KD) or that expressed variants of PCDH24. PCDH24-KD cells were transfected with knockdown-insensitive constructs, PCDH24–EGFP or ΔEC1–PCDH24–EGFP, to assess the ability of these constructs to rescue PCDH24 function.

For the treatment of CACO-2$_{BEE}$ cells with the calcium chelator BAPTA, cells were washed with buffer and then incubated in the presence of 1 mM BAPTA for 2 min. Cells were then immediately fixed for electron microscopy. For proteinase-K treatment, CACO-2$_{BEE}$ cells were grown on filters, washed once with buffer, and incubated in buffer containing 25 U/mL of proteinase-K for 5 min. The sample was washed twice with buffer and then fixed for examination by scanning electron microscopy. Glycosidase treatment followed the same protocol with the exception that 100 μL of a protein deglycosylation mixture was added to the cells followed by incubation for 4 h.

Microscopy

Cells and tissue sections were imaged using a laser-scanning confocal microscope, transmission electron microscopy, or a scanning electron microscope.

Cells were washed with buffer then fixed using 4% paraformaldehyde. For immunofluorescence, cells were washed after fixation with buffer and permeabilized in 0.1% Triton X-100, washed, and incubated with antibodies to F-actin or PCDH24 for 2 h at 37°C. The cells were washed again and then reacted with fluorescent secondary antibody. For scanning electron microscopy, the fixed tissue or cells were rinsed with buffer and then imaged without further processing using an environmental scanning electron microscope.

Mouse intestine was removed by dissection, fixed, and processed for scanning electron microscopy or freeze-etch electron microscopy. Tissue was fixed with 4% glutaraldehyde, washed with water, and then rapidly frozen by contact with the surface of a sapphire block cooled to $-186°C$ with liquid N_2. Samples were etched by raising the temperature of the sample to $-100°C$ for 8–15 min. Replicas were generated by platinum rotary shadowing and viewed using a transmission electron microscope.

Analysis of microvillar clustering

Cells from a 12-day postconfluency (12DPC) monolayer were stained for F-actin and PCDH24. Individual cells were scored according to whether they exhibited robust microvillar clustering (possessing one or more distinct clusters of five or more microvilli) or nonclustering microvilli. Cells were then scored for PCDH24 expression levels (negative, low, or high).

Bead aggregation assay

The extracellular adhesion domains of PCDH24, ΔEC1–PCDH24–-EGFP, MLPCDH-S, MLPCDH-L, or E-cadherin, were expressed as fusion proteins with an Fc domain, which promotes binding to the surface of fluorescent protein A-coated beads. Fusion proteins were incubated with the beads for 1 h with gentle agitation at 4°C. The beads were washed extensively in buffer and then either incubated on their own or mixed with beads carrying a different fusion protein. Beads were allowed to aggregate for 1 h in a glass depression slide in a humidified chamber and images were collected every 15 min. For aggregation assays in the absence of calcium, the beads were washed in buffer supplemented with 2 mM EGTA. Assays were performed at least three times using three independent transfections.

- Conduct a search to learn more about CACO-2$_{BEE}$ cells.
- Research/review how RNA interference functions to knock down expression of a specific protein.
- Explain how an RNAi knockdown cell can be "rescued."
- Compare the methods used for scanning electron microscopy and freeze etch electron microscopy.

- Predict what the results of a bead aggregation assay would look like if:
  - The proteins on the surface of the bead form homophilic interactions.
  - The proteins on the surface of the bead form heterophilic interactions.
  - The proteins on the surface of the bead do not participate in adhesion.

RESULTS

- Describe how the microvilli of CACO-2$_{BEE}$ cells develop over time with reference to their length and organization (Figure 10.3.1).
- Characterize the pattern of attachment that you observe in the 20-days postconfluency (20DPC) CACO-2$_{BEE}$ monolayer. How does this pattern compare with the attachment between microvilli of the mouse intestinal epithelium?
- Estimate the width of a brush border microvillus in a mouse.
- Debate whether similarity in the appearance of links in tissue culture cells and mouse epithelium indicates that they are the same structure.
- Summarize the conclusion that is supported by the data in Figure 10.3.2.
- What type of molecule would be affected by a glycosidase?
- Explain how cadherin proteins function to promote cell adhesion.
- Describe what you observe in each panel in the top row of Figure 10.3.3a.
- Why is there no staining in the anti-PCDH24 panel in the second row of Figure 10.3.3a?
- Propose an explanation for why percent microvilli clustering in nontransduced and control cells is less than in rescued PCDH24-KD cells.
- Discuss whether the difference in sample size between the experiments in Figure 10.3.3b influences your interpretation of the data.
- What effect does the absence of PCDH24 have on the microvilli of a 20DPC CACO-2$_{BEE}$ cell?
- Select the experiment from Figure 10.3.3 that provides the strongest evidence that microvilli clustering require PCDH24. Explain your choice.
- Predict what might happen if CACO-2$_{BEE}$ cells overexpressed PCDH24.
- Develop an argument for or against the conclusion that PCDH24 does not interact by homophilic binding using the data from Figure 10.3.4a.
- Does the size of the extracellular domain of MLPCDH affect its ability to bind? Justify your answer.
- Discuss how the results from the bead aggregation assay compare with similar experiments performed on cells. Use this comparison to argue for or against the validity of *in vitro* aggregation assay.
- Create a model of the interaction between PCDH24 and MLPCDH based on the results in Figure 10.3.4b.
- Estimate the size of the adhesion linkage formed between the two types of beads in Figure 10.3.4c. How does this structure compare with the links found in CACO-2$_{BEE}$ cells and in mouse epithelial tissue?

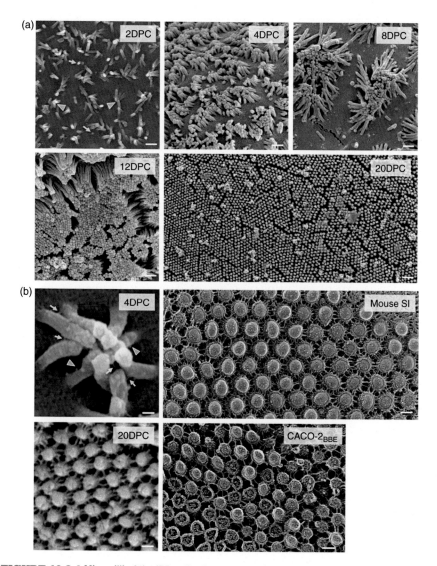

FIGURE 10.3.1 Microvilli of the IBB cells cluster during differentiation and are connected by thread-like links.

a. Scanning electron micrographs of CACO-2$_{BEE}$ tissue culture cells at increasing days postconfluency (DPC). Yellow arrows point to initial microvillar membrane buds and arrowheads indicate points of contact between the distal tips of longer microvilli. Scale bar = 500 nm. b. High-magnification images of CACO-2$_{BEE}$ cells and native intestinal tissue. Scanning electron micrograph of a microvillar cluster from a 4DPC CACO-2$_{BEE}$ cell is shown in the left, top panel. Yellow arrows point to intact intermicrovillar adhesion links and arrowheads indicate unpaired or broken links. Microvillar clusters from a 20DPC CACO-2$_{BEE}$ monolayer are shown in the left, bottom panel. Adhesion links between microvilli are more apparent. Intestinal epithelium from a mouse (right, top panel) or 20DPC CACO-2$_{BEE}$ monolayer (right, bottom panel) were examined using freeze etch electron microscopy. Scale bars = 100 nm.

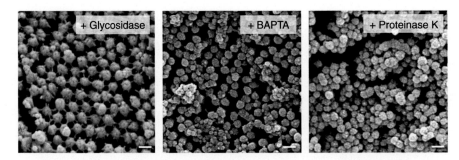

FIGURE 10.3.2 Intermicrovillar links are composed of protein and are sensitive to calcium levels.

20DPC CACO-2$_{BEE}$ cells were treated with glycosidase, BAPTA (a calcium chelator), and proteinase K and visualized by scanning electron microscope.

- Design an experiment using protein-coated beads to further test the nature of the adhesion linkage between PCDH24 and MLPCDH.
- Cadherin proteins are known to interact with a cell's cytoskeleton. Design an experiment to test whether PCDH24 binds to the microvillar cytoskeleton.
- Develop a hypothesis, based on all the data presented in this case, to explain the spacing and organization of the cell adhesion links seen in Figure 10.3.1b.

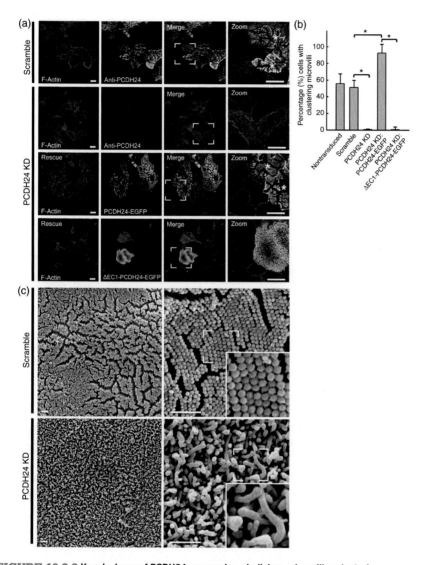

FIGURE 10.3.3 Knock-down of PCDH24 expression abolishes microvillar clustering.
a. Confocal fluorescence images of microvilli of control (SC) or PCDH24-knockdown CACO-2$_{BEE}$ cells. Cells were stained for F-actin (red) and PCDH24 (green). Areas of colocalization appear yellow. PCDH24-knockdown cells expressing the knockdown-insensitive construct PCDH24–EGFP rescued clustering; however, deletion of the first extracellular cadherin domain from the rescue construct (ΔEC1–PCDH24–EGFP) fails to restore clustering. Asterisks indicate regions of microvillar clustering. Scale bar = 15 μm. b. Quantitation of clustering for each of the constructs in (a). For quantification of rescue cell lines only EGFP-positive cells were scored (mean ± SD); nontransduced cells, $n = 2639$; Sc control cells, $n = 1019$; PCDH24-knockdown cells, $n = 1056$; PCDH24-KD rescued with PCDH24–EGFP, $n = 115$; and PCDH24-KD rescued with ΔEC1–PCDH24–EGFP, $n = 160$. *$p < 0.0001$ using a t-test. c. Scanning electron micrographs of the microvilli from 20DPC CACO-2$_{BEE}$ cells stably expressing the control (Sc) or KD constructs. Scale bars = 1 μm.

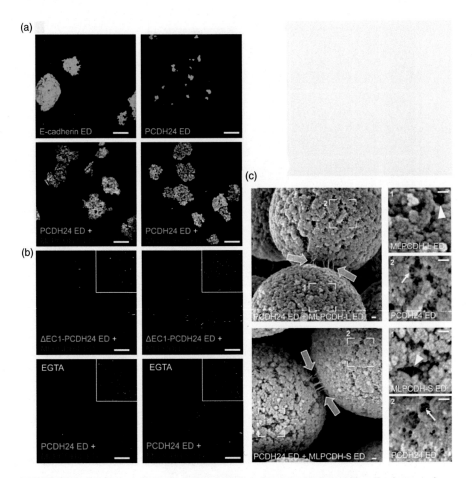

FIGURE 10.3.4 Protocadherin is part of a calcium-dependent heterophilic adhesion complex.
The ability of protocadherin to interact with either itself or MLPCDH, the other major cell adhesion protein found on microvilli, was examined using a bead aggregation assay. The extracellular domains (ED) of either PCDH24 or MLPCDH were fused to the surface of fluorescent beads. Two MLPCDH constructs of differing length (L = long; S = short) were tested. Beads coated with the cell adhesion protein E-cadherin were used as a positive control. a. Confocal images of ED-coated fluorescent beads after 60 min incubation. Inset images are magnified to show individual beads. Protein composition of the beads is indicated. Scale bar = 250 μm. b. Bead aggregation assays that paired MLPCDH with a PCDH24 construct lacking the first extracellular cadherin domain (ΔEC1–PCDH24–ED) or normal PCDH24, but in the presence of the calcium chelator EGTA. c. Scanning electron micrographs of *in vitro*–reconstituted *trans*-heterophilic adhesion complexes between beads coated with the extracellular domain of PCDH24 and either MLPCDH-L (top) or MLPCDH-S (bottom). White arrows point to adhesion links at bead–bead interfaces. Scale bar = 100 nm.

References

[1] Mah T-F, Pitts B, Pellock B, Walker GC, Stewart PS, O'Toole GA. A genetic basis for *Pseudomonas aeruginosa* biofilm antibiotic resistance. Nature 2003;426:306–10.

[2] Sung BH, Zhu X, Kaverina I, Weaver AM. Cortactin controls cell motility and lamellipodial dynamics by regulating ECM secretion. Curr Biol 2011;21:1460–9.

[3] Crawley SW, Shifrin DA Jr, Grega-Larson NE, McConnell RE, et al. Intestinal brush border assembly driven by protocadherin-based intermicrovillar adhesion. Cell 2014;157:433–46.

CHAPTER 11

Cell Metabolism

When Glucose is Low, Something Must Go [1]

INTRODUCTION

Cellular life depends upon a constant supply of energy. Energy, in the form of ATP, depends upon the availability of glucose. **Glycolysis** is a series of 10 chemical reactions that convert the energy stored in a molecule of glucose into four molecules of ATP and two molecules of NADH. In the presence of oxygen, the two pyruvates produced as a result of glycolysis can be further catabolized into three molecules of CO_2 and a molecule of GTP. The resulting NADH and $FADH_2$ molecules then carry their electrons to the electron transport chain, driving the production of more ATP through oxidative phosphorylation.

Glucose levels can be variable, especially for unicellular heterotrophs such as yeast. When glucose levels are depleted, yeast cells temporarily shut down energy-intensive activities such as translation or pumping hydrogen ions into the vacuole. Once glucose (and ATP) levels return to normal, these activities are restored. Protein synthesis and acidification of vacuoles are not the only cellular activities that require large amounts of energy. The impact of glucose depletion on the intracellular transport of organelles is the focus of the experiments described in this case study.

- Where do nonphotosynthetic cells get the glucose they need in order to produce ATP?
- Where is the majority of ATP produced in a eukaryotic cell?
- Research/review the steps involved in glycolysis.
- Discuss whether glucose levels would be variable for a unicellular autotroph.

- Explain why ATP energy is required to transport hydrogen ions into the lumen of the vacuole.
- Predict how the shutdown of translation and vacuolar-ATPase will influence the cell.

BACKGROUND

The budding yeast *Saccharomyces cerevisiae* is a very well established model system for the study of cell biology. Experiments described in this case study made use of the ability to alter the yeast genome, generating strains of yeasts that express fluorescent fusion proteins. The gene for **green fluorescent protein (GFP)** was spliced on the gene for the yeast motor protein **myosin 2 (Myo2)**. Myo2 is an example of an **unconventional myosin** that is involved in the movement of membrane-bound organelles from the "mother cell" to the "daughter cell " or bud. Just like the conventional myosin motor that is associated with muscle contraction, Myo2 has a "motor domain" that binds and hydrolyzes ATP to drive a conformation change in the myosin motor that creates the force required to allow the motor protein to "walk" along an actin filament. However, unlike muscle myosin, Myo 2 has a short "tail" domain that binds to membranes. Since organelle transport requires ATP energy, will the depletion of glucose cause transport to stop?

One of the techniques used in this work was **fluorescence recovery after photobleaching (FRAP)**. The FRAP technique relies on the ability of a high-intensity laser to permanently prevent a small, defined population of fluorescent molecules in a sample from fluorescing. The destruction of a fluorophore is called "photobleaching." The unique feature of this technique is that the fluorescence intensity of the photobleached area is then measured over time. If surrounding fluorescently labeled proteins are able to diffuse into the photobleached zone, then fluorescence will "recover."

- Research/review what GFP is and why it is a handy tool for scientists.
- Research/review the myosin-crossbridging cycle characterized for striated muscle contraction.
- Predict what types of membrane-bound organelles might need to be transported from the mother cell to the daughter yeast cell.
- Draw or describe the relationship between an unconventional myosin, a membrane-bound organelle, and an actin filament.
- List some of the variables that would influence the ability of fluorescently labeled proteins to diffuse within a cell.

METHODS

Yeast growth and carbon depletion

S. cerevisiae were grown in synthetic complete (SC) media overnight, diluted the next morning into fresh media and allowed to grow to mid-log phase.

Cells were added to a 35 mm glass-bottomed dish pretreated with 0.5 mg/mL concanavalin A to promote cell adhesion and quickly transferred into the indicated conditions by three washes with the appropriate media. For galactose depletion, cells were grown to mid-log phase in media supplemented with 2% galactose instead of glucose and washed into medium lacking any carbon source. Immediately after changing the medium, cells were subjected to fluorescence microscopy.

Measurement of ATP levels
ATP concentrations were measured using a bioluminescence assay based on the ability of luciferase to produce light in the presence of its substrate luciferin and ATP. There is a linear relationship between the amount of ATP present in the sample and the amount of light produced. Luminescence of a sample is compared with a standard curve of known ATP concentrations. At each time point, 12.5 μL of the cell sample was added to an equal volume of 10% trichloroacetic acid and vortexed vigorously for 1 min to extract ATP. The mixtures were then neutralized with 1 mL of neutralization buffer, and 10 μL of sample was reacted with 100 μL of the luciferin/luciferase mixture. ATP concentrations were normalized and expressed as the ratio of ATP levels in prestarved cells.

Fluorescence recovery after photobleaching
Yeast cells were attached to a glass-bottomed dish coated with concanavalin A to promote attachment of the cell. An argon laser was used to photobleach a medium-sized bud for 1000 ms. Recovery of Myo2-GFP fluorescence intensity in the bud was imaged every 2 s afterwards. The prebleach GFP intensity of the daughter cell was normalized for comparison between cells. Experiments with glucose-depleted cells were always performed within 10 min after glucose withdrawal.

Yeast cell permeabilization
Mid-log phase yeast cells were attached to concanavalin A–coated glass dishes. The cells were changed into media supplemented with 15 mM ATP and 0.01% digitonin for 5–10 min. Immediately after permeabilization, the cells were transferred into the indicated media for subsequent experiments.

- What advantage might there be to using a synthetic media to grow the yeast cells?
- Define the following:
  - Carbon source
  - Log phase growth
  - Bioluminescence
  - Permeabilization
- Predict whether the dilutions described in the protocol for the measurement of the ATP levels in cells would have a negative impact on the values obtained from the assay.

RESULTS

- Predict why the majority of Myo2 staining is localized at one end of yeast cells in the presence of glucose in Figure 11.1.1a.
- Describe the pattern of Myo2 distribution in the cell in the absence of glucose in Figure 11.1.1a.
- Explain the significance of the yellow color in the bottom right panel of Figure 11.1.1b.
- Summarize the conclusion supported by the results shown in Figure 11.1.1b.
- How does the loss of actin microfilaments affect the distribution of Myo2 in the presence and absence of glucose?
- Conduct a search for an image of the chemical structure of galactose. How is galactose different from glucose?

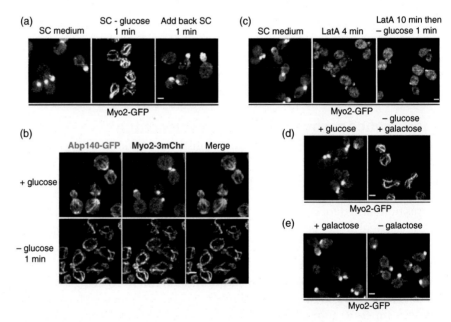

FIGURE 11.1.1 Myo2 rapidly relocalizes to actin cables in the absence of glucose.
a. Cells expressing the Myo2-GFP fusion protein were grown to mid-log phase in culture media and then transferred to fresh culture media with (SC medium) or without (SC−glucose) glucose. Glucose was added back to the −glucose cultures after 1 min (add back SC). The pattern of myosin distribution in the cell was followed using expression of Myo2:GFP. b. Cells coexpressing Myo2 fused to the red fluorescent protein 3mCherry (Myo2-3mCherry) and actin-binding protein 140 fused with GFP (Abp140-GFP) were transferred to fresh media with (+glucose) or without (−glucose) for 1 min. c. Cells as in (a) were transferred into fresh media (SC medium) or media containing 120 μM latrunculin A, an actin-depolymerizing agent. d. Cells as in (a) were transferred to fresh media with glucose (+glucose) or without glucose, but with 2% galactose (−glucose/+galactose) for 1 min. e. Cells were grown overnight on media with 2% galactose and then transferred to fresh media with (+galactose) or without (−galactose) galactose.

- Research/review how galactose enters the glycolytic pathway.
- Are the yeast cells shown in Figure 11.1.1d able to quickly use galactose to produce ATP? Justify your answer.
- Propose an explanation for the differences in the images in Figure 11.1.1d and e.
- Relate your review of the steps involved in glycolysis to the schematic diagram shown in Figure 11.1.2a.

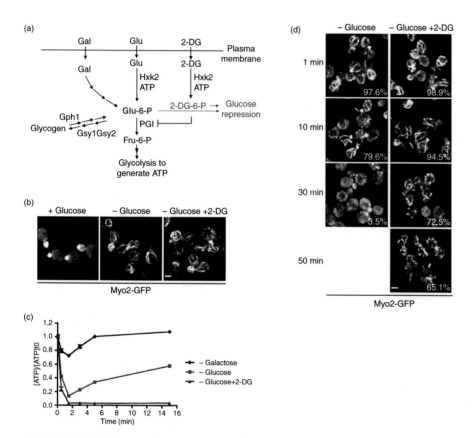

FIGURE 11.1.2 Myo2 localization correlates with intracellular ATP levels.
a. Schematic diagram of the entry of galactose, glucose, and the glucose analog 2-deoxyglucose (2-DG) into the glycolytic pathway of the yeast cell. b. Cells expressing Myo2-GFP protein in normal culture media were switched to fresh media either with glucose (+glucose), without glucose (−glucose), or to media with 2% 2-DG, but without glucose (−glucose + 2-DG). c. ATP levels in cells after glucose depletion (blue), galactose depletion (black), or glucose depletion and 2-DG addition (red). ATP concentrations were normalized to prestarved cells. Error bars represent standard deviation. d. Myo2-GFP distribution in cells depleted of glucose without (−glucose) or with 2-DG (−glucose + 2-DG) after prolonged incubation. The percentage of cells containing Myo2 fibers is shown in the lower right corner. At least 200 cells were scored for each condition.

- Conduct a search for an image of 2-deoxyglucose. How does 2-DG differ from glucose? Connect the structure of 2-DG with its inability to be used in glycolysis.
- 2-DG is phosphorylated by hexokinase and is a competitive inhibitor of the enzyme phosphoglucose isomerase (PGI). How do these two facts connect to the observation that intracellular ATP levels are reduced in the presence of 2-DG?
- How does the addition of 2-DG influence Myo-2 distribution in the cell in the absence of glucose (Figure 11.1.2b)?
- Describe the results shown in the graph in Figure 11.1.2c.
- Speculate on where the ATP is coming from in the glucose-depleted and galactose-depleted cells in Figure 11.1.2c?
- How do the results in Figure 11.1.2d relate to the information in the graph in Figure 11.1.2c? Why is Myo2 staining after 30 min of incubation different between the −glucose and −glucose + 2-DG?
- What effect does the addition of 15 mM ATP have on Myo2 distribution in the permeabilized cells in Figure 11.1.3?
- Research/review the rigor state of actin:myosin binding. Discuss how the results shown in Figure 11.1.3 relate to this concept.
- Under which of the conditions shown in the graph in Figure 11.1.4a is there recovery after photobleaching?
- Explain what must be happening in a cell to account for the recovery of Myo2 fluorescence in a photobleached area.
- Describe what is happening to the cell in Figure 11.1.4b.
- Create a graph like the one in Figure 11.1.4a. Graph your predictions for the following experiments:
  - −glucose + galactose
  - −glucose + 2-DG
  - −glucose with glucose add back after 60 s

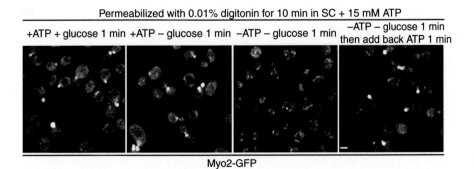

Permeabilized with 0.01% digitonin for 10 min in SC + 15 mM ATP

+ATP + glucose 1 min   +ATP − glucose 1 min   −ATP − glucose 1 min   −ATP − glucose 1 min then add back ATP 1 min

Myo2-GFP

FIGURE 11.1.3 ATP dependence of Myo2 distribution can be replicated *in vitro*.
Cells expressing Myo2-GFP were permeabilized using 0.01% digitonin and transferred to the culture conditions indicated.

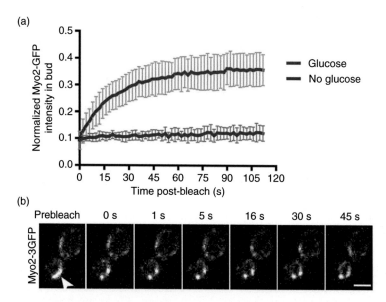

(a)

(b)

FIGURE 11.1.4 Myo2 forms a rigor complex with actin microfilaments in the absence of glucose. a. Normalized Myo2-GFP intensity after photobleaching in the bud of cells incubated with fresh media (red, *n* = 17) or less than 10 min after glucose depletion (blue, *n* = 16). Error bars represent standard deviation. b. Frames from a movie showing photobleaching of a small region of cable-bound Myo2-GFP in a glucose-depleted cell. Arrow indicates the bleached area.

■ Discuss whether the Myo2 response to ATP depletion seen in unicellular yeast would also occur in the cells of a multicellular organism.

■ What challenges would you face in designing an experiment to test Myo2 transport in a mammalian cell?

■ Predict how depletion of O_2 might influence the distribution of Myo2 in the yeast cell. Explain the basis of your prediction.

Do Plants Really Need Two Photosystems? [2]

INTRODUCTION

During photosynthesis light energy is used to split water, generating O_2 and electrons that are then used to produce the ATP and NADPH required for **carbon fixation**. Photosystem II (PSII) functions to capture light energy and transfer it to plastoquinone, the first molecule in an electron transport chain that leads to the production of ATP. The oxidized reaction center pigment P680 returns to a reduced state by stripping electrons from water in a process known as **photolysis**, which ultimately results in the production of O_2. **Photosystem I (PSI)** is also capable of absorbing light energy. Electrons from its reaction center pigment **P700** are transferred to the protein ferredoxin, which can then donate the electrons to the electron carrier $NADP^+$ to form NADPH or to the electron transport chain resulting in the production of additional ATP.

ATP and NADPH produced by the light-dependent reactions of the photosystems are used by the **Calvin cycle** in the stroma of the chloroplast. Molecules of CO_2 gas are **fixed** into molecules of 3-phosphoglycerate in a reaction catalyzed by the enzyme **Rubisco**. Subsequent reactions convert molecules of 3-phosphoglycerate into molecules of glyceraldehyde-3-phosphate, some of which will ultimately be converted into glucose in the cytoplasm of the plant cell.

- Where in a eukaryotic cell do photosynthesis and the Calvin cycle occur?
- Research/review the reactions associated with noncyclic photosynthesis. Identify where O_2 and NADPH production occurs.
- Research/review the reactions of the Calvin cycle. Identify where CO_2 is consumed and glyceraldehyde-3-phosphate is produced.
- Describe how a plant cell would use a molecule of glucose.

BACKGROUND

The unicellular green algae *Chlamydomonas reinhardtii* are **photoautotrophs**. *Chlamydomonas* cells have a single large chloroplast. Mutations in *Chlamydomonas* have been identified that affect the ability of these cells to undergo photosynthesis. The genes *PasA* and *PasB* code for proteins required for P700 function. *Chlamydomonas* mutants **B4** and **F8** block processing of *PasA* mRNA while the mutants ***pasA*Δ** and ***pasB*Δ** have deleted exons. How will the loss of P700 function impact these cells?

- Define the term "photoautroph" in your own words.
- The *PasA* gene is located in the nucleus of *Chlamydomonas*. Outline the steps involved in the transcription of a gene and the transport of mRNA into the cytoplasm of a cell.
- What is an exon? How would the loss of an exon affect a protein's function?
- Predict whether a photoautotroph can survive in the absence of PSI. Explain your answer.

METHODS

Algal strains and growth conditions

The *psaA* and *psaB* mutants (referred to as *psaA*Δ and *psaB*Δ) were constructed by deletion of an exon. Mutant strains F8 and B4 were obtained from the laboratories of E. Greenbaum and L. Mets, respectively. Algae stocks were grown heterotrophically at 23°C in rich liquid media under low-light conditions. Photoautotrophic growth was tested by spotting 12 μL of each log phase culture onto agar plates of either rich media or minimal media. Plates were incubated in the presence of low, medium, or high-intensity light for 4 days.

Gas exchange measurements

Algal cultures were harvested by low-speed centrifugation, washed, and resuspended in minimal media. 1.5 mL of the suspension was placed in the measuring chamber of a mass spectrometer. O_2 and CO_2 exchange of the algae was measured by withdrawing gases directly from the liquid cultures. The use of ^{18}O-enriched O_2 allowed for *in vivo* determination of the O_2 produced by PSII from photolysis in the presence of O_2-consuming processes. CO_2 exchange measurements were performed following $NaH^{13}CO_3$ addition in the dark before recording CO_2 consumption. Light was supplied by a fiber optic illuminator. All gas measurements were performed at 25°C.

Immunoblotting

Liquid cultures were harvested, washed in media, and sonicated to disrupt the cells. The sonicated suspension was centrifuged at $2000g$ for 1 min. The pellet was discarded and the supernatant was centrifuged again at $15,000g$ for 10 min at 4°C. Membranes were carefully collected from above the starch granule pellet. The membrane sample was prepared for SDS-PAGE and immunoblotting using a polyclonal antibody raised against the amino terminal domain of PsaA.

- Exchange of research organisms and reagents is an important part of science. Discuss the circumstances that would justify a lab <u>not</u> sharing resources.
- How are the *Chlamydomonas* obtaining carbon when they are grown heterotrophically?
- What are the advantages of a polyclonal antibody?

RESULTS

- O_2 production is a measure of which cellular process?
- Propose an explanation for why O_2 production in Figure 11.2.1a differs between the wild type and the *psaAΔ* mutant.
- O_2 consumption is a measure of which cellular process?
- Compare the pattern of O_2 consumption with increasing light for the wild-type and mutant cells in Figure 11.2.1b. What conclusion is supported by the similarity in the two lines?
- Suggest a reason the mutant cell is consuming less O_2 than the wild type.
- Why is there no loss in the amount of CO_2 during the dark phase of the experiment in Figure 11.2.2?
- Explain why the slope of the line for wild-type cells differs with the intensity of light.
- Suggest an explanation for the line for the similarities in CO_2 consumption in F8 and wild-type cells.
- Predict what the graph in Figure 11.2.2 would look like if you were measuring CO_2 production.

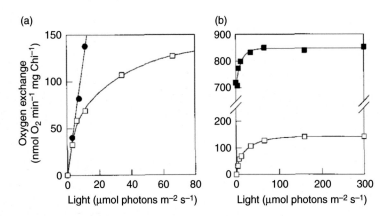

FIGURE 11.2.1 Effect of light intensity on O_2 production and consumption.
^{18}O-enriched O_2 was used to measure the rates of O_2 production (a) and consumption (b) of wild-type (filled symbols) and *psaAΔ* (open symbols) cells with increasing intensity of light.

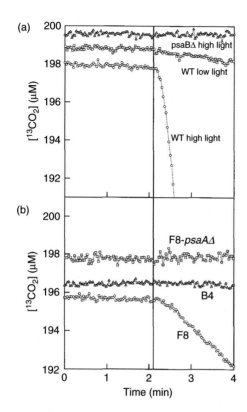

FIGURE 11.2.2 CO₂ consumption differs between wild-type and mutant *Chlamydomonas.*
CO₂ consumption was measured using ^{13}C-enriched CO₂ in a dark/light assay. Cells were exposed to
light at the time indicated by the vertical line. a. CO₂ consumption by wild-type (WT) and *psaB∆* mutant
cells following exposure to the indicated intensities of light. b. Same as above but with the mutants B4,
F8, and the double mutant F8-*psaA∆*.

- How does the combination of media and light intensity influence the growth of wild-type *Chlamydomonas*? (Figure 11.2.3)
- Relate the ability of the various strains to grow on rich media with their specific mutations. What must the rich media be providing to support cell growth?
- Suggest a reason the F8[WT] strain was able to grow on minimal media.
- What is meant by "anaerobic"? How would this condition impact the ability of these cells to grow?
- Can carbon fixation occur in the absence of P700 function? Justify your answer using the results from this case study.

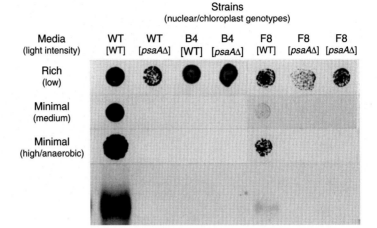

FIGURE 11.2.3 Growth of PSI mutant cell lines relies on exogenous carbon sources.
Wild-type and *Chlamydomonas* mutant strains were grown on different types of solid media in the presence of different intensities of light. Two independent F8-*psaA*Δ constructs are shown. Rich media provide all nutrients required for growth even in low light. Minimal media requires that the cells are capable of fixing carbon. Membrane proteins (25 μg per lane) were prepared from the cultures and immunoblotted against an antibody to PsaA protein (lower panel).

FREX: Opening a Window Into Cellular Metabolism [3]

INTRODUCTION

Nicotinamide adenine dinucleotide (NAD) plays a very critical role in a wide range of cellular reactions. The conversion of NAD from its oxidized form (NAD^+) to its reduced form (**NADH**), and back, provides the cell with a mechanism for accepting and donating electrons. NAD^+/NADH plays a significant role in the reactions associated with glycolysis, oxidative phosphorylation, and fermentation. Given its importance to cell function, it would be useful if there were a means of visualizing NADH in living cells. The work presented in this case study introduces a new tool for research in cell metabolism – a NADH fluorescent sensor.

- Conduct a search for an image of NAD^+. What part of the molecule is reduced in the conversion to NADH?
- Outline the involvement of NAD^+/NADH in the following reactions:
  - Glycolysis
  - Fermentation
  - Citric acid cycle – Krebs cycle; TCA cycle
  - Mitochondrial electron transport
- How does NADH move from the cytoplasm into the mitochondrial matrix?

BACKGROUND

A **biosensor** combines a biological component with a component that will respond to a physiological or chemical cue. The fluorescent biosensor developed for the study of intracellular NADH is a fusion of a NADH-sensitive protein to a fluorescent protein. The **Rex protein** is a NADH-sensitive transcription factor from the bacteria *Bacillus subtilis*. The Rex protein only responds to NADH and not to NAD^+ or NADPH. The fluorescent component of the fusion protein is a **circularly permuted yellow fluorescent protein (cpYFP)**. Like all fluorescent molecules, cpYFP absorbs light at one wavelength, the **excitation wavelength**, and emits light at a second lower wavelength, the **emission wavelength**. Linking cpYFP to a "sensor" protein means that its excitation/emission properties will change in response to changes in its environment; namely, binding of

NADH by Rex. The researchers who developed this biosensor named it **Frex** (fluorescent Rex).

- Suggest a reason a transcription factor would be sensitive to NADH levels.
- Explain why a bacterial protein can be expressed in a eukaryotic cell.
- Conduct a search to find the excitation/emission spectra for YFP.
- Why is the emission wavelength always lower than the excitation wavelength for a fluorescent molecule?

METHODS

Charactertization of Frex *in vitro*

Purified Frex protein was stored at $-20\,^{\circ}C$ until use. Initial fluorescence levels were determined prior to the addition of any reagents. Reactions were initiated by the addition of the indicated nucleotides, and spectra were immediately recorded. For excitation experiments spectra were recorded with an excitation range from 300 nm to 515 nm and an emission wavelength of 530 nm. Readings were taken every 1 nm. For emission spectra the emission range was 510–600 nm with an excitation wavelength of 498 nm.

Cell culture

293FT cells were maintained in tissue culture media supplemented with 10% fetal bovine serum at $37\,^{\circ}C$ with 5% CO_2. Cells were plated in antibiotic-free high-glucose media 16 h before transfection.

Fluorescence microscopy

For confocal fluorescence microscopy, cells were plated on 35 mm glass-bottomed dishes in growth media and observed for 24–30 h post transfection. Frex was expressed in different subcellular compartments by tagging it with organelle-specific signal peptides. Images were acquired using a laser scanning confocal microscope. For pseudocolor analysis the pixel-by-pixel ratio for a 488 nm excitation image by 405 nm excitation image of the same cell was determined. The purple (RGB value 255, 0, 255) was defined as the lowest ratio and red (RGB value 255, 0, 0) was defined as the highest ratio.

Live cell fluorescence measurements

293FT cells were harvested, trypsinized, and counted 24–48 h after transfection. Cells were washed and aliquots of cells were incubated at $37\,^{\circ}C$ with different drugs. Fluorescence excited at 485 nm was measured using a microplate reader. Dual excitation ratios were obtained using excitation wavelengths of 410 and 500 nm and an emission wavelength of 528 nm. Fluorescence values were corrected by subtracting background fluorescence of cells not expressing Frex.

- Conduct a search to learn more about 293FT cells.
- How does a signal peptide work to target a protein at a particular organelle?

RESULTS

- Suggest a reason binding of NADH causes a conformational change in Rex.
- Use the model in Figure 11.3.1b to explain how the Frex biosensor works.
- Explain the differences in fluorescence intensity in Figure 11.3.1c using the model of Frex activity in Figure 11.3.1b.
- Subsequent experiments using the Frex biosensor measured fluorescence of a sample after excitation with 410 and 500 nm wavelength light. What is being measured for each of the two excitation wavelengths?

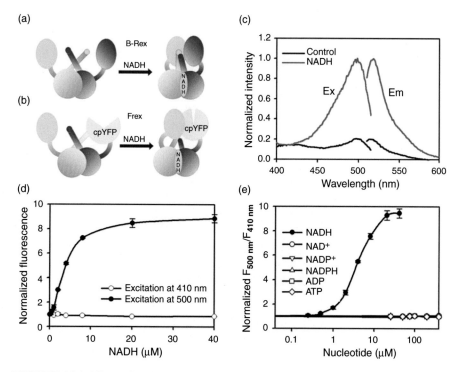

FIGURE 11.3.1 Properties of an NADH fluorescent biosensor.
a. Model for the NADH-sensing properties of Rex. NADH binding induces conformational change bringing the amino terminal DNA-binding domains (spheres) together. b. Design of the Rex-cpYFP fusion protein Frex. c. Fluorescence excitation spectra of purified Frex is altered by NADH binding. In the absence of NADH (control), two excitation peaks are present at 420 and 500 nm with an emission peak at 530 nm. In the presence of NADH, Frex fluorescence increases in intensity with an excitation peak at 498 nm and an emission peak at 518 nm. d. Fluorescence intensities at an emission wavelength of 528 nm were measured for excitation with 410 and 500 nm light in the presence of increasing concentrations of NADH. Fluorescence was normalized relative to the initial value. e. Fluorescence intensity generated by an excitation wavelength of 500 nm was divided by the intensity of the emission generated at 410 nm in the presence of different concentrations of NADH or its analogs. Error bars represent the standard error of the mean.

- Describe the graph in Figure 11.3.1d in the context of the information presented in Figure 11.3.1c.
- What is determining the specificity of the reaction shown in Figure 11.3.1e?
- Is the fluorescence of cpYFP-Mit influenced by NADH concentrations? Explain your answer.
- Which of the treatment conditions in Figure 11.3.2a reduced the amount of fluorescence produced by the mitochondria? What does the loss of fluorescence indicate about the availability of NADH?

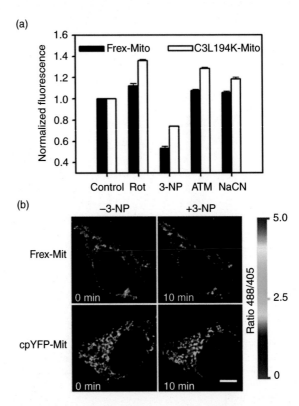

FIGURE 11.3.2 Inhibition of the electron transport chain oxidizes mitochondrial NADH.
a. Tissue culture cells were incubated in the presence of different inhibitors of mitochondrial electron transport for 30 min and imaged using an excitation wavelength of 485 nm. Frex-Mit (high affinity) and C3L194K-Mit (low affinity) fluorescence was normalized to cpYFP-Mit fluorescence. Abbreviations: Rot, rotenone (a mitochondrial complex I inhibitor); 3-NP, 3-nitropropionic acid (a mitochondrial complex II inhibitor); ATM, mitochondrial inhibitor; NaCN, sodium cyanide (depolarized mitochondria). Error bars represent the standard error of the mean. b. Confocal fluorescence images of Frex-Mit or cpYFP-Mit expressing cells before (−3-NP) and 10 min after (+3-NP) addition of the mitochondrial complex II inhibitor 3-NP. Images are pseudocolored based on the ratio of emission at 488/405 as shown to the right of the panels. Scale bar = 10 μm.

- Research/review the relationship between NADH and the electron transport chain mitochondrial complex I. Predict what would happen to NADH levels in the mitochondria if complex I was inhibited. Do the data in Figure 11.3.2a support your prediction?
- Explain what the colors in the images in Figure 11.3.2b represent.
- What is the basis for the cpYFP-Mit staining in Figure 11.3.2b?
- Why did the staining pattern for cpYFP-Mit not change in the presence of 3-NP?
- Describe the basis for the difference in Frex-Mit staining in Figure 11.3.2b.
- Define the pools of NADH that are being measured in the graphs in Figure 11.3.3a and b.
- Explain why Frex-Mit fluorescence is increasing in the graph in Figure 11.3.3a.

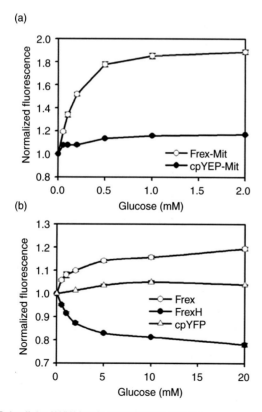

FIGURE 11.3.3 Subcellular NADH levels depend upon glucose.
a. Dose-dependent fluorescence response of Frex-Mit and cpYFP-Mit to different concentrations of glucose supplementation. b. Dose-dependent fluorescence response of cytosolic Frex, high-affinity Frex (FrexH), and cpYFP to different concentrations of glucose.

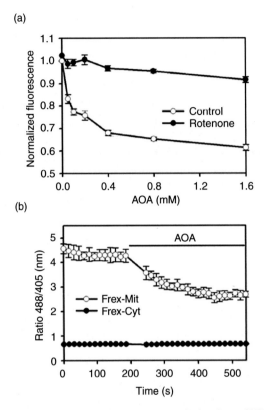

(a)

(b)

FIGURE 11.3.4 Inhibition of mitochondrial transport results in a loss of NADH.
The malate–aspartate shuttle is responsible for the transport of NADH from the cytoplasm and into the matrix of the mitochondria. The shuttle is inhibited by the compound AOA. a. Frex-Mit fluorescence was measured in increasing concentrations of AOA in the absence (control; open symbol) or presence of the mitochondrial complex II inhibitor rotenone (filled symbol). b. Time course of the average fluorescence of the mitochondrial (Frex-Mit) and cytoplasmic Frex biosensor (Frex-Cyt) in the absence ($t = 0$ to $t = 200$ s) or presence ($t = 250$ to $t = 550$) of AOA.

- What metabolic reaction is responsible for the increase in Frex fluorescence in Figure 11.3.3b?
- Which graph, Figure 11.3.3a or b, would reflect a change if the experiment was repeated in the absence of O_2?
- Propose a reason for the loss of FrexH fluorescence in Figure 11.3.3b.
- What is occurring in the control cells of the experiment in Figure 11.3.4a?
- How does treatment with the inhibitor rotenone rescue NADH fluorescence?
- Summarize the results of the experiment in Figure 11.3.4b.

- If NADH is being prevented from transporting from the cytoplasm into the mitochondria where is it going? Does the graph in Figure 11.3.4b provide any answers? Design an experiment to address this question.
- In your opinion, which of the experiments in the case study highlight the usefulness of Frex as a "powerful tool for investigating the effect of various stimuli on NADH in living cells"?
- Rotenone is a common ingredient in insecticides. Predict whether exposure to rotenone-containing insecticides runs a risk to human health. Conduct a search to check your prediction.
- Funding of basic research (research without immediate obvious benefit to humankind) is often a controversial topic. Debate the value of basic research in the context of the work presented in this case study.

References

[1] Xu L, Bretscher A. Rapid glucose depletion immobilizes active myosin V on stabilized actin cables. Curr Biol 2014;24:2471–9.

[2] Cournac L, Redding K, Bennoun P, Peltier G. Limited photosynthetic electron flow but no CO_2 fixation in *Chlamydomonas* mutants lacking photosystem I. FEBS Lett 1997;416:65–8.

[3] Zhao Y, Jin J, Hu Q, Zhou H-M, Yi J, Yu Z, Xu L, Wang X, Yang Y, Loscalzo J. Genetically encoded fluorescent sensors for intracellular NADH detection. Cell Metab 2011;14:555–66.

CHAPTER 12

Cell Signaling

How Cells Know When It's Time to Go [1]

INTRODUCTION

The ability to sense and respond to the environment is one of the properties of life. For cells, this behavior takes the form of protein–protein interactions that occur in a sequence known as a **signal cascade**. A signal from the environment, a **ligand**, binds to a **receptor protein** located either on the surface of the plasma membrane or in the cytoplasm of the cell. Ligand binding triggers a conformational change in the receptor that causes a change in a second protein. Like a row of falling dominos, conformational changes in one protein in the **signal pathway** results in a change in the next until the signal results in a cellular **response**. A response could be as simple as opening or closing an ion channel or it could be as complicated as altering the gene expression of the cell.

One of the important signal pathways in a cell makes use of a class of receptor proteins known as **G protein-coupled receptors** (**GPCRs**). Two characteristics that distinguish GPCRs are the presence of seven transmembrane domains that insert the receptor into the cell's plasma membrane and the ability of the receptor to bind to a collection of heterotrimeric **G proteins** (G_α, G_β, and G_γ). Small G proteins bind and hydrolyze the nucleotide GTP. The activity of the G_α proteins depends upon GTP. With GTP attached, the G_α protein is active. It dissociates from the G_β and G_γ subunits and interacts with **effector** proteins, causing conformational changes and propagating a signal. In some cases, the G_β and G_γ subunits also play a role in activating effector proteins. Eventually, the GTPase activity of the G_α converts the GTP to guanosine diphosphate (GDP),

283

at which point G_α is inactive. At this point, it reassociates with G_β and G_γ, and rebinds to the cytoplasmic domain of the GPCR. Should another ligand bind to the receptor, a conformational change in GPCR will trigger the exchange of GTP for GDP on the G_α subunit, and the process begins again.

In the GTP-bound state, G_α subunits activate the effector protein adenylyl cyclase, an enzyme that catalyzes the synthesis of **cyclic AMP (cAMP)** from ATP. Cyclic AMP is a **second messenger** molecule that can influence the activity of a host of other proteins in the cell. Activation of adenylyl cyclase leads to the production of multiple molecules of cAMP, effectively amplifying the signal from the original interaction between receptor and ligand.

- What type of environmental signals would be important for a cell to be able to sense?
- What determines the specificity between a ligand and its receptor?
- List the steps involved in gene expression. Where in your list could a protein:protein interaction from a signal pathway influence gene expression?
- Define the term "heterotrimeric" in your own words.
- How might cAMP contribute to the change in gene expression described previously?
- Conduct a search for images or videos that illustrate the activity of GPCRs and G proteins.

BACKGROUND

The protozoan *Dictyostelium discoideum* spends most of its life as a single-cell amoeba. However, if the number of *Dictyostelium* in a population exceeds their food supply, then individual amoebae will **aggregate** to form a multicellular **slug** that moves through the environment until it finds a favorable spot, at which point it transforms into a structure known as a **fruiting body**. The fruiting body is made up of a stalk and a head. The head of the fruiting body is a mass of spores that are released and germinate into new amoebae.

The ability of the amoebae to sense and respond to starvation conditions requires signal pathways that ultimately alter gene expression in individual cells, resulting in the formation of the slug and fruiting body. Aggregation of cells into the slug form, occurs in response to gradients of extracellular cAMP, produced by individual cells within the population. Other soluble factors released by *Dictyostelium* cells into the surrounding environment, **conditioned medium factors (CMFs)**, in combination with cAMP, have been shown to influence changes in the expression of genes for cAMP receptors and cell adhesion molecules. Binding of CMFs to *Dictyostelium* cells activates a G protein–independent signaling pathway and a second, G protein–dependent pathway that involves the heterotrimeric G proteins, $G_\alpha 1$, G_β, and G_γ. Activation of the CMF receptor activates $G_\alpha 1$, freeing G_β and G_γ to bind to **phospholipase C (PLC)**, triggering an increase in inositol 1,4,5-triphosphate (IP_3).

The protein **receptor phosphoinositol kinase A (RpkA)** is a seven-membrane-spanning domain protein. Given the apparent role of a G protein in the signaling pathway associated with the development of *Dictyostelium* from a single-cell to a multicellular organism it is reasonable to investigate whether RpkA plays a part in the starvation signal pathway.

- Conduct a search for images of the life cycle of *D. discoideum*.
- Discuss whether the extracellular cAMP produced by *Dictyostelium* in response to starvation would be considered a second messenger.
- Propose a model to explain how a cell can "sense" a concentration gradient of cAMP.
- What characteristic defines RpkA as a GPCR?

METHODS

Cell culture
Wild-type or mutant *D. discoideum* were grown in either liquid nutrient broth with shaking or on nutrient agar at 21 °C in the presence of the bacteria *Klebsiella aerogenes* provided as a food source. Development of fruiting bodies was induced by starvation. Cells were grown to a density of 3×10^6 cells/mL, washed with buffer, plated onto agar plates at various cell densities, and allowed to develop at 21 °C. The RpkA gene was removed (knocked out) of the *Dictyostelium* genome using a homologous recombination. Removal of the gene was confirmed using PCR.

Assays
Streaming assays were performed by starving the cells at 1×10^6 cells/mL in 1 mL of buffer. Conditioned medium and recombinant CMF (1 ng/mL) were added to the buffer, and cells were observed using a stereomicroscope equipped with a video camera.

Production of IP_3 was determined using a commercially available kit (GE Healthcare). The assay was performed in the presence or absence of recombinant CMF (1 ng/mL).

Northern and immunoblot analysis
Total RNA was extracted using the Qiagen RNeasy kit. RNA (10 µg per lane) was subjected to electrophoresis in 1.2% agarose/formaldehyde gels and transferred onto a nylon membrane. The cDNA fragments that were used as probes were radiolabeled with $[\alpha^{32}P]ATP$.

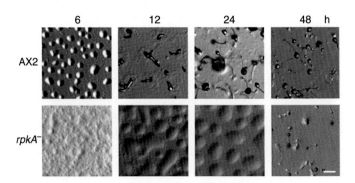

FIGURE 12.1.1 RpkA plays a role in early development of *Dictyostelium*.
Cells with (AX2) or without (*rpkA⁻*) the RpkA gene were grown on agar at a cell density of 5×10^6 cells/cm^2 and allowed to develop at 21°C. Images were captured at the indicated time points.

Total cell extracts were prepared by lysis of *Dictyostelium* cells in SDS-PAGE loading buffer. Homogenates of equal amounts of cells were separated on a 10 or 12% polyacrylamide gel, and blotted onto a nitrocellulose membrane. Proteins were visualized using the indicated primary antibodies and chemiluminescent secondary antibody.

- Predict whether starvation might occur more quickly for cells in a liquid or a solid media.
- Compare and differentiate the protocols for a Northern blot and a Western blot.
- Debate whether lysis of an equivalent number of cells, is an acceptable standard if a researcher wants to claim that the same amount of total protein has been loaded on a gel.

RESULTS

- Estimate how long it took for the cells in Figure 12.1.1 to deplete their food supply.
- Label the AX2 panels in Figure 12.1.1 to indicate, which stage in the life cycle of *Dictyostelium* is represented.
- Describe how the response of the AX2 and *rpkA⁻* cells to starvation differed.
- Summarize the phenotype of the *rpkA⁻* cells.
- What is meant by "total RNA?" What is the relationship between a cDNA and an mRNA?
- Why was cyclase-associated protein (CAP) expression included in the results for Figure 12.1.2?
- What is the implication of the following result from Figure 12.1.2a cDNA specific to adenylyl cyclase binds to the RNA sample from wild-type AX2 cells after 4 and 8 h in culture?

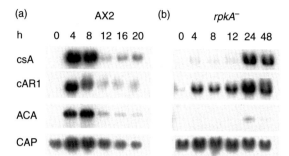

FIGURE 12.1.2 Expression of developmentally regulated genes is altered in *rpkA⁻* cells. Northern blot analysis of total RNA prepared from; a. wild-type (AX2) and b. *rpkA⁻* cells at the indicated time points after starvation. cDNA probes were used to identify RNAs for specific proteins required for early development. These proteins included the cell adhesion molecule, contact site A (**csA**), the cAMP receptor (**cAR1**), and adenylyl cyclase (**ACA**). Cyclase-associated protein (**CAP**) cDNA was used as the control.

- Connect the functions of the three genes listed in Figure 12.1.2 with the activities associated with the aggregation phase of *Dictyostelium* early development.
- Relate the pattern of gene expression in Figure 12.1.2 with the pattern of behavior in Figure 12.1.1. What can you conclude?
- Develop an argument for or against the statement, "proper expression of early developmental genes does not take place in *rpkA⁻* cells."
- Summarize the results presented in Figure 12.1.3a.
- Does activation of PLC require RpkA? Explain your answer.
- How do the data from Figure 12.1.3 support the conclusion that cells lacking the RpkA protein can produce, but cannot respond to, factors required for aggregation?
- It has been hypothesized that activation of the G protein $G_\alpha 1$ releases G_β and G_γ to bind and activate PLC. Explain the data in Figure 12.1.3b in the context of this hypothesis.
- Compare the phenotype of the wild-type AX2 cells and the $g_\alpha 1^-$ cells that are missing from the G protein $G_\alpha 1$.
- How are the data in Figure 12.1.4b different from those in Figure 12.1.2?
- Complete the following table based on the results presented in Figures 12.1.2 and 12.1.4.

Strain	RpkA Expressed?	$G_\alpha 1$ Expressed?	csA Expressed? (by 8 h starvation)	Fruiting Body?
AX2				
rpkA⁻				
$g_\alpha 1^-$				
$g_\alpha 1^-/rpkA^-$				

(a)

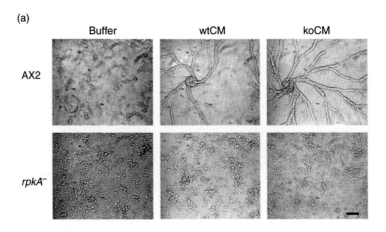

(b)

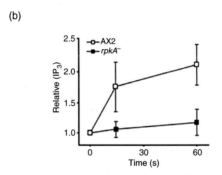

FIGURE 12.1.3 The rpkA⁻cells produce, but cannot respond to, conditioned medium factors.
a. Conditioned media was collected from wild-type AX2 and RpkA knockout (*rpkA⁻*) cells. AX2 cells
(top row) were starved in the presence of media supplemented with buffer, media from wild-type cells
(wtCM), or media from *rpkA⁻* cells (koCM). The parallel experiment was performed by starving *rpkA⁻*
cells (bottom row). b. IP_3 production was measured in AX2 and *rpkA⁻* cells in the presence or absence
of CMF (1 ng/mL). This graph illustrates the relative difference in IP_3 levels of the two cell types. Data are
presented as the mean ± standard deviation of the two independent experiments.

- What can you conclude from the aforementioned table?
- Restate the following excerpt using your own words, "RpkA seems to be
  needed for enabling signaling downstream of CMF, such as production of
  IP_3 and establishment of full responsiveness to cAMP. Inactivation of $G_\alpha 1$,
  a negative regulator of CMF signaling, rescued the developmental defect
  of the *rpkA⁻* cells, suggesting that RpkA actions are mediated by this G
  protein."

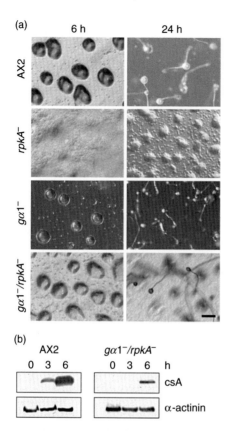

FIGURE 12.1.4 G$_\alpha$1 regulated CMF signaling through RpkA.

a. Development of wild-type AX2, *rpkA$^-$*, *g$_\alpha$1$^-$*, and *g$_\alpha$1$^-$/rpkA$^-$* double mutants on agar plates at a cell density of 5×10^6 cells/cm^2. Cells were allowed to develop at 21°C and images were captured at the times indicated. b. Wild-type and *g$_\alpha$1$^-$/rpkA$^-$* double mutants were starved in a suspension culture at a density of 1×10^7 cells/cm^2. Cells were collected at the indicated times and lysed in SDS-PAGE sample buffer. Homogenates from equal amounts of cells were analyzed by SDS-PAGE. The cell adhesion molecule csA, was detected using a monoclonal antibody. The protein α-actinin was used as the loading control and was also detected using a monoclonal antibody.

Can You "Ad" Hear Me Now? Signaling and Intraflagellar Transport [2]

INTRODUCTION

A **signal transduction pathway** represents the flow of information in a cell. The pathway starts when a signaling molecule or **ligand** binds to its **receptor** protein. The interaction triggers a change in the receptor that is carried or **transduced** to the next protein in the pathway. In the example of a **protein tyrosine kinase pathway**, activation of the kinase function of the receptor by ligand binding results in **phosphorylation** of an **adaptor** or other protein. If the next protein in the sequence is an inactive kinase, then the addition of a phosphate group will alter its conformation, converting it into an active kinase. The resulting **phosphorylation cascade** will eventually result in a change in the cell's behavior either by modifying the activity of proteins already present within the cytoplasm or by altering gene expression.

Receptor proteins are located in the cell's plasma membrane and, in some cases, within the cytoplasm of the cell. However, receptor proteins and their associated signal pathways not necessarily are evenly distributed over the surface of the cell. The importance of a specialized region of a eukaryotic cell known as the **primary cilium** in cell signaling has recently been established. Like other cilia and flagella, the primary cilium is formed by a microtubule cytoskeleton or **axoneme**, surrounded by plasma membrane, which extends out into the extracellular space. However, unlike other cilia and flagella, the primary cilium is nonmotile. Instead, these cilia appear to acts as sensors of the chemical and mechanical environment. Loss of signaling ability of the primary cilia found on the epithelial cells lining the kidney, is linked to **polycystic kidney disease**. Mutations that impact the function of primary cilia can lead to a range of development defects including blindness and mental retardation.

- Explain what determines the specificity between ligand and receptor binding.
- Conduct a search for an image of a MAP kinase pathway. Identify the ligand, receptor, kinase phosphorylation events, and cellular outcome in the pathway.

- Define the term "phosphorylation cascade" using your own words.
- Research/review the structure of a microtubule axoneme. Which part(s) of the axoneme would you predict to be missing in the nonmotile primary cilium?
- Conduct a search to learn more about polycystic kidney disease.

BACKGROUND

Chlamydomonas reinhardtii is a biflagellate green alga that is used as a model system for the study of cilia and flagella. *Chlamydomonas* uses its flagella both for swimming and mating. In the life cycle of *Chlamydomonas*, vegetative haploid cells divide by mitosis. However, if nitrogen is depleted from the environment the vegetative cells will differentiate into gametes. Differentiation involves expression of a collection of **mating-type (MT)** genes. *Chlamydomonas* are either MT^+ or MT^-. The proteins expressed by the MT^+ cell allow it to mate with an MT^- cell. Some of those genes encode cell adhesion proteins called **agglutinins** that become localized along the length of the flagella. Agglutinins promote adhesion between the flagella of the two mating types and initiate a signal transduction pathway. Flagellar adhesion activates a flagellar protein tyrosine kinase (**PTK**) that phosphorylates a 105 kDa **cyclic GMP-dependent protein kinase (PKG)**, resulting in an increase in the intracellular concentrations of cAMP, leading to the loss of the cell walls and the eventual fusion of the two cell bodies to form a zygote. Once environmental conditions improve, the zygote will divide to form new vegetative cells.

Another interesting feature that the flagella of *Chlamydomonas* shares with other cilia and flagella, including the primary cilium, is **intraflagellar transport (IFT)**. IFT is required for the growth and maintenance of cilia and flagella. Proteins targeted for delivery to the flagella associate with intraflagellar transport-specific proteins to form an **intraflagellar transport particle**. The particle is then transported along the microtubule axoneme through the activity of microtubule motor proteins. Intraflagellar transport uses the microtubule motor proteins kinesin and cytoplasmic dynein to move the material forward toward the tip of the flagella (kinesin) or backward toward the cell body (cytoplasmic dynein). The role of intraflagellar transport in signaling within the flagella is the topic of this case study.

- Explain why a haploid vegetative cell can divide by mitosis?
- Predict how agglutinin proteins work to promote cell adhesion between the two *Chlamydomonas* mating types.
- Create an outline of the steps involved in the signal transduction pathway associated with flagellar adhesion.
- What type of cell division is required in order to produce vegetative cells from the zygote?
- Predict the ratio of MT^+ and MT^- daughter cells formed from the zygote.

METHODS

Cells and cell culture

Chlamydomonas reinhardtii (*MT*⁺ and *MT*⁻ strains) were cultured vegetatively in a liquid medium with aeration provided over a 13 h light and 11 h dark cycle. To induce the formation of gametes, vegetative cells were switched into a nitrogen-free medium and were cultured for 15–18 h with the same light:dark cycle. Cell adhesion was measured using an electronic particle counter. Cell fusion was determined by fixing a sample in 0.5% glutaraldehyde and determining the number of cells with two or four flagella.

Fractionation of cells and flagella

Flagella were isolated by lowering the pH to 4.5 using 0.5 M acetic acid with constant stirring. Deflagellation usually occurred within 15 s, and samples were neutralized to pH 7.2 with 0.5 M KOH. Cell bodies were collected by centrifugation through a 25% sucrose cushion. Flagella were collected from the overlaying supernatant. Flagellar membranes were disrupted by freezing and thawing. Frozen flagella were thawed and then centrifuged at 13,000 rpm for 20 min at 4°C to obtain a low-speed supernatant (S1) and pellet (P1). In some experiments the low-speed supernatant was centrifuged for 50,000 rpm for 1 h at 4°C to yield a high-speed supernatant (S2) and pellet (P2). Equal volumes of each fraction were used for SDS-PAGE and immunoblotting.

To obtain the flagellar membrane matrix, freshly prepared flagella were isolated and suspended in a buffer solution containing 1% NP40, a detergent, and incubated for 30 min at room temperature. Samples were then centrifuged at 13,000 rpm for 20 min at 4°C to yield a membrane/matrix fraction and an axonemal fraction.

Immunoblotting and immunofluorescence

Samples were analyzed using SDS-PAGE and immunoblotting with the indicated antibodies. Blots were incubated overnight at 4°C, washed extensively, and reacted with a HRP-conjugated secondary antibody and detected using chemiluminescence.

For immunofluorescence, cells were fixed with 2% paraformaldehyde and 0.1% glutaraldehyde, washed with buffer, and permeabilized with 0.2% Triton X-100. Samples were incubated in preimmune serum or with the indicated antibodies overnight at 4°C. Samples were washed with buffer and incubated with a fluorescent secondary antibody and examined using a compound fluorescence microscope.

In vitro protein kinase assay

Flagella from wild-type adhering gametes were isolated as described, and immunoprecipitation was accomplished using control IgG or anti-PKG antibodies.

Samples were incubated with the indicated antibody and protein A–Sepharose beads overnight at 4°C. The beads were washed three times with buffer. The immunoprecipitated protein complexes were incubated in the presence of 20 μCi [γ-^{32}P] ATP, 1 mM ATP, and 0.5 μg myelin basic protein. At the end of incubation, samples were boiled in SDS-PAGE sample buffer and resolved by electrophoresis. Gels were dried and subjected to autoradiography.

- Why was it necessary to culture these cells in the presence of light?
- Calculate the percentage of zygotes in a sample given the following data: 439 biflagellate cells; 754 quadraflagellate cells.
- Outline the steps involved in the freeze–thaw protocol. Include the low-speed and high-speed centrifugation steps.
- Suggest a reason why the disruption of flagella by freezing and thawing is different from the use of a detergent such as NP40.
- What assumptions are being made in the use of *in vitro* assay for kinase activity?

RESULTS

- Refer back to your outline of the flagellar signal transduction pathway for mating *Chlamydomonas*. Which protein in the pathway, is capable of phosphorylating tyrosine amino acids?
- Conduct a search to learn more about cGMP-dependent protein kinases. Add the information you find to your outline of the signal pathway.
- How many flagellar proteins are phosphorylated as a result of the adhesion signal (Figure 12.2.1a)?
- Predict whether an immunoblot of whole cell lysate reacted with the α-pTyr antibody would look the same or different than the blot in Figure 12.2.1a. Explain your answer.
- Describe the relative concentration of proteins in the different samples in Figure 12.2.1b. Explain whether the differences make sense in the context of the samples being analyzed.
- What is the significance of the immunoblot in Figure 12.2.1b?
- What change was made to the protocol for the experiment in Figure 12.2.1c? How did that change influence your interpretation of the results?
- Summarize the results in Figure 12.2.1d. Is the staining of the cell body by the α-PKG antibody meaningful? Why or why not?
- Where in the flagella would you expect to find PKG? Use the results from the experiments in Figure 12.2.2 to defend your answer.
- *Chlamydomonas* PKG protein does *not* have a transmembrane domain. Propose a model for the presence of PKG in the flagella that is consistent with this fact and the results from Figure 12.2.2.

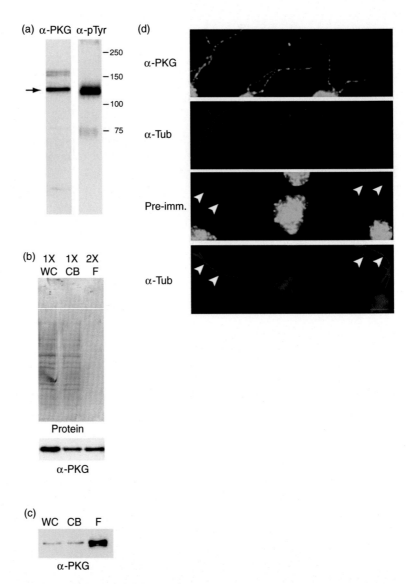

FIGURE 12.2.1 *Chlamydomonas* cGMP-dependent protein kinase is a flagellar protein and is the substrate for adhesion-activated flagellar PTK.

a. Antibody specific to *Chlamydomonas* PKG (α-PKG) recognizes a protein from *Chlamydomonas* whole cell lysate that comigrates with a protein recognized by an antibody specific to phosphorylated tyrosine (α-pTyr) present in a sample prepared from flagella isolated from adhering gametes. Arrow indicates PKG. b. Flagella were removed from cells by pH shock. Whole cells (WC), cell bodies (CB), and flagella (F) were analyzed by SDS-PAGE (top) and immunoblot for PKG (bottom). The amount of sample loaded in the flagellar lane was twice that of WC or CB lanes. c. Equal amounts of protein (5 μg) from WC, CB, or F were analyzed by SDS-PAGE, immunoblotted, and reacted with the α-PKG antibody. d. *Chlamydomonas* gamete cells were fixed and stained with an antibody to tubulin (α-Tub; red) and either α-PKG or preimmune serum (green).

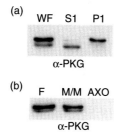

FIGURE 12.2.2 *Chlamydomonas* cGMP-dependent protein kinase associates with the membranes of the flagella.

a. Whole flagella isolated from non-adhering gamete cells (WF) were frozen, thawed, and centrifuged to separate soluble proteins (S1) from proteins associated with the axoneme or flagellar plasma membrane (P1). Samples were analyzed by SDS-PAGE and immunoblotting with the α-PKG antibody. b. Freshly isolated flagella from non-adhering gamete cells (F) were incubated with buffer containing 1% NP-40 followed by centrifugation to separate the detergent soluble membrane–matrix (M/M) from the insoluble axoneme (AXO). Samples were processed as described earlier.

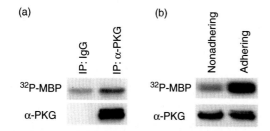

FIGURE 12.2.3 Flagellar adhesion signals the activation of PKG.

a. Control (IP:IgG) and anti-PKG antibodies (IP:α-PKG) were used to immunoprecipitate proteins from flagella isolated from nonadherent cells. The kinase activity of the immunoprecipitated protein was determined by an *in vitro* assay using [^{32}P]ATP and the substrate myelin basic protein. Phosphorylated myelin basic protein (^{32}p-MBP) was visualized using autoradiography. b. The *in vitro* kinase described earlier was used to measure the relative activity of PKG immunoprecipitated from flagella isolated from nonadhering and adhering cells. The upper panel shows the autoradiograph of MBP, and the lower panel shows the immunoblot with α-PKG antibody.

- Would you expect to find myelin basic protein in flagella?
- Defend the following statement using data from Figure 12.2.3. "Flagellar PKG is part of a signal transduction cascade that starts with agglutinin binding between mating types."
- Summarize the conclusions supported by the results from the experiments in Figure 12.2.4a.
- Describe the differences in the results shown in Figure 12.2.4b compared to Figure 12.2.4a.

(a)

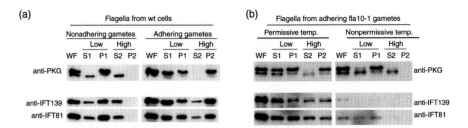

(b)

FIGURE 12.2.4 Flagellar adhesion promotes the association of *Chlamydomonas* PKG with IFT particles.

a. Flagella isolated from wild-type adherent and nonadherent gamete cells were frozen, thawed, and fractionated by differential centrifugation. Whole flagella (WF), low-speed supernatant (S1) and pellet (P1), and high-speed supernatant (S2) and pellet (P2), were analyzed by SDS-PAGE and immunoblotting with antibodies to PKG (anti-PKG) and the intraflagellar transport proteins IFT 139 (anti-IFT139) and 81 (anti-IFT81). b. *Fla10-1* is a temperature-sensitive *Chlamydomonas* kinesin mutation. *fla10-1 MT$^+$ and MT$^-$* gametes were mixed together for 5 min at the permissive temperature (25°C) or were preincubated for 40 min at the nonpermissive temperature (32°C), and then mixed together for 10 min at the nonpermissive temperature. Flagella were isolated and treated as described earlier.

(a)

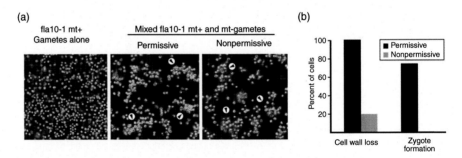

(b)

FIGURE 12.2.5 Flagellar adhesion–induced signaling requires active IFT.

a. *Chlamydomonas fla10-1* kinesin mutants were incubated alone (*fla10-1 MT$^+$*) or mixed (*fla10-1 MT$^+$ and MT$^-$*) at permissive or nonpermissive temperatures. Arrowheads indicate clumps of adhering gametes. b. *Fla10-1 MT$^+$ and MT$^-$* cells that had been preincubated separately at nonpermissive and permissive temperatures for 40 min were mixed together and assessed for cell wall loss and cell fusion 40 min after incubation at the nonpermissive temperature.

- What happened to the IFT proteins at the nonpermissive temperature?
- Predict whether the flagellar adhesion signal pathway would be functional in *fla10-1* cells at the nonpermissive temperature. Design an experiment to test your prediction.
- How did the loss of kinesin influence cell adhesion, as shown in the experiment in Figure 12.2.5a?

- Develop an argument for or against the following: "PKG is necessary for flagellar adhesion-induced signaling."
- Refer to your outline of the signal transduction pathway for flagellar adhesion. What steps occur prior to the loss of cell wall and cell fusion?
- Construct a model of the flagellar adhesion signaling pathway that is consistent with the results presented in this case study.

References

[1] Bakthavatsalam D, Brazill D, Gomer RN, Eichinger L, Rivero F, Noegel AA. A G protein-coupled receptor with a lipid kinase domain is involved in cell-density sensing. Curr Biol 2007;17:892–7.

[2] Wang Q, Pan J, Snell WJ. Intraflagellar transport particles participate directly in cilium-generated signaling in *Chlamydomonas*. Cell 2006;125:549–62.

CHAPTER 13

Cell Cycle

Now You See It, Now You Don't: The Discovery of Cyclin [1]

INTRODUCTION

The life of a cell can be divided into two distinct stages: interphase and M-phase. M-phase includes all the steps involved in mitotic cell division while interphase is divided into G1, S, and G2 stages that represent every other aspect of a cell's life. The progression of cells from interphase to M-phase and back to interphase is controlled by the activity of a collection of proteins known as **cyclins** and **cyclin-dependent protein kinases** (Cdks). As its name implies, a cyclin-dependent protein kinase is a type of enzyme that phosphorylates proteins, but only in the presence of cyclin. Cyclin proteins act as allosteric activators for Cdks. In the presence of cyclin, Cdk adds phosphate groups to serine and threonine amino acids located within specific domains on its target protein or proteins, altering the activity of phosphorylated proteins. Cdk is inactive in the absence of cyclin.

Each of the stages of the cell cycle is regulated by a distinct set of cyclins and Cdk proteins. For example, the transition from G1 to S phase is controlled by activation of Cdk2 by cyclin E while the transition from G2 to M phase requires the binding of cyclin B to Cdk 1. Once the cell has successfully entered the next phase in the cell cycle Cdk activity is "turned off" by destruction of its partner cyclin protein. Cdk proteins remain inactive until cyclin becomes available in the cytoplasm. Transcription and translation of each of the cyclin proteins must occur before the cell can repeat the cycle.

299

Tracking specific proteins is a challenge given the number and diversity of proteins in a cell. The standard technique of SDS-polyacrylamide gel electrophoresis (SDS-PAGE) followed by Coomassie Blue staining would produce a complicated smear of hundreds of protein bands. **Autoradiography** is a technique that helps reduce this complexity by detecting only those proteins that have been radioactively labeled. Cells incubated with the labeled amino acid ^{35}S-Met, will incorporate radioactive methionine into any proteins that were translated during the incubation period. Radioactivity does not affect the function of the protein. Following SDS-PAGE, the gel is imaged using a sheet of X-ray film. Radioactive particles emitted by the labeled proteins expose the film, creating a black-and-white image of the protein bands.

- Sketch and label all of the stages of the cell cycle
- Use your knowledge of enzymes to describe the structure, function, and regulation of Cdk.
- What is the likely source of the phosphate groups used by Cdk?
- Explain why changing the activity of a protein could lead to progression through the cell cycle.
- Why is methionine a good choice as a labeled amino acid for the technique of autoradiography?

BACKGROUND

All cells go through the cell cycle at some point in their life. Developing embryos go through rapid rounds of cell division as the embryo progresses from a fertilized egg to a multicellular organism. Eggs from fish, frogs, and marine invertebrates such as sea urchins and clams have long been valued as model systems to study various problems in cell biology. Unfertilized eggs can be obtained in large quantities, fertilized in the lab, and experimentally manipulated.

The work presented in this case study made use of eggs from the sea urchins, *A. punctulata* and *Lytechinus pictus*. Fertilization triggers an increase in the rate of protein synthesis within the newly formed zygote by stimulating translation of a collection of **maternal mRNAs** that are stored in the cytoplasm of the egg. When eggs are fertilized in the presence of ^{35}S-Met, radioactive amino acids get into the proteins that are translated from those maternal mRNAs. Exposing eggs to chemical agents such as the ionophore A23817 or ammonia (NH_4Cl) can similarly result in activation of protein synthesis. Chemically activated eggs are not fertilized and will not go through mitotic cell division.

- Conduct an online search for basic information about *A. punctulata* and *L. pictus*.
- What are the advantages of storing mRNAs in an egg?

- Use your knowledge of the process of mitosis to develop an explanation for why chemical activation of an egg cannot lead to normal cell division.

METHODS

Preparation and incubation of eggs and embryos

Eggs were collected from *A. punctulata* or *L. pictus* and maintained in 20°C seawater. The eggs were rinsed using several changes of filtered seawater and suspended to a final density of 15,000–20,000 eggs/mL. Labeled methionine was added to a final concentration of one volume of 5 mCi/mL stock to 200 volumes of egg suspension. The eggs were incubated for 5 min and then divided into four treatment groups. Sperm were diluted with filtered seawater and allowed to stand for 5 min before being added to eggs in the fertilization treatment group. Ionophore A23187 (10 μM) or NH_4Cl (10 mM) were added to the other two treatment groups, respectively. The final group received no additional treatment.

Sample preparation and analysis

50 μL of egg suspension were removed at 10 min intervals starting 25 min after fertilization/activation. The samples were transferred into a microcentrifuge tube containing 100 μL of 25% trichloroacetic acid (TCA). Proteins were precipitated out of solution in the presence of TCA and were collected by centrifugation. The pelleted proteins were washed with acetone to remove residual TCA and then dissolved in 30 μL of SDS gel sample buffer. Protein samples were run on a SDS-polyacrylamide gel and stained with Coomassie Blue. The gels were then dried and autoradiographed using X-ray film. The autoradiographs were scanned and the density of the radioactive protein bands was measured.

Determination of the cleavage index

A small sample of the embryos from each of the time points examined earlier was fixed using 1% glutaraldehyde in filtered seawater. Fixed embryos were examined by light microscopy and the number of cells dividing was counted. The percentage of cells undergoing mitosis at a given time represents the cleavage index. The cell cycle of fertilized eggs is synchronous during the first three rounds of cell division.

- Calculate the volume of ^{35}S-Met you would add to a 1 mL sample of egg suspension using the protocol described earlier.
- Which of the treatment groups is the control? Explain your answer.

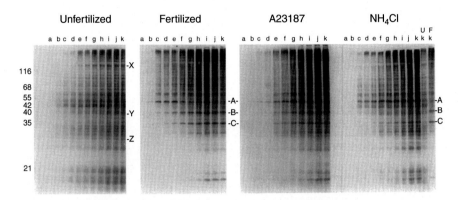

FIGURE 13.1.1 Patterns of egg protein synthesis before and after activation.
Unfertilized *Arbacia* eggs were incubated in the presence of ^{35}S-Met. After 5 min, the eggs were divided into four treatment groups that were either left untreated (unfertilized), fertilized, or activated by addition of 10 μM A23187 or 10 mM NH$_4$Cl. Samples were collected every 10 min starting at 25 min after fertilization/activation (lanes a) until 115 min (lanes j). Samples were also collected at 127 min (lanes k). Autoradiographs of the 127 min samples from unfertilized and fertilized eggs are shown with a short exposure time (lanes Uk and Fk, respectively). Bands identified as X, Y, and Z represent proteins that become less abundant following fertilization. Bands identified as A, B, and C represent nonhistone proteins whose synthesis is increased in response to fertilization.

- Explain the relationship between the density of a band on an autoradiograph and the amount of that specific protein in the egg/embryo.
- Why would it be important for fertilized eggs to go through synchronous rounds of cell division?

RESULTS

- Does protein synthesis occur in unfertilized eggs? Support your answer with evidence from Figure 13.1.1.
- Review/research histone proteins. Explain why the authors emphasize the point that bands A, B, and C are nonhistone proteins.
- Summarize the differences you observe in the pattern of protein synthesis between the unfertilized and fertilized treatment groups in Figure 13.1.1.
- Describe why the authors gave the name *cyclin* to protein band A.
- How does the pattern of protein synthesis in the chemically activated eggs compare with that of fertilized eggs in Figure 13.1.1?
- Estimate the size (molecular weight) of bands A, B, and C in Figure 13.1.1.

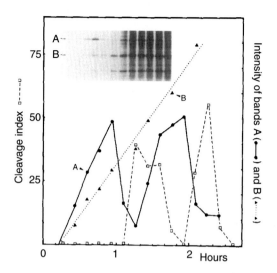

FIGURE 13.1.2 Correlation of the level of cyclin with the cell division cycle.
A suspension of *Arbacia* eggs was fertilized. After 6 min, ^{35}S-Met was added to the suspension. Samples were collected for SDS-PAGE autoradiography at 10 min intervals starting 16 min after fertilization. Within 20–30 s a second sample was collected and fixed using 1% glutaraldehyde. The fixed sample was later examined under light microscopy to determine the cleavage index (□) corresponding to each time point. The intensities of protein bands A (●) and B (▲) were measured by scanning the autoradiograph (shown as inset).

- What units apply to the numbers on the cleavage index axis?
- Estimate the time between cell divisions in *Arbacia* based on the data in Figure 13.1.2.
- Assume that you started this experiment with 400 μL of a 15,000 eggs/mL suspension. How many cells would you expect to be present in the sample 2 h after fertilization?
- Propose an explanation for the pattern of intensity versus time for band B.
- Describe the relationship between the pattern of intensity versus time for cyclin (band A) and the cleavage index.
- Using the data in Figure 13.1.2, predict where in the cell cycle this cyclin (protein A) is active.
- Generate a list of possible explanations for the loss in intensity of band A.
- Summarize the changes in protein synthesis found in a fertilized *Lytechinus* egg in Figure 13.1.3.
- Which of the proteins in *Lytechinus* are cyclins?
- How does activation with NH_4Cl affect egg protein synthesis compared with fertilization?

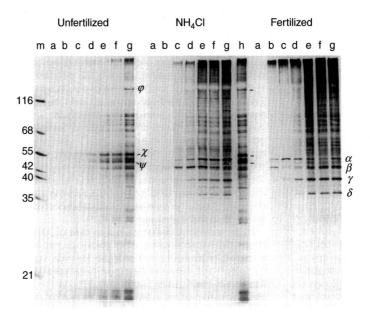

FIGURE 13.1.3 *Lytechinus* **eggs contain more than one cyclin.**
Lytechinus eggs were labeled with ^{35}S-Met and divided into three treatment groups that were either left untreated (unfertilized), chemically activated (NH$_4$Cl), or fertilized. Samples were collected every 10 min starting 15 min after fertilization (lane a). First cleavage occurred in samples corresponding to lane f. Bands labeled φ, χ, and ψ are proteins whose synthesis declines after fertilization. Bands α, β, γ, and δ appear strongly labeled after fertilization/activation. Lane h is a sample of unfertilized eggs taken 10 min after the sample shown in unfertilized lane g. Lane m is molecular weight markers.

- Predict whether band A in *Arbacia* and band α in *Lytechinus* are the same protein. Support your answer with data.
- Construct a graph like the one in Figure 13.1.2 using the data from Figure 13.1.3. Estimate relative values for each of the time points.
- Where in the cell cycle would band β possibly be active?

Sorting Out Cyclins [2]

INTRODUCTION

Each stage of the cell cycle is controlled by the activity of a unique combination of cyclins and cyclin dependent kinases (Cdks). Cyclin proteins bind to and activate their partner Cdk. Active kinase then phosphorylates a host of protein substrates within the cell. Phosphorylation can alter the activity of a protein. The phosphorylation of a specific set of proteins by Cdk triggers the transition from one stage of the cell cycle to the next.

Cyclin proteins, as their name implies, undergo cyclic changes in abundance within the cytoplasm of a cell. Regulation of the timing of the transcription and translation of cyclin mRNAs controls the accumulation of specific cyclins within the cytoplasm. Similarly, cells control the destruction of cyclin proteins through the activity of the ubiquitin–proteasome pathway. The net effect is a series of biological "switches" turning the activity of the various Cdk proteins "on" or "off" and thereby moving the cell from one stage to the next in the cell cycle.

While the cell cycle includes all aspects of cellular life, it can also be viewed simply as a cyclic pattern of DNA replication and cell division. From this perspective, the cell is either 2N or 4N. In the normal diploid (2N) state chromosomes exist in pairs with one chromosome coming from each parent. A *tetraploid* (4N) cell has copied all of its genetic material, doubling the number of chromosomes. The classic "X" shape of a chromosome reflects this duplication. Only after a cell has gone through mitotic cell division will it restore itself to the 2N state.

Cell ploidy can be measured using an instrument called a **flow cytometer**. A flow cytometer uses laser light to detect a variety of properties of a cell. In the work described here, cells were stained with the DNA-specific fluorescent molecule **propidium iodide**. The cells are suspended in buffer and passed single file through the laser beam. Propidium iodide emits light in response to excitation by the laser. The intensity of the light emitted is proportional to the amount of DNA in the cell. The flow cytometer counts the number of cells in each of the populations being measured.

- Review/research the combinations of cyclins and Cdks associated with each stage of the cell cycle.
- Propose a model to explain how the activity of one cyclin could lead to the transcription of the next cyclin in the cell cycle.
- Which stage of the cell cycle is associated with DNA replication?
- Sketch and label the stages of the cell cycle. Indicate on your drawing which stages are 2N and which are 4N.
- Predict how the staining intensity of propidium iodide would change with the cell cycle.

BACKGROUND

After DNA replication, several cellular events must occur before a cell can go through mitosis. Replicated centrosomes must migrate to opposite sides of the nucleus. The membrane surrounding the nucleus, the **nuclear envelope**, disassembles in a process known as nuclear envelope breakdown (**NEB**) in order to allow the microtubules growing out from those centrosomes to form a mitotic spindle and connect with the chromosomes. Cyclin proteins A2, B1, and B2 are all thought to be involved with these processes.

One way to investigate the role of each of the cyclins is to alter their expression in the cell. The technique of RNA interference (RNAi) allows researchers to **knock down** or reduce the level of expression of a protein of interest by selectively targeting its mRNA for destruction by the cell. Small interfering RNAs (**siRNAs**) are short double-stranded RNA sequences that interact with a collection of *ribonucleases,* proteins that function to degrade RNA molecules. Complementary base pairing between siRNA and mRNA results in the degradation and removal of mRNA from the cytoplasm.

Knockdown experiments require that researchers be able to transfect cells with the gene sequence for the specific siRNA. Transfected cells transcribe foreign DNA to make multiple copies of the siRNA. While many kinds of cells can be transfected, tissue culture cells are especially useful for these types of experiments. In this work, the researchers used **HeLa cells**, a very well researched human tissue culture cell line. HeLa cells were isolated from a woman named Henrietta Lacks who died of cancer in 1951. Because HeLa cells were originally tumor cells they divide rapidly in culture with a total cell cycle time of approximately 24 h.

- Sketch and label the stages of mitosis. Identify when nuclear envelope breakdown occurs.
- Explain how reducing the levels of cyclin mRNA would affect the activity of a Cdk.

■ Research/review the steps involved in RNAi.
■ Learn more about the story of Henrietta Lacks.

METHODS

Synchronization of HeLa cells

Incubation of tissue culture cells in the presence of excess amounts of thymidine inhibits DNA synthesis causing the cells to become stuck at the transition from G1 to S phase. HeLa cells were treated with culture media containing 2 mM thymidine for a total of 35 h to insure that all the cells were blocked in S phase. Cells were released from the block by washing out the thymidine using fresh culture media lacking thymidine. All the cells simultaneously entered S phase and continued through the rest of the cell cycle as a synchronized population.

Tracking nuclear envelope breakdown using a fluorescent mitosis biosensor

Fluorescent mitosis biosensor (MBS) is a genetically engineered protein that is designed to change its localization inside the cell in response to nuclear envelope breakdown. A plasma membrane–targeting domain was added to one end of yellow fluorescent protein (YFP) while a nuclear localization sequence was added to the other end. YFP accumulates in the nucleus when the nuclear membrane is intact and travels to the plasma membrane when it is not.

Knockdown of cyclins by siRNA

Pools of double-stranded siRNA (d-siRNA) molecules complementary to portions of the mRNA sequences for cyclins A2, B1, and B2 were generated using a recombinant version of the protein dicer. HeLa cells were transfected with the d-siRNAs prior to their release from the thymidine block. A pool of d-siRNAs specific to firefly luciferase (GL3) were also generated and transfected into HeLa cells.

Determination of ploidy by flow cytometry

HeLa cells treated with various d-siRNAs were collected 20 h after thymidine release. The cells were fixed in 2% paraformaldehyde followed by permeabilization with ice-cold methanol and stained with a solution of 50 $\mu g/\mu L$ propidium iodide. Cell numbers for each population of fluorescence intensity were measured at various times after release from the thymidine block.

Immunoblot analysis

HeLa cells were lysed and centrifuged to collect soluble proteins released from the cell. Total protein concentration was measured, and 8 μg of each protein sample was loaded onto an SDS polyacrylamide gel. Proteins were transferred from the gel onto a PVDF membrane and reacted with primary antibodies specific to cyclins A2, B1, and B2. Primary antibody binding was detected using an HRP-labeled secondary antibody followed by incubation with the HRP substrate.

- Excess thymidine accumulates in cells in the form of deoxythymidine triphosphate (dTTP). Predict the chemical structure of this molecule. How does this molecule compare with ATP?
- dTTP acts as an allosteric inhibitor of the enzyme required to synthesize cytosine. Explain why inhibiting this enzyme would result in blocking progression of the cell through the cell cycle.
- Research/review the function of the nuclear localization sequence.
- How does the protein dicer contribute to the story of RNAi?
- What is *luciferase*? Suggest a reason the researchers included GL3 in their experiments?

RESULTS

- Summarize the results shown in Figure 13.2.1a.
- Does the absence of a band on the immunoblot mean that there is no cyclin protein in the cell?
- How much time is required before NEB in HeLa cells?
- Use the data presented in Figure 13.2.1c to determine which cyclin (or cyclins) is involved in regulating NEB.
- Propose an explanation for the difference in the results shown in Figure 13.2.1c for A2 alone (red line), B1 alone (green line), and A2 + B1 (blue line).
- Considering all the data presented in Figure 13.2.1, suggest a reason the researchers included immunoblot data for 5 and 19 h after release from the thymidine block?
- Approximately how much time is required for the majority of HeLa cells to complete S phase under control conditions?
- Approximately how much time is required for the majority of HeLa cells to complete mitosis under control conditions?
- Describe the evidence provided in Figure 13.2.2 that supports the conclusion that cyclin A2 regulates mitosis and not S phase.
- Describe the pattern of cyclin B1 fluorescence in the GL3 knockdown cell 45 min after removal of the thymidine block.
- Approximately how long does it take the centrosomes in the GL3 knockdown cell to orient themselves at opposite ends of the nucleus?
- Propose an explanation for why cyclin B1–YFP fluorescence is lost 80 min after centrosome separation.

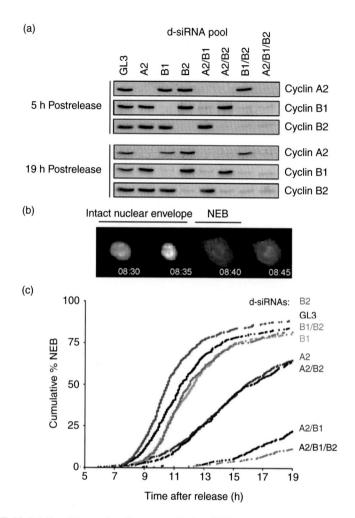

FIGURE 13.2.1 Knockdown of cyclin expression by siRNA inhibits nuclear envelope breakdown.

a. Each of the d-siRNA pools was specific in its ability to reduce the level of cyclin present in the cytoplasm as detected by immunoblotting. b. Fluorescence micrograph of a HeLa cell following release from the thymidine block. The time stamp shows hours and minutes after release. NEB is easily detected by the change in the fluorescent signal generated by the MBS. c. Timing of NEB following thymidine release for cells transfected with various d-siRNAs. Approximately 500 cells were counted for each treatment.

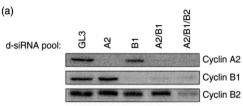

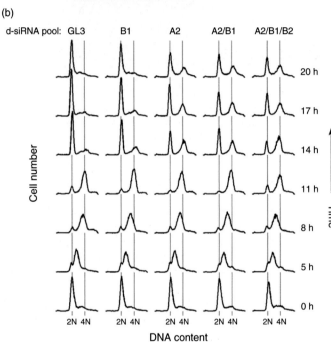

FIGURE 13.2.2 Cyclin A2 regulates G2/M phase, not S phase.
a. Cyclin levels for each of the d-siRNA treatment groups remains low 20 h after release from the thymidine block as shown by immunoblotting. b. DNA content of HeLa cells changes with time as the cells complete the cell cycle. Silencing of cyclin A2 inhibits transition from the G2/M phase back to G1 phase.

- Summarize the effects of cyclin A2 knockdown on the cell shown in Figure 13.2.3.
- What is the interval of time between the accumulation of cyclin B1 in the nucleus and NEB?
- Can you conclude that cyclin B1 causes NEB based on the data presented in Figure 13.2.3?
- Cyclin B1 must be phosphorylated at it N-terminus before it can accumulate in the nucleus. Develop a model to explain how cyclin A2 might regulate the ability of cyclin B1 to move into the nucleus.

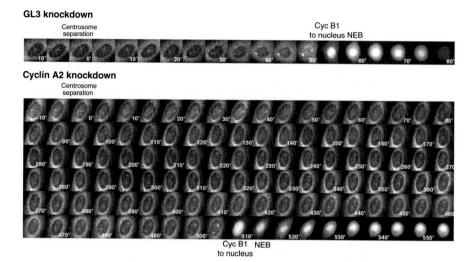

FIGURE 13.2.3 Centrosome separation and cyclin B1 behavior is influenced by cyclin A2.
Still images from a video showing accumulation of cyclin B1 at the centrosomes, centrosome separation, and finally movement of cyclin B1 into the nucleus in a control (GL3 knockdown) and cyclin A2 knockdown HeLa cell. Fluorescently labeled cyclin B1 was expressed in cells transfected with the gene for cyclin B1-YFP. Control cells underwent centrosome separation 11.8 h after release from the thymidine block. Cyclin A2 knockdown cells underwent centrosome separation 6.5 h after release from thymidine. Images were captured at a rate of one frame per 5 min starting 10 min prior to centrosome separation. NEB was determined using the MBS (data not shown).

Of Centriole Separation and Cyclins [3]

INTRODUCTION

Microtubules are an important part of a cell's cytoskeleton. Microtubules are responsible for maintaining a cell's shape, transporting organelles within the cytoplasm, and, in the case of cilia or flagella, moving the whole cell. Microtubules also play a critical role in the cell cycle by forming the mitotic spindle. All of these important cellular functions depend on polymerization of alpha (α) and beta (β) tubulin protein dimers into a microtubule. The organelle responsible for nucleating the polymerization of microtubules is called a **centrosome**.

Centrosomes are made up of two **centrioles** surrounded by a mass of **pericentriolar material** (PCM). Centrioles are cylindrical structures composed of multiple sets of three short microtubules organized into a pinwheel-like pattern. The two centrioles are usually oriented at right angles to one another. The PCM is a matrix of protein and fibrous material. Nucleation of microtubule assembly occurs within the PCM. Microtubule growth is "seeded" by ring structures composed of the protein gamma tubulin (**γ tubulin**). γ tubulin rings are believed to attach to the fibers within the PCM.

A cell in the G1 phase of the cell cycle has a single centrosome containing a pair of centrioles. By M phase, however, a cell needs two centrosomes, each with its own pair of centrioles, which will function as the spindle poles required to build a mitotic spindle. This means that the centrioles must **replicate** as part of the cell cycle. Centriole replication begins at the G1/S phase transition. Replication is semiconservative with each centrosome containing one centriole from the original parent centrosome, and one newly formed centriole. Replicated centrosomes remain together until the cell enters prophase of mitosis.

- Compare and contrast the following cell biology terms: centrosome, basal body, microtubule organizing center (MTOC).
- Look up images of centrioles and centrosomes to gain a better understanding of the organization of this organelle.
- Identify the spindle poles/centrosomes in images of mitotic spindles.
- Describe another example of semiconservative replication in biology.
- Describe the cellular changes associated with prophase of mitosis. What role do centrosomes play in prophase?
- How does a mitotic cell restore the normal ratio of "one centrosome per cell?"

BACKGROUND

Centrosome duplication is an essential part of the cell cycle. Centrosome duplication requires that paired centrioles located at the heart of the centrosome replicate. The process of replication begins at the G1/S transition when the two original centrioles physically separate from each other and new centrioles begin to form from the side of the original. PCM accumulates around the two centrosomes during S phase, and the new centrioles finish growing by the G2/M transition. The two centrosomes remain in close proximity until the start of prophase.

Eggs from the frog *Xenopus laevis* are a very useful model system for studies related to the cell cycle and mitosis. Fertilized eggs undergo synchronous and rapid cell divisions as the embryo begins to develop. These early cell divisions happen so quickly that the G1 stage of the cell cycle is essentially skipped as the cells replicate their DNA and then divide. What makes this model system especially powerful is the ability of extracts of soluble cytoplasmic contents of the early embryo cells to support the replication of DNA and other aspects of the cell cycle *in vitro*. Being able to reproduce progression of the cell cycle outside a cell allows researchers the opportunity to investigate the regulation and mechanics involved in the process.

Embryo extracts can be tested using a variety of experimental manipulations. The role of protein synthesis can be investigated through the use of a protein synthesis inhibitor such as **cycloheximide**. The activity of a protein such as **Cdk** can be examined either by the addition of an inhibitor protein or by the removal or **depletion** of the Cdk from the extract. Proteins such as **p21** or **p27** bind and inhibit Cdks as a normal part of the cell cycle. Since Cdk function is dependent on the presence of cyclins, depletion of these proteins also serves to inhibit Cdk function. Finally, Cdks can be selectively removed from the embryo extract through the use of chemically modified beads. The protein **p13^{suc1}** is known to bind specifically to Cdk2. When p13^{suc1} is attached to the surface of a bead it can "capture" Cdk2 by binding to it. Centrifugation of the sample will cause the beads to pellet, along with any proteins attached to their surface. The same methodology is used for the depletion of cyclins; however, in this case cyclin-specific antibodies are used in place of p13^{suc1}.

- Why would the lack of two centrosomes inhibit cell division?
- What normally happens during the G1 phase of the cell cycle? How can embryonic cells skip this phase?
- How does protein synthesis contribute to progression of the cell cycle?
- Cycloheximide is a bacterial compound that specifically targets the ribosomes of eukaryotic cells. How do prokaryotic and eukaryotic ribosomes differ?

- Cycloheximide inhibits protein synthesis by blocking the E-site on a eukaryotic ribosome. How would blocking this site affect protein synthesis?
- Predict whether the use of p13^{suc1} or antibody-coated beads could completely remove all of the Cdk2 or cyclin proteins from an embryo extract.

METHODS

Frog embryo injections and fluorescent labeling

Cells from the animal pole of a 16- to 64-cell stage frog embryo were coinjected with various proteins and a fluorescein isothiocyanate (FITC)-labeled dextran. Fluorescent dextran served as a marker to identify which cells had been injected. After injection, the embryos were incubated in a saline solution that contained 1 mg/mL cycloheximide for 4 h. Cells were then fixed and processed for incubation with primary antibodies specific to either α or γ tubulin. Excess primary antibody was removed by washing with PBS followed by incubation in the presence of a rhodamine-labeled secondary antibody.

In vitro centriole separation assay

Centrosomes were isolated from *Xenopus* fibroblast tissue culture (XTC) cells. Isolated centrosomes consisted of a single pair of centrioles and the attached PCM. Fertilized frog eggs were incubated with 1 mg/mL cycloheximide for 45 min after the first mitotic cell division. The fertilized eggs were lysed, releasing the cytoplasmic contents. The lysate was then centrifuged at 100,000g and the supernatant containing soluble proteins was used as the embryo extract. Centriole separation was initiated by adding 100 μL of the embryo extract to 10 μL of purified XTC centrosomes and incubating at 25°C. The reaction was stopped by incubating the mixture on ice in the presence of the microtubule-depolymerizing agent *nocodazole*. Centrosomes were collected onto coverslips and processed for immunofluorescence using primary antibodies specific to α and γ tubulin.

Embryo extract modifications

Frog embryo extracts were manipulated in various ways to test for the effects on centriole separation. Extracts were incubated with beads that had the protein p13^{suc1} attached to their surface. Centrifugation of the extract causes the beads and all the attached proteins to pellet, depleting the extract. Mock depletions used beads coated with bovine serum albumin (BSA). Antibodies specific to cyclin E, cyclin A, or Cdk2 were attached to beads coated with protein A. The embryo extract was incubated with the antibody-coated bead mixture and then centrifuged to pellet the beads and attached proteins. Frog embryo extract was pushed into mitosis through the addition of human cyclin B. The status of the extract was confirmed by the addition of sperm nuclei.

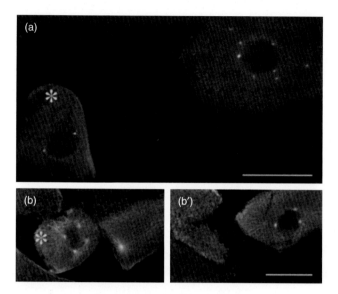

FIGURE 13.3.1 Inhibition of Cdk2 blocks centrosome replication.
Frog embryo cells were injected with the Cdk2 inhibitor p21 and treated with cycloheximide for 4 h to inhibit protein synthesis. Asterisks indicate injected cells as determined by coinjection with FITC dextran. a. Injection with p21 followed by staining for γ tubulin. b and b′. Embryo cells injected with p21 and excess cyclin E followed by staining for α tubulin.

- Find an image of the embryonic staging system for *Xenopus*. Estimate the stage numbers of the 16- to 64-cell embryos used in this experiment.
- Predict which stage of the cell cycle the frog embryos were in after incubation with cycloheximide.
- Compare the excitation/emission spectra of FITC and rhodamine. Explain why these two fluorophores could be used in a single cell.
- Brainstorm a list of proteins that you would expect to find in the embryo extract.
- Why would the addition of human cyclin B cause the frog embryo extract to be "pushed" into mitosis?
- Predict how sperm nuclei would behave in the presence of mitotic cytoplasm.

RESULTS

- Describe the differences you observe between the two cells shown in Figure 13.3.1a.
- Explain why inhibition of protein synthesis would stop cell division.
- Does inhibition of protein synthesis block centrosome replication?

Table 13.3.1 Summary of *In Vivo* Injection Experiments

| | Average Number of Centrosomes* | | |
Solution Injected	Injected Cells	Uninjected Cells	Ratio (I/U)**
Control[†]	6.4 ± 0.8 $n = 18$	7.3 ± 0.6 $n = 31$	0.88
p21 (3 μM)	1.9 ± 0.3 $n = 20$	6.0 ± 0.6 $n = 20$	0.32[‡]
p21 (2 μM)	2.1 ± 0.2 $n = 36$	10.4 ± 0.6 $n = 38$	0.20[‡]
p21 (1.5 μM)	2.5 ± 0.2 $n = 32$	5.5 ± 0.4 $n = 23$	0.45[‡]
p21 (1 μM)	3.1 ± 0.3 $n = 52$	5.6 ± 0.5 $n = 51$	0.55[‡]
p21 + cyclin E[§]	7.7 ± 0.5 $n = 40$	7.5 ± 0.5 $n = 34$	1.03
p21 N-terminus (10 μM)	0.9 ± 0.3 $n = 11$	5.8 ± 0.4 $n = 15$	0.16[‡]
p21 N-terminus (5 μM)	1.2 ± 0.2 $n = 29$	7.6 ± 0.8 $n = 19$	0.16[‡]
p21 C-terminus (8 μM)	5.9 ± 0.3 $n = 33$	5.4 ± 0.3 $n = 36$	1.09
p27 N-terminus (16 μM)	0.9 ± 0.2 $n = 35$	5.8 ± 0.7 $n = 29$	0.16[‡]
p27 N-terminus (8 μM)	2.8 ± 0.2 $n = 30$	7.5 ± 0.5 $n = 30$	0.37[‡]
p27 C-terminus (10 pM)	4.6 ± 0.3 $n = 40$	5.3 ± 0.4 $n = 40$	0.87
p27 C-terminus (5 μM)	6.9 ± 0.5 $n = 29$	6.9 ± 0.5 $n = 30$	1.00

n, *number of cells counted for each condition.*
**The average number of centrosomes is shown ± SEM.*
***The ratio is centrosome number for injected cells divided by centrosome number for uninjected cells.*
[†]*The control solution used was PBS.*
[‡]*For these experiments, the average number of centrosomes in injected cells versus. uninjected cells is significantly different within a 99% confidence level.*
[§]*Cyclin E was added at twice the molarity of p21.*

- What conclusion is supported by the data shown in Figure 13.3.1?
- Define the following terms based on information provided in Table 13.3.1:
    - n
    - Ratio I/U
    - Control
- What would be the predicted value of the ratio I/U for the control? Propose an explanation for why the reported value does not match the predicted value.

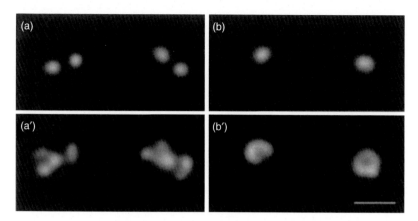

FIGURE 13.3.2 *In vitro* centriole separation assay.
Centrosomes isolated from *Xenopus* XTC cells were incubated with frog embryo extract. XTC centrosomes were examined using an α tubulin antibody at the start of the assay (a) and 1 h after incubation with the embryo extract (b). These same centrosomes were stained with a γ tubulin antibody at the start of the assay (a') and after 1 h (b'). Scale bar = 1 μm.

- Defend the statement, "the effect of p21 on centrosome duplication is dose dependent" using the data provided in Table 13.3.1.
- Proteins p21 and p27 can inhibit the activity of Cdk2. Which region of each of these proteins is involved with Cdk2 inhibition?
- Provide an explanation for the difference in staining in Figure 13.3.2a and Figure 13.3.2a'.
- Why was it necessary to check for the presence of γ tubulin in this assay?
- What is "mock depletion?" How do these data contribute to your interpretation of the results of this experiment?
- What proteins are being depleted by incubation of the embryo extract with p13^{suc1} beads in Figure 13.3.3a?
- Discuss whether the "rescue" of conversion shown in Figure 13.3.3a should be considered "significant."
- Does the depletion of cyclin E or A influence centriole separation relative to control conditions? Explain your answer.
- Propose a model to explain why centriole separation could continue in the absence of either cyclin E or A, but not both.
- What are the implications of the results shown in Figure 13.3.3c?
- Explain which experiment(s) you feel best supports the conclusion that centrosome replication is regulated by cyclin E/Cdk2.

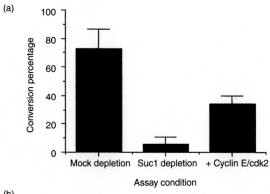

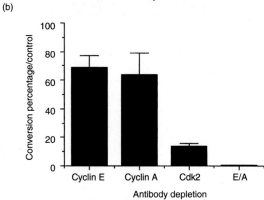

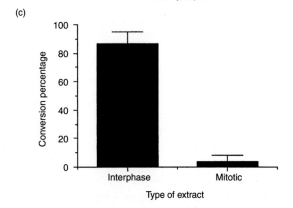

FIGURE 13.3.3 Centriole separation requires cyclin-dependent kinases.
Frog embryo extracts were modified prior to incubation with XTC centrosomes to identify which factors
are important for centriole separation. Centriole separation is defined as the conversion of centriole
doublets (Figure 13.3.2a) into singlets (Figure 13.3.2b). a. Pretreatment of embryo extracts with p13[suc1]
beads, but not BSA-coated beads, inhibits centriole conversion. The activity of the extract can be rescued
by the addition of cyclin E/Cdk2. b. Embyro extracts were depleted of either cyclin E, cyclin A, Cdk2, or a
combination of cyclins E and A. c. Mitotic embryo extract inhibits centriole separation.

The Path to S Phase is Paved With Phosphorylation [4]

INTRODUCTION

Transition from the G1 phase of the cell cycle to S phase represents a commitment by a cell to undergo cell division. Progression through G1 is controlled by two different sets of cyclins and cyclin-dependent kinases. A combination of cyclins D1, D2, and D3 is present in high concentrations early in G1 where they bind to their partner kinases, Cdk4 and Cdk6. The cytoplasmic concentration of the D cyclins diminishes with time in G1 and is replaced by increasing levels of cyclin E. Cyclin E activates Cdk 2. G1/S phase transition is one of the checkpoints in the cell cycle.

Progression from G1 to S is accompanied by changes in gene expression in the cell. Genes for S phase–specific proteins must be transcribed and translated. Transcription factors are proteins that bind to the promoter region of a gene and facilitate the binding of RNA polymerase II, leading to the production of a molecule of mRNA. The protein **E2F** is a transcription factor that controls the expression of S phase–specific proteins. The activity of E2F is controlled by another protein called **Rb**. During G1, Rb attaches to E2F forming a repressor complex that binds to the promoter regions of S phase–specific genes, but functions to block their transcription. As long as Rb is attached, E2F is **inactive**. Rb must "let go" of E2F in order for the transcription factor to become active and allow for the production of the proteins required for S phase.

Protein–protein binding, like the association between Rb and E2F, depends on the conformation (shape or folding) of the proteins. One way to temporarily alter a protein's conformation is through phosphorylation. The addition of clusters of negative charge to certain amino acids within the sequence of a protein can cause it to adopt a slightly different shape. Kinase proteins catalyze the addition of phosphate groups to amino acids within a specific sequence known as a consensus sequence. Cyclin-dependent kinases phosphorylate the amino acids serine (S) or threonine (T) if they are followed by the amino acids proline and lysine (or arginine).

- Explain why S phase commits a cell to going through cell division.
- What is being "checked" at the G1/S checkpoint?
- Predict what types of proteins would be required for a cell to go through S phase.

- How do the concepts of protein phosphorylation and regulation of the cell cycle connect?
- Create a diagram showing E2F in an active state and an inactive state.
- Why would activation of a transcription factor represent an irreversible step in the cell cycle?
- The name Rb stands for retinoblastoma, an inherited type of cancer associated with a mutation in the gene for the Rb protein. Propose an explanation why a loss in function of the Rb protein would lead to cancer.

BACKGROUND

E2F binds to a region of the Rb protein known as the central pocket domain. The central pocket domain is located from amino acid 379 to amino acid 792 in the primary sequence of the Rb protein. Amino acids 793–928 in the primary sequence make up the carboxy-terminal (C-terminal) domain of the protein. While the central pocket domain is sufficient to bind E2F and form the repressor complex, the C-terminal domain is required for the regulation of Rb binding. The C-terminal domain contains several consensus motifs for phosphorylation by Cdks.

Using molecular techniques it is possible to construct genetically engineered versions of Rb that allow scientists to investigate the interactions that occur between different regions of the protein and how those interactions are modified by phosphorylation. In the work presented in this case study, the researchers compared the activity of various constructs of Rb including Rb with only the central pocket domain (Rb(A + B)), central pocket domain plus C-terminal domain (Rb(A + B) + C)), truncated C-terminal domain (Rb882), and C-terminal domain without four of the consensus phosphorylation sites (RbCΔ4). Each of these variations of the Rb protein were linked to the DNA-binding domain of the yeast cell transcription factor *Gal4*. The Rb-Gal4 fusion protein functions to repress gene expression in the same manner as Rb-E2F. Phosphorylation of the Rb portion of the fusion protein inhibits repressor activity and restores gene expression. DNA sequences for the various constructs were transfected into a human cancer cell line that lacked the gene for Rb protein (Rb(−)C33a).

C33a cells were also transfected with a DNA sequence for the reporter chloramphenicol acetyltransferase (**CAT**). CAT is a bacterial protein that, as its name implies, catalyzes the addition of an acetyl group to the antibiotic chloramphenicol, neutralizing the antibiotic. By constructing a DNA sequence that links the CAT gene to the promoter of interest it is possible to measure the activity of that promoter under various experimental conditions. In this work, the CAT gene was linked to an SV40 promoter/enhancer that was located "downstream" of a Gal4-binding site. CAT expression is high in the absence and low in the presence of the Rb-Gal4 fusion protein.

- Use online resources to look up the sequence of human Rb protein. Identify one or more examples of the Cdk phosphorylation consensus sequence within the C-terminal domain.
- Use online resources to learn more about the C33a tissue culture cell line.
- Suggest a reason the researchers used Gal4 instead of E2F in the design of their Rb and reporter sequences.
- Connect expression of CAT in bacteria to the concept of antibiotic resistance.
- Create a diagram of the DNA sequence for the CAT reporter gene (include the SV40 promoter and the Gal4-binding site).

METHODS

Plasmid constructs and transfection

Constructs of the gene sequence for the retinoblastoma protein Rb were linked to the DNA-binding domain of Gal4 and inserted into plasmids and transfected into a human cancer cell line that does not express Rb (Rb(−)C33a). The following is a list of the constructs used in the experiments discussed in this case:

Construct Name	Protein Domain	Amino Acids
Rb	Central pocket + C-terminal	379–928
Rb(A + B)	Central pocket	389–792
RbC or C	C-terminal domain	767–928
RbCΔ4	Mutation of serine and threonine to alanine	S-807, S-811, T-821, and T-826

The gene sequence for the reporter protein CAT was inserted into a plasmid that contained a Gal4-binding site upstream of an enhancer domain and TATA box. For other experiments, the plasmid construct used an E2F binding and TATA box linked to the CAT gene. Cells were transfected with a combination of 0.2 μg of reporter plasmid and 0.5 μg of the Rb construct plasmid DNA. "Empty" plasmids lacking an Rb gene sequence were used as controls.

For some experiments, Rb(−)C33a cells were cotransfected with plasmids for Rb-Gal4 fusion proteins and plasmids carrying the gene for either cyclin D or cyclin E. In other experiments cotransfections included a plasmid carrying a dominant negative mutant form of Cdk2 (**dn-K2**). Expression of this mutant protein interferes with the activity of the cell's endogenous, nonmutated Cdk2.

CAT assay

Expression of the reporter plasmid leads to the accumulation of CAT as a soluble protein in the cytoplasm of transfected cells. Cell extracts containing soluble cytoplasmic proteins were isolated by lysing the transfected cells. The level of CAT activity has been shown to be directly proportional to the amount of CAT present in the cell extract. CAT activity was measured by reacting cell extracts with radioactively labeled chloramphenicol (^{3}H-chloramphenicol) in the presence of acetyl-CoA. The solution was then separated using the technique of thin-layer chromatography. Chloramphenicol migrates different distances on the chromatograph depending on the number of acetyl groups that have been added to it. CAT is capable of adding a maximum of two acetyl groups to ^{3}H-chloramphenicol. Because ^{3}H-chloramphenicol is the only radioactive protein in the solution it can be detected by exposing the chromatographs to X-ray film. The location and intensity of spots on the chromatograph were measured to determine the relative activity of CAT.

Immunoprecipitation and Western blot

Cytoplasmic extracts from transfected cells were isolated. Ten percent of the extract was used for Western blotting and the remainder was incubated with a monoclonal antibody to Gal4. Antibody was precipitated by incubation with protein A–coated beads. All protein samples were run on SDS-PAGE and blotted to nitrocellulose. Blots were reacted with a polyclonal antibody to DNA-binding domains (pAb–anti-LexA).

- Why would changing a serine or threonine to alanine influence the ability of Cdk to phosphorylate Rb?
- How would transfection with a plasmid carrying the gene for cyclin D or E affect a cell?
- What are the functions of the enhancer domain and TATA box in the CAT plasmid?
- How does acetyl-CoA contribute to CAT assay?
- Research/review the principles behind the technique of thin-layer chromatography.

RESULTS

- What effect does expression of the Rb-Gal4 fusion protein have on CAT gene expression based on the data in Figure 13.4.1a?
- Explain why coexpression of the Rb-Gal4 fusion protein with either cyclin D or E rescued CAT expression.
- Suggest a possible explanation for why the presence of cyclin D had a greater effect on CAT gene expression than cyclin E.
- Why did expression of the Cdk2 dominant negative mutate (dn-K2) not affect CAT gene expression in the presence of cyclin D?

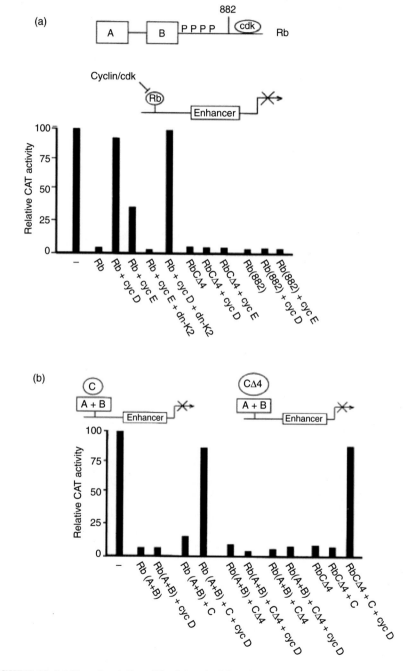

FIGURE 13.4.1 Phosphorylation of the C-terminal domain regulates repression of gene expression by Rb. (*Continued*)

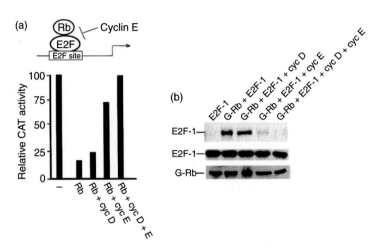

FIGURE 13.4.2 Rb inactivation of E2F depends on cyclin E–Cdk4/6, not cyclin D–Cdk2.
Rb(−)C33a cells were cotransfected with plasmids for the Rb-Gal4 fusion protein, cyclin proteins, and a reporter plasmid containing an E2F-binding site. a. Schematic diagram and CAT assay data for E2F inactivation by Rb-Gal4. b. Western blot of proteins that coimmunoprecipitated in the presence of an antibody to Gal4 (top blot). Total E2F and Rb-Gal4 fusion protein (G-Rb) present in the extract prior to immunoprecipitation is shown in the lower two blots. Blots are stained using a polyclonal anti-LexA antibody.

- Summarize the conclusion supported by the data for the RbCΔ4 experiments.
- How do the results from the Rb(882) experiments add to your summary?
- Defend the following statement using the data in Figure 13.4.1b, "The central pocket domain of Rb is sufficient to inactivate E2F; however, the C-terminal domain is required to regulate that inactivation."
- Which data in Figure 13.4.2a are evidence that the Rb-Gal4 fusion protein can bind to E2F?
- Look closely at Figure 13.4.2b. What is wrong with this figure?

▶ **FIGURE 13.4.1 (*cont.*)** CAT activity was measured in Rb(−)C33a cells transfected with various constructs of the Rb-Gal4 fusion protein. a. Schematic diagrams of the Rb portion of the fusion protein and its interaction with the reporter gene. The central pocket domain (A + B) is attached to the C-terminal domain that has a site for Cdk binding (cdk) and four consensus sequences for Cdk phosphorylation (P). Relative CAT activity is shown for each of the indicated constructs. (−), empty plasmid; Rb, Rb-Gal4 expression; cyc, coexpression of cyclin; dn-K2, expression of a double-negative gene for Cdk2; CΔ4, Rb lacking four consensus sequences for Cdk phosphorylation; Rb(882), Rb missing amino acids 883–928. b. The Rb central pocket (Rb(A + B)) and C-terminal domains (C) or C-terminal domain lacking consensus phosphorylation sites (CΔ4 or CΔ2) were coexpressed as separate proteins in Rb(−)C33a cells with or without coexpression of cyclin D (cyc D). RbCΔ4 was coexpressed with the intact C-terminal domain (C), with or without cyclin D (cyc D).

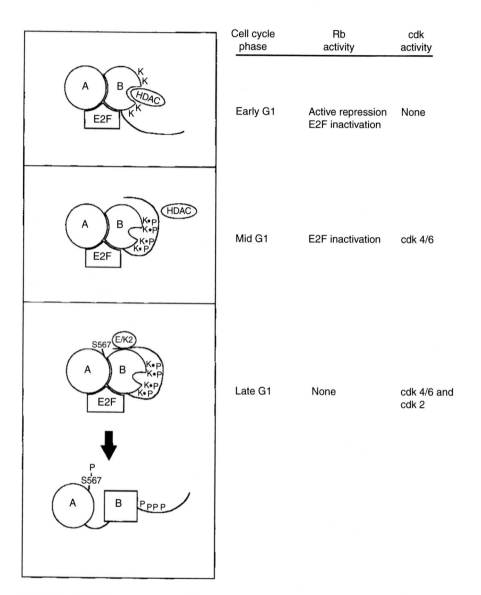

Cell cycle phase	Rb activity	cdk activity
Early G1	Active repression E2F inactivation	None
Mid G1	E2F inactivation	cdk 4/6
Late G1	None	cdk 4/6 and cdk 2

FIGURE 13.4.3 Proposed model of the regulation of E2F by Rb.
The central pocket of Rb(A + B) binds E2F and the protein histone deacetylase (HDAC), a protein involved in chromatin structure. Phosphorylation of serine and threonine residues in the C-terminal domain of Rb displaces HDAC. Binding of cyclin E/Cdk2 triggers the release of E2F.

- Summarize the conclusion supported by the data in Figure 13.4.2.
- Discuss how the error in Figure 13.4.2b affects your opinion of the validity of this research.
- What is the substrate for cyclin E/Cdk2 based on the model shown in Figure 13.4.3?

- Which of the images in Figure 13.4.3 are supported by the data presented in this case study?
- Histone deacetylase (HDAC) is an enzyme that removes acetyl groups from histone proteins causing a tight association between histone proteins and the DNA. Predict how the presence of HDAC would influence gene expression.

References

[1] Evans T, Rosenthal ET, Youngblom J, Distel D, Hunt T. Cyclin: a protein specified by maternal mRNA in sea urchin eggs that is destroyed at each cleavage division. Cell 1983;33:389–96.

[2] Gong D, Pomerening JR, Myers JW, et al. Cyclin A2 regulates nuclear-envelope breakdown and the nuclear accumulation of cyclin B1. Curr Biol 2007;17:85–91.

[3] Lacey KR, Jackson PK, Stearns T. Cyclin-dependent kinase control of centrosome duplication. Proc Natl Acad Sci USA 1999;96:2817–22.

[4] Harbour JW, Luo RX, DeiSanti A, Postigo AA, Dean DC. Cdk phosphorylation triggers sequential intramolecular interactions that progressively block Rb functions as cells move through G1. Cell 1999;98:859–69.

CHAPTER 14

Cell Division

Push and Pull: How Motor Proteins Help Build a Spindle [1]

INTRODUCTION

Proper distribution of a cell's genetic material in mitosis relies on the presence of a **bipolar spindle**. The classic shape of the mitotic spindle is a reflection of the mechanics required for the alignment of chromosomes at metaphase, the separation of sister chromatids at anaphase, and the establishment of physically distinct daughter cells following telophase. A spindle is composed primarily of **microtubules (MTs)** that grow out from the **spindle poles** or **centrosomes**. At the core of the centrosome is a pair of centrioles. Centrioles replicate during S phase and separate to form two centrosomes at the beginning of mitosis. As MTs polymerize off the centrosome they form a star-like structure called an **aster**. Spindle formation requires that a subset of the MTs emerging from the two centrosomes interact, forming overlapping, antiparallel MTs and causing final separation of the centrosomes to opposite poles of the spindle. Spindle MTs that bind to the kinetochores located at the centromere of each chromosome are responsible for the movement of chromosomes and chromatids. Nonkinetochore MTs that overlap along the midline of the spindle contribute to the movement of the spindle itself. Errors in spindle assembly result when centrosomes fail to separate leading to the formation of **monopolar spindles**.

■ Discuss how the shape and function of a spindle contributes to the process of meiosis.

- Research/review MT polymerization. Predict the location of the plus (fast-growing) end and minus (slow-growing) ends of an MT growing off a centrosome.
- Draw an image of a mitotic spindle with at least one chromosome. Label the centrosome, centromere, kinetochore MT, and nonkinetochore MTs. Indicate the plus and minus ends of spindle MTs.
- Explain what is meant by description of "antiparallel MTs" in the context of the spindle.
- Predict whether a monopolar spindle could separate sister chromatids. Explain your answer.

BACKGROUND

A **push–pull model** has been proposed to explain assembly of the MT spindle. According to this model, forces generated by **MT motor proteins** are responsible for the separation of centrosomes and the final bipolar shape of the spindle. MT motor proteins fall into two general categories; **dyneins** and **kinesins**. Dynein is a minus end–directed motor protein while kinesins, typically but not always, are plus end–directed motors. Multiple motor proteins are known to function in the spindle. It has been proposed that **antagonistic** force production by the different motor proteins is required for spindle assembly and function. This case study examines the interplay between the plus end–directed motor **Eg5** (kinesin-5) and **dynein**.

- List other examples of cellular activities that rely on MT motor proteins.
- Define the term *antagonistic* using your own words.
- Refer back to your drawing of a spindle. Use arrows to indicate the *push* or outward direction of *force* on the spindle and the *pull* or inward *force* direction. Predict which MT motor proteins are exerting these forces.
- Force production by MT motor proteins is thought to occur in areas where antiparallel MTs overlap. Create a diagram that illustrates the interaction of antiparallel MTs and MT motors.

METHODS

Cell culture and nocodazole treatment

LLC-PK1 cells expressing GFP-tubulin were maintained in tissue culture media supplemented with 10% fetal calf serum in an atmosphere of 5% CO_2 at 37°C. Cells were plated on glass coverslips 2 days prior to observation. Nocodazole at a concentration of 3.3 μM was added to the medium 3–5 h before imaging of live cells. Nocodazole washout was performed by repeated removal of media and replacement with fresh media without nocodazole.

Inhibitors

Monastrol was used at a concentration of 200 μM. p150-CC1 plasmid expressing a protein fragment known to inhibit dynein function was prepared and
injected in cells at a concentration of 25 μM.

Time lapse microscopy

Cells were plated on glass coverslips and images were acquired between 5 s and
2 min intervals with exposure times between 400 and 800 ms.

- What are LLC-PK1 cells? What was their origin?
- Conduct a search to learn more about monastrol and nocodazole.

RESULTS

- How did the experiments in Figure 14.1.1a and b alter the shape of the
  spindle?

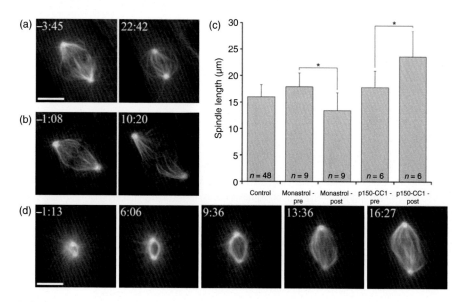

**FIGURE 14.1.1 Antagonistic forces are produced by the spindle motor proteins Eg5
and dynein.**
Cells expressing GFP-tubulin were treated with the Eg5 inhibitor monastrol (a) or the dynein inhibitor
p150-CC1 (b). Images represent spindle length before (left) and after (right) treatment. c. Average and
standard deviation of spindle length before and after each treatment. Data are significant (asterisk) at
$p = 0.01$ for monastrol and $p = 0.05$ for p150-CC1. d. Selected images from a time lapse series of a
monastrol-treated cell containing a monopolar spindle. All times are relative to motor inhibition (0:00) and
are displayed as min:s. Scale bars = 10 μm.

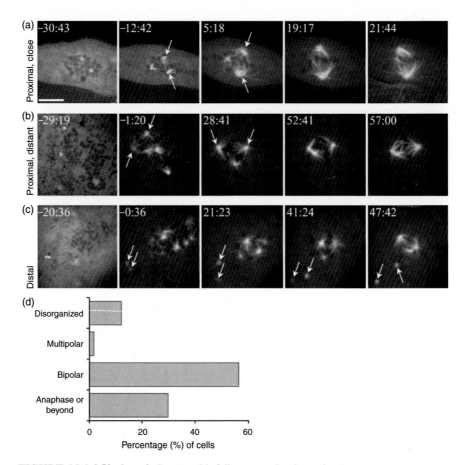

FIGURE 14.1.2 Bipolar spindle assembly follows nocodazole washout.

Selected images from time lapse sequences of cells treated with and released from nocodazole illustrating three centrosomal configurations. Centrosomes can be seen as white dots in the first panel showing the cell prior to nocodazole washout. Arrows indicate the location of the centrosomes in subsequent images. a. Centrosomes are in close proximity to the chromosomes (proximal) and each other (close). b. Centrosomes are in proximity to chromosomes (proximal) but are separated from each other (distant). c. Centrosomes are separated from chromosomes (distal). All times are relative to final nocodazole washout (0:00) and are displayed as min:s. d. Percentage of fixed cells at the indicated mitotic stage after 60 min post nocodazole washout.

- Refer to your drawing of a spindle. Add labels to indicate which MT motor protein, is responsible for the *push* (outward force) and *pull* (inward force) based on the data in Figure 14.1.1a and b.
- Explain the mechanics behind the different directions of force production based on your knowledge of these motor proteins.
- Are the pretreatment spindle lengths in Figure 14.1.1c equivalent to the control spindle length? Justify your answer.
- Develop an argument based on the data in Figure 14.1.1d to support the conclusion that dynein is responsible for the formation of monopolar spindles.

- What effect does nocodazole have on the spindles in Figure 14.1.2 (first panels)?
- Describe the pattern of spindle recovery in the experiments in Figure 14.1.2a and b.
- Locate the centrosomes and chromosomes in Figure 14.1.2c. Where are the MTs located in these images?
- What can you conclude from the experiment in Figure 14.1.2c?
- Use the data in Figure 14.1.2d to develop an argument for or against the statement that the nocodazole treatment protocol does not adversely affect spindle function.
- How are the images in Figure 14.1.3 different from those in Figure 14.1.2? What is different between the two experimental protocols?
- Which motor protein is responsible for the formation of monopolar spindles? Explain your answer.
- If MT overlap is important for motor function, what can you conclude about the arrangement of MTs in the three centrosome configurations shown in Figures 14.1.2 and 14.1.3?
- Predict what the spindles in Figure 14.1.3 would look like, if the experiment inhibited dynein and not Eg5?

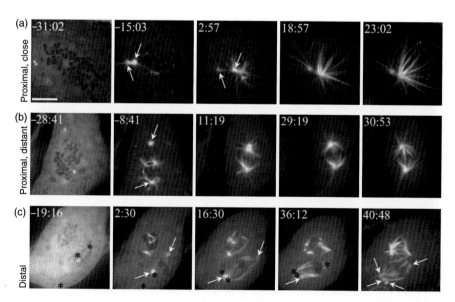

FIGURE 14.1.3 Distance between centrosomes is a predictor of monopolar spindle formation. Selected images from time lapse sequences of cells treated with nocodazole and monastrol, and then released into monastrol-containing media. a. Monopolar spindle formation from proximal, close centrosomes. b. Bipolar spindle formation from proximal, distant centrosomes. c. Bipolar spindle formation from distal centrosomes. Two cells have fused in (c). The upper spindle has formed without centrosomes. Arrows indicate the location of centrosomes. Asterisks indicate out-of-focus centrosomes.

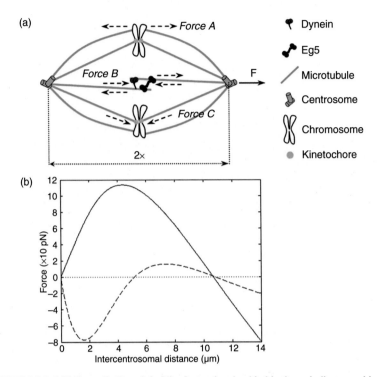

FIGURE 14.1.4 Mathematical model of the forces involved in bipolar spindle assembly.
a. Schematic of the mathematical model. The total force *F* acting on the centrosomes is a function of forces *A*, *B*, and *C*. *Force A* is generated by MTs attached to chromosome arms where polymerization or kinesin motors produce an outward, repulsive force. *Force B* is the product of Eg5 and dynein working antagonistically in areas where antiparallel MTs overlap. *Force C* is an inward force generated by the pull created by kinetochores. Assuming all chromosomes align at the midpoint of the spindle, outward force *F* can be computed as a function of the length of the half-spindle *x* and the average MT length *L*, using the formula $F(x) = (Ae^{-x/L} - C) - 2Bxe^{-2x/L}$. b. Predications of force versus intercentrosomal distance based on the model for an uninhibited or coinhibited spindle (solid line; $L = 2$, $A = 1$, $B = 0$, and $C = 0.03$) or an Eg5-inhibited spindle (dashed line; $L = 2$, $A = 1$, $B = 2$, and $C = 0.03$). Positive values for *F* reflect outward force while negative values reflect inward force.

- Describe how the direction of force predicted by the mathematical model in Figure 14.1.4 changes with intercentrosomal distance for the uninhibited spindle.
- What is the predicted length of a spindle in which all the antagonistic forces are balanced?
- How does the loss of Eg5 activity alter predicted force dynamics in the spindle?
- What sign would you use for the value *B* to predict the effect of inhibition of dynein, not Eg5?
- Discuss which of the experiments presented in this case study provides the most compelling evidence of the *push–pull* model of spindle assembly.

Ready, Set, Anaphase! [2]

INTRODUCTION

Throughout the entire cell cycle, there are three **checkpoints** that insure that the cell and its DNA are undamaged and able to progress through mitosis. Each of these checkpoints is regulated by its own set of signaling molecules. One of the most dramatic checkpoints, the **mitotic or metaphase checkpoint**, occurs at the transition between metaphase and anaphase. Proteins involved in the mitotic checkpoint monitor the attachment of spindle microtubules (MTs) to the kinetochores located at the centromere of each chromosome. Signals created by the presence of an unattached kinetochore inhibit anaphase. Only after every kinetochore has captured an MT will the checkpoint be satisfied, allowing separation of the sister chromatids and progression of anaphase to continue.

- Create a diagram showing where the three checkpoints occur in the cell cycle.
- Describe what happens at the transition from metaphase to anaphase.
- Predict what would occur if one kinetochore remained unattached to a spindle MT and there was no mitotic checkpoint.

BACKGROUND

The mitotic checkpoint is regulated by a series of proteins centered around the activity of the **anaphase-promoting complex (APC)**. APC is a **ubiquitin ligase** that is located at the kinetochore. Active APC adds ubiquitin to a set of specific substrate proteins. One of the key protein substrates of APC is **securin**, the protein responsible for cohesion between sister chromatids. Proteins that have ubiquitin attached to them are targeted for destruction by the **proteasome**. However, in kinetochores that have not attached to a spindle microtubule the proteins **Mad1** and **Mad2** bind to APC, inhibiting its activity. Only after a MT has attached to the kinetochore will Mad2 release APC. How does Mad2 "know" that a MT has formed at attachment? What role, if any, does the kinetochore-associated MT motor protein, **CENP-E** play in the signaling pathway that leads to activation of APC?

Xenopus frog egg extracts are a very useful model system for *in vitro* studies of cell division including kinetochore-dependent activation of the mitotic checkpoint. The mitotic checkpoint can be established *in vitro* by assembling spindles in the presence of a high concentration of sperm nuclei, an endogenous **cytostatic factor (CSF)** which blocks spindles at metaphase, and the MT-depolymerizing agent **nocodazole**. Addition of calcium to the reaction inhibits CSF, allowing spindles to proceed through anaphase. However, spindles affected by the mitotic checkpoint will remain in metaphase.

- Discuss the various ways enzyme activity can be regulated. Speculate on the way Mad2 may be able to inhibit APC function.
- Why are sperm nuclei added to egg extract in this *in vitro* reaction?
- Sketch a diagram illustrating the relationship between MTs, Mad2, APC, and the kinetochore.
- Conduct a search for images of the structure of a kinetochore. Localize Mad2, APC, and CENP proteins.

METHODS

Xenopus egg extracts

CSF-arrested egg extracts were prepared and the mitotic checkpoint was activated by the addition of approximately 9000 sperm nuclei/μL and 10 μg/mL nocodazole 30 min before release from CSF arrest with the addition of 0.4 mM calcium chloride. MTs were visualized by the addition of fluorescently labeled tubulin. CENP-E was depleted from CSF extracts by adding anti-CENP-E antibody and protein A-Sepharose beads to immunoprecipitated CENP-E. For the antibody addition experiment, 100 μg/mL of purified CENP-E was added to the reactions.

Histone H1 phosphorylation assay

At indicated time points after calcium chloride addition, 1 μL of extract was frozen in liquid nitrogen. Extract samples were thawed by the addition of 9 μL of histone H1 mix that included [γ^{32}P]-ATP and incubated for 10 min at room temperature. The reactions were terminated by the addition of 10 μL of SDS-PAGE sample-loading buffer and were analyzed by SDS-PAGE and autoradiography.

Securin degradation

A securin-encoding plasmid was used to *in vitro* transcribe and translate securin using a reticulocyte lysate system. Translation occurred in the presence of [γ^{35}S]-methionine. 1 μL of the *in vitro* translated protein was incubated with 20 μL of extract, and 2 μL were taken at the indicated time points after calcium chloride addition. The reactions were terminated by the addition of 10 μL of SDS-PAGE sample-loading buffer and were analyzed by SDS-PAGE and autoradiography.

Immunofluorescence

Extracts containing mitotic spindles were diluted in buffer and sedimented onto a coverslip through a sucrose cushion. Coverslips were fixed in –20 °C methanol, rehydrated in buffer, blocked using 1% BSA, and reacted with primary antibodies. Antibody binding was visualized using fluorescent-conjugated secondary antibodies. Chromatin was visualized using DAPI.

- Speculate why the addition of excess sperm nuclei and depolymerization of MTs with nocodazole, would activate the mitotic checkpoint.
- Explain how the addition of calcium could alter the behavior of a protein.
- Predict whether the phosphorylation assay would work if myelin basic protein was used in place of histone H1.

RESULTS

- What is "mock depletion"? Why would this treatment be considered a control?
- Identify the control conditions in Figure 14.2.1b. What combination of conditions triggers the mitotic checkpoint? Explain your answer.

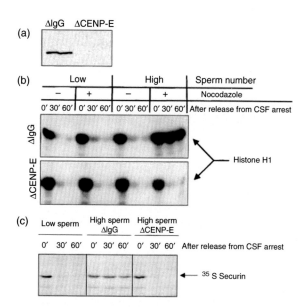

FIGURE 14.2.1 CENP-E depletion prevents activation of the checkpoint in _Xenopus_ egg extracts.
a. CSF-arrested egg extracts were immunodepleted of CENP-E (ΔCENP-E) or mock-depleted (ΔIgG) and analyzed by immunoblotting with an antibody to CENP-E. b. Activation of the mitotic checkpoint is demonstrated by continued phosphorylation of histone H1. c. _In vitro_ translated ^{35}S-radiolabeled securin was incubated in CENP-E-depleted and mock-depleted extracts for 20 min before release of CSF arrest. At the indicated times aliquots were removed and analyzed by SDS-PAGE and autoradiography.

- What type of protein is involved in phosphorylation reactions? How do you explain the difference in phosphorylation of histone H1 in Figure 14.2.1b?
- Activation of APC is known to lead to the destruction of securin. Under what conditions is APC becoming active?
- Synthesize the results presented in Figure 14.2.1 to develop an argument in support of the statement, "CENP-E is required for checkpoint-mediated inhibition of sister chromatid separation in response to nocodazole-induced spindle damage."
- How do we know that CENP-E depletion of the extracts in Figure 14.2.2 was successful?
- Can either of the Mad proteins bind to DNA in the absence of CENP-E?
- How would you respond to the criticism that the Mad1 and Mad2, shown in Figure 14.2.2, are nonspecific and artifacts?
- The top row of Figure 14.2.3a (+IgG) is the control for this experiment. Explain what is happening in each of the panels.

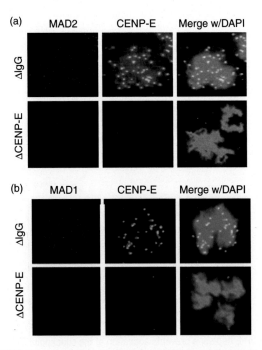

FIGURE 14.2.2 CENP-E is required for association of Mad2 and Mad1 with kinetochores.
CSF-arrested extracts were mock-depleted (ΔIgG) or CENP-E-depleted (ΔCENP-E) and mixed with a high concentration of sperm nuclei and nocodazole to induce the mitotic checkpoint. The colocalization of Mad2 (a) or Mad1 (b) with CENP-E was examined using immunofluorescence microscopy. DNA was stained using DAPI. The three fluorescent images were merged in the right column.

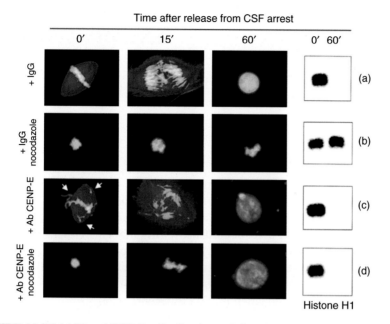

Time after release from CSF arrest

FIGURE 14.2.3 Addition of CENP-E antibodies does not disrupt centromere function but still prevents activation of the mitotic checkpoint.

CSF extracts containing high levels of sperm nuclei were allowed to progress through one cell cycle and were blocked at the next metaphase by readdition of CSF. Affinity purified CENP-E antibody or control IgG was incubated with the extract for 30 min in the presence of absence of nocodazole followed by release of CSF arrest. Aliquots were collected at the indicated times for examination by fluorescence microscopy and were assayed by phosphorylation of histone H1. Arrows indicate misaligned chromosomes.

- Describe how and why the images in Figure 14.2.3b are different from those in Figure 14.2.3a.
- Is CENP-E required for kinetochore/centromere function? Explain your answer.
- Look again at your diagram of the possible relationship between a kinetochore, Mad2, APC, and a MT. Use the results from this case study to add CENP-E to your diagram. Are there any missing pieces to the story of kinetochore attachment and the mitotic checkpoint?

Building Cell Walls – Cytokinesis in a Plant Cell [3]

INTRODUCTION

The final critical step in cell division is separation of the two new daughter nuclei into two daughter cells, a process known as **cytokinesis**. In animal cells, cytokinesis involves the cytoskeletal proteins actin and myosin. Actin microfilaments create a "belt" or **contractile ring** that encircles the cell and attaches to the plasma membrane (PM). Myosin motor proteins use ATP energy to crossbridge and exert force on actin filaments, causing the contractile ring to tighten. A **cleavage furrow** is created as the PM of the cell is drawn inward. Daughter cells are formed when the PM fuses with itself.

Cytokinesis in plants requires a different mechanism. Daughter cells are separated by the formation of a new cell wall. A **cell plate** is assembled along the midline of the cell, perpendicular to the spindle. The cell plate consists of Golgi-derived vesicles that carry cell wall precursor material. The vesicles travel on a specialized cytoskeletal structure known as the **phragmoplast**, which is made up of MTs and microfilaments. Similar to a mitotic spindle, the bundles of phragmoplast MTs extend from two sides of the cell, overlapping at the midline. The MTs are organized with their *plus* ends oriented toward the middle. Golgi vesicles accumulate at the *plus* ends of these MTs and eventually fuse to form a larger and larger membrane network. Pectin and other cell wall precursor proteins are contained within the vesicles and will form the middle lamella of the new cell wall.

- Predict what might happen to a cell that did not go through cytokinesis.
- Describe another example of force production between actin and myosin.
- Explain why actin/myosin-based cytokinesis does not occur in a plant cell.
- What is the source of the new PM that forms between the daughter cells in a plant?
- Research/review the mechanics of vesicle transport along an MT.
- Budding yeast (*Saccharomyces cerevisiae*) have cells walls, but use actin and myosin during cytokinesis. Speculate the differences between yeast and plants, which might account for this.

BACKGROUND

The protein composition of the phragmoplast includes more than just MTs. Proteins that share homology with the MT motor protein kinesin, **kinesin-like proteins**, are concentrated within the phragmoplast, as are a collection of protein kinases. The kinase **nucleus- and phragmoplast-localized protein kinase 1 (NPK1)** has been shown to become active during late M phase. NPK1 is associated with lateral expansion of the cell plate. What controls the activation of NPK1? Preliminary evidence points to a role for the kinesin-like protein **NACK1**. Interaction between the kinesin-like motor protein NACK1 and the protein kinase NPK1 was examined in this case study using the yeast *S. cerevisiae* and the tobacco plant *Nicotiana tabacum*.

- Explain the difference between a "kinase" and "kinesin".
- What direction does kinesin move along an MT?
- Why is the timing of the activation of NPK1 significant?
- Predict what would happen to a cell if NPK1 did not function.

METHODS

Plant and cell material

Yeast (*S. cerevisiae*) was used to express various combinations of NPK1 and NACK1 for immunoprecipitation. The tobacco cell line BY-2 was used for immunofluorescence studies of the localization of the two proteins. BY-2 cells were maintained as a suspension in culture media at 26°C in the dark. Transgenic *N. tabacum* were generated using *Agrobacterium*-mediated transformation with a NACK1 construct lacking a motor domain. Seeds were germinated on solid media supplemented with or without 0.1 µM DEX.

Immunoprecipitation

Yeasts expressing the constructs of interest were washed twice, resuspended into buffer, and disrupted using glass beads. Cell lysates containing 400 µg of protein were diluted into buffer and incubated with 1 µg of NPK1-specific antibody on ice for 90 min, followed by 60 min incubation at 4°C in the presence of protein A-Sepharose beads. Beads were washed several times in buffer and prepared for SDS-PAGE and immunoblotting.

Kinase assay

NPK1 immunoprecipitates were prepared as described earlier. Sepharose beads with the attached NPK1:NACK1 complex, were washed with buffer and then incubated with a reaction cocktail that included 1 µg myelin basic protein, 10 µCi [γ^{32}P]ATP and 50 µM ATP. Kinase reactions were allowed to proceed for 30 min at 25°C, after which samples were prepared for SDS-PAGE and gels were subjected to autoradiography to visualize phosphorylated proteins.

Immunofluorescence and light microscopy

BY-2 cells were fixed with 3.7% formaldehyde in buffer for 1 h. Cells were washed with buffer, and cell walls were briefly digested with 0.5% cellulose followed by permeabilization with 0.5% Triton X-100. Cells were incubated overnight with the indicated antibodies, washed, and incubated with the appropriate fluorescent-conjugated secondary antibody. Cells were washed, incubated with DAPI, and prepared for observation using a fluorescence microscope. For staining of the cell wall, cells were fixed with 3.7% formaldehyde in buffer for 1 h and stained with 0.005% calcofluor, 1 μg/mL propidium iodide, and 10 μg/mL RNase A.

To observe nuclei in cotyledons, the plant tissue was soaked overnight in a solution of ethanol and acetic acid, and then stained with 1% orcein for 4 h before observation.

- Debate whether expression of a tobacco plant protein in a yeast cell would have an influence on the results of an experiment.
- Outline the basic protocol of an immunoprecipitation experiment.
- What extra challenges exist for performing immunofluorescence on a plant cell? How were these challenges addressed in the methods described here?

RESULTS

- Which lane or lanes in Figure 14.3.1a demonstrate that NACK1 and NPK1 interact? Explain your answer.
- Why is there no NACK1 band in lanes 5 and 8 of Figure 14.3.1a?
- Summarize the evidence in Figure 14.3.1b that supports the conclusion that NACK1 binding activates NPK1.
- Provide an explanation for the appearance of two bands on "Autorad H" but only one on "Autorad L" in Figure 14.3.1b.
- Suggest a reason why there are bands in lanes 1, 2, 4, and 5 of the autoradiographs in Figure 14.3.1b.
- DAPI is a fluorescent stain specific to DNA. Explain why that statement makes sense in the context of the images in Figure 14.3.2a.
- Describe what happens to the MT cytoskeleton of the BY-2 cells during the stages of mitosis (Figure 14.3.2a).
- How does the distribution of NACK1 on the MTs in Figure 14.3.2a change during mitosis?

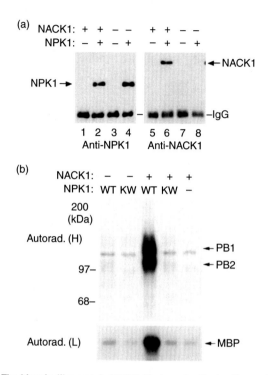

FIGURE 14.3.1 The kinesin-like protein NACK1 binds and activates the protein kinase NPK1.
a. Coimmunoprecipitation of NACK1 and NPK1 from cell extracts isolated from yeast cells expressing cDNAs for NACK1 and NPK1 in different combinations. The expression of cDNAs is indicated by a (+). Cell extracts from each of the combinations were immunoprecipitated with an NPK1-specific antibody. Samples were analyzed by SDS-PAGE and immunoblotted with antibody to NPK1 (lanes 1–4) or NACK1 (lanes 5–8). b. Proteins were extracted from yeast expressing either wild-type NPK1 (WT; lanes 1 and 3), a kinase mutant NPK1 (KW; lanes 2 and 4), or no NPK1 (lane 5) in the absence (lanes 1 and 2) or presence (lanes 3–5) of NACK1. Extracts were immunoprecipitated with NPK1-specific antibodies, and kinase assay was performed in the presence of ^{32}P-ATP with myelin basic protein as a substrate. Samples were analyzed by SDS-PAGE (8% polyacrylamide, H; 12% polyacrylamide, L), and phosphorylated proteins were detected by autoradiography (Autorad.).

- Connect your observations of the change in NACK1 distribution with its potential function as a kinesin-like protein.
- Which of the images in Figure 14.3.2b demonstrates a specific interaction between NPK1 and NACK1?
- Develop an argument for or against the following: "The presence of yellow staining in the merged image of the GFP expressing BY2 cell is an indication of a specific association between GFP and NACK1."
- To what extent does the loss of NACK1 motor function impact cell plate formation in Figure 14.3.3a and b?

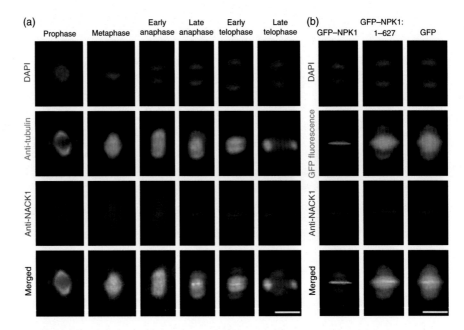

FIGURE 14.3.2 Subcellular localization of NACK1 and NPK1 in BY-2 tobacco tissue culture cells.

a. Localization of NACK1 in BY-2 cells at differing stages of mitotic cell division. Cells were triple-stained with NACK1-specific antibodies (red), α-tubulin–specific antibodies (green), and DAPI (blue). Merged images are shown in the bottom row. Overlap of the red and green fluorophores appears yellow. White indicates regions of overlap of all three fluorophores. b. Colocalization of NACK1 and NPK1 at telophase in BY-2 cells expressing GFP–NPK1, NPK1 missing its carboxy-terminal domain (GFP–NPK1:1-627), or GFP alone. Cells were stained with the NACK1-specific antibody (red) or DAPI (blue). Merged images are shown in the bottom row. Scale bar = 20 μm.

- Does NACK1 support vesicle transport along the phragmoplast MTs. Describe the evidence that supports your answer.
- Are the effects of overexpression of a mutant NACK1 the same in tissue culture and whole plant cells? Explain your answer.
- What affect did cell plate defects associated with the NACK1 mutation have on the growth of the germinated seedling? How would this change if cell plate formation was eliminated?
- Design an experiment to test whether NACK1 motor function is required for the telophase localization of NPK1 seen in Figure 14.3.2b.

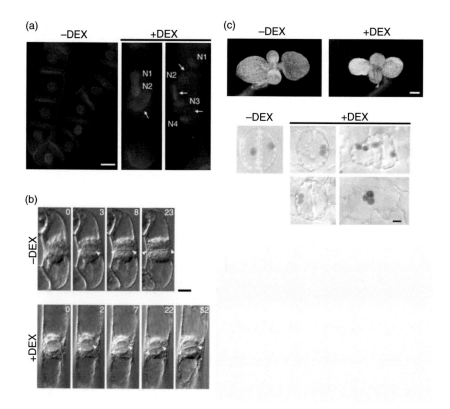

FIGURE 14.3.3 Overexpression of a truncated from of NACK1 results in defects in cytokinesis.
a. BY-2 cells transfected with a DEX-inducible construct of NACK1 missing its motor domain were cultured for 3 days in the presence (+) or absence (−) of 0.1 μM DEX. Cells were fixed and stained with a cell wall–specific stain (blue) or the DNA-specific stain propidium iodide (red). Arrows indicate incomplete cell plates. Scale bar = 50 μm. b. Time lapse micrographs of cell plate formation in living BY-2 cells described earlier cultured with or without 0.1 μM DEX for 2 days. Images were recorded at the indicated times (min) after the start of anaphase. c. *Nicotinana* seeds, transgenic for the DEX-inducible constructs shown previously, were allowed to germinate on solid media with or without DEX. Surface view of the cotyledons of 10-day-old seedlings (top; scale bar = 1 mm). Light micrograph images of guard cells in orcein-stained cotyledons (bottom; scale bar = 10 μm).

References

[1] Ferenz NP, Paul R, Fagerstrom C, Mogilner A, Wadsworth P. Dynein antagonizes Eg5 by cross-linking and sliding antiparallel microtubules. Curr. Biol. 2009;19:1833–8.

[2] Abrieu A, Kahana JA, Wood KW, Cleveland DW. CENP-E as an essential component of the mitotic checkpoint *in vitro*. Cell 2000;102:817–26.

[3] Nishihama R, Soyano T, Ishikawa M, Arakai S, et al. Expansion of the cell plate in plant cyto-kinesis requires a kinesin-like protein/MAPKKK complex. Cell 2002;109:87–99.

CHAPTER 15

Cell Systems

Do Bumblebees Have B cells? A Case of Insect Immunity [1]

INTRODUCTION

Your immune system is made up of two parts that work together to limit the damage caused when a foreign cell, virus, or other pathogen enters your body. The **innate immune system** is your first line of defense. It is made up of a class of cells called **leukocytes** (neutrophils, macrophages, and mast cells) that control infections by physically removing the invading pathogen through the process of **phagocytosis**. Should the infection overwhelm the leukocytes, a different class of immune cells, the **lymphocytes** (B cells and T cells), will become activated. Lymphocytes are part of the **adaptive immune system**.

The adaptive immune system functions differently than the innate immune system. One way the adaptive immune system is different, is the ability of the lymphocytes to produce **antibody proteins**. Antibody proteins recognize and bind with high specificity to **antigens** on the surface of foreign cells. The molecules that make up the surface features of a cell (proteins, glycoproteins, polysaccharides, etc.) are all potential antigens. Antibodies bind to specific antigens. Antibodies act as a kind of cellular "red flag," which signals other immune cells to remove the pathogen either by phagocytosis or cell-mediated killing. This means that only cells carrying that antigen will be targeted for destruction. Another important difference is the ability of the cells of the adaptive immune system to remember. The ability of the immune system to "remember" previous encounters with pathogens means that the immune response to subsequent

345

exposures can occur more quickly and more efficiently, saving the organism from damage or disease.

For humans, vaccinations (immunizations) provide protection from diseases by introducing nonlethal versions of a pathogen to an individual's immune system. Let us use the example of a vaccination against influenza virus (the flu). Your body reacts to the presence of the injected flu virus as if it were capable of causing the disease. Lymphocytes that can produce antibodies specific to the surface antigens of the virus are activated and will begin to remove the virus from your body. A subpopulation of these activated lymphocytes will not participate in the removal of the virus. Instead, these cells are stored in your body as **memory cells**. Later, if you again get exposed to the actual flu virus, these memory cells will rapidly produce the antibodies needed to quickly target and destroy the virus, protecting you from getting sick.

Although it is easy to think of this sophisticated type of immune response as an exclusive feature of higher animals like humans, the pathogens of the world do not discriminate between phyla. However, controversy exists as to whether invertebrates possess a type of adaptive immune system that provides the same specificity and long-term protection that are the hallmarks of your immune system.

- What happens to a bacterial cell after it is internalized by a macrophage through the process of phagocytosis?
- How can an antibody protein "recognize" and bind specifically to a particular antigen?
- Outline the roles and relationships between each of the following cells in the adaptive immune response: *dendritic cells, T helper cells, cytotoxic T cells, B cells, plasma cells, and memory cells.*
- Propose an explanation for why you need annual flu vaccines, but not annual polio vaccines.

BACKGROUND

Bumblebees (*Bombus terrestris*) are social insects that live in colonies. This means that bees are in constant, close proximity to other bees and are therefore exposed to any pathogens carried by other members of the colony. Pathogens known to infect bees include the Gram-negative bacterium *Pseudomonas fluorescens* and the Gram-positive bacteria *Paenibacillus alvei* and *Paenibacillus larvae*. In a high concentration, these pathogens are capable of killing the bees.

Bumblebees, like other insects, have an **open circulatory system**, which transports **hemolymph** through the bee's body, where it can come into direct contact with tissues and any invading pathogens. In addition to the oxygen-carrying molecule hemocyanin, hemolymph contains proteins and cells called

granulocytes and **plasmatocytes** that function as part of an innate immune response by encapsulating, phagocytosing, and killing pathogens. Insects are not known to produce antibodies. The following experiment sought to determine whether prior exposure to a pathogen provided specific, long-term protection in bumblebees.

- What types of antigens might be present on the surface of a bacterium?
- How would the surface features of a bacterium vary if it is Gram-negative versus Gram-positive?
- Outline the steps involved in phagocytosis of a pathogen.
- Would it be possible for an organism to have an adaptive immune response in the absence of antibodies? Explain your answer.

METHODS

Insects
All animals used in this study were of the species *B. terrestris*. Worker bumblebees were obtained from healthy colonies set up from queens collected in northwestern Switzerland in the spring of 2005. Immature workers were removed from the colony, maintained in isolation, and randomly assigned to one of the experimental groups 5 days after emergence from their pupal case. For each experiment, treatments were fully repeated in each of the colonies used. All colonies and individual bees were kept at $28 \pm 2°C$ under red light with pollen and sugar water provided *ad libitum*. All injections took place between 2 p.m. and 6 p.m. and treatments carried out on a particular day were randomly assigned.

Bacteria
P. fluorescens (DSM 50090), *P. alvei* (DSM 29), and *P. larvae* (DSM3615) were used. These bacteria were cultured in the appropriate media at $30°C$ prior to use. Immediately before use, bacterial cells were washed three times in Ringer saline solution by centrifugation. Cell concentrations were determined and adjusted to the appropriate dosage level; 5×10^4 cells/mL for initial injections and 2.5×10^6 cells/mL for lethal injections. Bacterial clearance from the hemolymph was tested using an intermediate dose of 1.5×10^6 cells/mL. Bacteria or saline solution were injected into the abdomen of worker bees that had been anesthetized by chilling.

Exposure combinations and survival
Individual bees received an initial dose of one of the three bacterial species or saline 5 days after emerging from their pupal case. At 8 or 22 days after initial exposure, the bees received a second higher dose injection, of one of three bacteria to achieve four possible exposure combinations. The "Ringer" group

contained individuals that had previously received an injection with saline. The "heterologous" group received combinations of *P. fluorescens* with either *P. alvei* or *P. larvae*. The "related heterologous" group received combinations of *P. alvei* with *P. larvae*. The "homologous" group contained individuals receiving the same bacterial species twice. After the second exposure, survival was recorded every 4 h for 7 days.

Exposure combinations and bacterial clearance

Injection combinations were carried out as described earlier, except that second bacterial challenges consisted of an intermediate dose of 1.5×10^6 cells/mL. Twenty four hours after second exposure, individuals were chilled on ice and all hemolymph was removed with a chilled glass microcapillary needle. The hemolymph was added to 100 μL of chilled bacterial media. The solution was further diluted by two sequential 10× dilutions. Ten microliters of each of the solutions, were spread onto plates containing agar media and incubated for 24 h. Bacterial colonies were counted and the number of bacteria per bee, was calculated taking into account the initial volume of hemolymph and the dilution factors. It was assumed that each differentiated bacterial colony originated from a single cell.

- Why were immature worker bumblebees maintained in isolation?
- What variables are being addressed in the description of the insects provided in the "Methods" section?
- Explain the difference between heterologous and related heterologous.
- Create a table/matrix outlining all the possible exposure combinations for the Ringer, heterologous, related heterologous, and homologous groups.
- Suggest a reason for using an intermediate dose of bacteria in the hemolymph clearance experiments.
- Assume 15 μL of hemolymph is collected from a bee and placed into 100 μL of bacterial media (tube 1). What volume would you need to transfer from tube 1 to tube 2 in order to make a 1:10 dilution if tube 2 contains 90 μL?
- You transfer 10 μL of media from tube 2 onto an agar plate. After a 24 h incubation you count 72 colonies on the plate. Calculate the original concentration of bacteria in the hemolymph sample in units of cells per milliliter.
- The "Methods" section of this chapter does not indicate the temperature used to incubate the hemolymph dilution plates. Based on your knowledge, what temperature would you guess would be appropriate. Justify your answer.
- Hemolymph is known to contain granulocytes and other eukaryotic cells. Predict what conditions would be required in order to grow insect cells in culture.

RESULTS

- Which of the bacterial strains shown in Table 15.1.1 are the most lethal for bumblebees?
- Interpret the data for the effect of a 5×10^4 cells/mL injection of *P. fluorescens* on the survival of a worker bumblebee.
- What biological process could account for the 0 values in the final column of Table 15.1.1?
- Which of the treatment groups had the highest mortality 8 days after initial exposure?
- Which of the treatment groups demonstrates the best survival 8 days after initial exposure?
- What was the purpose of injecting some individuals with Ringer saline in the first injection?
- Summarize how worker bumblebee survival changes with a 22 days lag between the first and second bacterial exposures (Figure 15.1.1b).
- Estimate the proportion of bees that survived 48 h following injection with the lethal dose (2.5×10^6 cells/mL) of homologous bacteria 8 and 22 days after initial, sublethal exposure. Compare your estimates with the values reported in Table 15.1.1 for single injections of each of the bacterial strains used in this case study. What can you conclude from the comparison?
- Explain what is meant by "clear hemolymph" in Figure 15.1.2.

Table 15.1.1 Preliminary Results of Pathogen Exposure to *Bombus terrestris* Workers

Injection Type	Dose of 2 μL Injection (Cells/mL)	Proportion of Bees Surviving 24 h (±95% Confidence Interval)	Proportion of Bees Surviving 48 h (±95% Confidence Interval)	Mean Counts of Bacteria in the Hemolymph 48 h after Injection ($n = 12$)
Ringer saline	NA	1.0	1.0	0
Bacillus thuringiensis	5×10^4	0.33 ± 0.13	0.11 ± 0.09	–
	2.5×10	0.11 ± 0.09	0	–
Serratia marcescens	5×10^4	0	0	–
	2.5×10^6	0	0	–
P. fluorescens	5×104	1.0	1.0	0
	2.5×10^6	0.91 ± 0.06	0	–
Paenibacillus alvei	5×10^4	1.0	1.0	0
	2.5×10^6	0.88 ± 0.09	0	–
P. larvae	5×10^4	1.0	1.0	0
	2.5×10^6	0.88 ± 0.09	0	–

A "low" dose is 5×10^4 cells/mL and a "high" dose is 2.5×10^6 cells/mL. If all bees survived to 48 h, hemolymph was checked to determine if the bacteria had been cleared. The results obtained determined the bacterial species used in the main study.

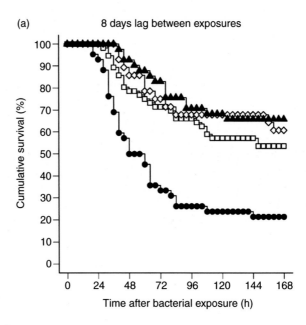

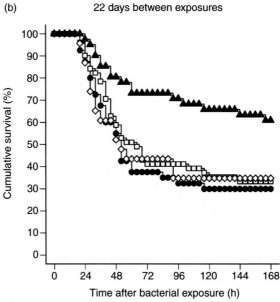

FIGURE 15.1.1 Survival of worker bumblebees after secondary exposure to bacteria.
Survival curves of workers after exposure to lethal bacterial doses either 8 (a) or 22 (b) days following an initial sublethal injection. Groups refer to the initial injection and its relatedness to secondary bacterial exposure: individuals receiving an initial injection of Ringer saline (circle, $n = 42$ for 8 days lag, $n = 40$ for 22 days lag) or individuals previously injected with a heterologous (square, $n = 56$, $n = 51$), related heterologous (diamond, $n = 28$, $n = 23$), or homologous (triangle, $n = 41$, $n = 41$) bacterial strain.

(a)

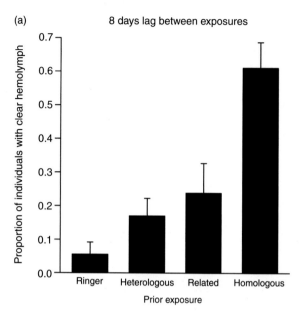

8 days lag between exposures

(b)

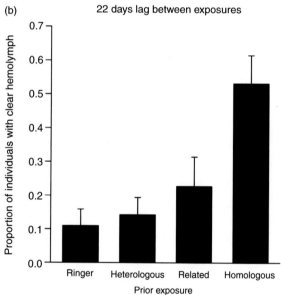

22 days lag between exposures

FIGURE 15.1.2 Clearance of bacterial cells 24 h after injection into the hemolymph of worker bumblebees.

Worker bumblebees were injected with an intermediate dose (1.5×10^6 cells/mL) of bacteria 8 days (a) or 22 days (b) following injection with Ringer saline or a sublethal concentration of bacteria. Groups refer to the relatedness of the two exposures. Worker bumblebees were injected with Ringer saline (Ringer; $n = 36$ for 8 days lag, $n = 36$ for 22 days lag), heterologous ($n = 47$, $n = 42$), related heterologous ($n = 21$, $n = 21$), or homologous ($n = 36$, $n = 32$) bacterial strain. Error bars refer to 95% confidence intervals for binomial distributions.

- Binomial distribution refers to an experiment where there are only two possible outcomes. Describe what the two outcomes are for the experiment in Figure 15.1.2.
- Summarize the conclusion supported by the data in Figure 15.1.2.
- Connect the data in Figure 15.1.2 to the data in Figure 15.1.1. How are these two experiments related?
- Why is Figure 15.1.2, a bar graph, while Figure 15.1.1 is a line graph?
- Suggest a reason the Ringer population of worker bumblebees had such a limited clearing of their hemolymph.
- Propose an explanation for why the ability of the bumblebee to "clear" a related pathogen (related heterologous) would be better than its ability to clear an unrelated pathogen (heterologous), but not as good as the response to the same pathogen (homologous)?
- The normal lifetime of a worker bumblebee is about 28 days. Based on your interpretation of the data, does prior exposure to a pathogen provide long-term and specific immune protection for bumblebees?
- Develop a hypothesis to explain the biological basis of long-term protection against pathogens observed in bumblebees.
- Propose an experiment that would test your hypothesis.

What Happens When the Endosymbionts "Bug Out"? [2]

INTRODUCTION

Endosymbionts are organisms that form a symbiotic relationship with another cell or organism. Some endosymbionts can be found either inside cells (intracellular), while others attach to the surface of cells (extracellular). Symbiotic relationships are ones in which both parties benefit. In the example of corals, the host animal is a multicellular cnidarian, and the endosymbiont is a unicellular alga called a dinoflagellate. The host animal serves as a protected environment, providing the alga with a constant supply of nutrients in the form of nitrogen, phosphorus, and sulfur. In turn, the alga provides the host with glucose. Another important example of a beneficial relationship between a host and an endosymbiont is found in the roots of bean and pea plants. Bacteria called rhizobia colonize the root cells of plants forming nodules. Rhizobia bacteria convert atmospheric nitrogen into ammonia that the plant can use to build biological molecules. The conversion of nitrogen gas into ammonium, requires large amounts of ATP. Plants supply the bacteria with carbohydrates to help drive the reaction.

- Explain how an alga endosymbiont could provide a host organism with glucose.
- Describe how a symbiotic relationship might alter an organism's phenotype.
- What biological molecules require nitrogen?
- Connect the activity of nitrogen-fixing rhizobia bacteria to the global nitrogen cycle.
- The *endosymbiotic theory* explains the origin of mitochondria and chloroplasts. Speculate on the interactions between a host protoeukaryote and a protomitochondria or protochloroplast endosymbiont.

BACKGROUND

Weevils are insects that feed on grains and other plant material. Like other insects, the life cycle of the weevil starts with an egg that is laid inside a grain or seed. When the larva hatches from the egg, it begins to feed on the material stored within the grain. After its pupal stage, the adult form of the insect emerges from the grain and begins the cycle again by mating and laying eggs.

Weevils are considered to be serious agricultural pests because of the damage that they can cause to stored grain and seeds. A single female can lay up to 300 eggs at a time, depositing each individual egg into its own grain. Weevils can also be household pests, showing up in flour and cereals.

The survival of an adult weevil is dependent, in part, on its exoskeleton or **cuticle**. The cuticle is composed of a mix of chitin and proteins that are organized into three layers, the **epicuticle**, the **exocuticle**, and the **endocuticle**. The cuticle goes through a hardening process known as *sclerotization* and *melanization* that requires the presence of the compound dihydroxyphenylalanine (**DOPA**). DOPA is synthesized from the amino acid tyrosine, which in turn, is synthesized from the amino acid phenylalanine. Phenylalanine is an essential amino acid.

The midgut of a weevil contains small pouches called **caeca** that function to increase the surface area of the gut, enhancing digestion and absorption of nutrients. At the tip of each caeca is a **bacteriome**, a specialized organ made up of cells called bacteriocytes, which protect endosymbiotic bacteria from exposure to the host's immune system. Bacteriocytes not only house the endosymbionts in their cytoplasm, but they also provide nutrients required to support bacterial growth. In turn, gut endosymbionts are thought to aid in the digestion of grain and plant material consumed by the weevil. Bacteriomes containing endosymbionts are also found in reproductive tissues of the female weevil. While the bacteriomes associated with eggs and ovaries are present throughout the life of the weevil, those associated with the caeca are lost soon after the weevil completes its final adult metamorphosis. The ability of these insects to modulate the population of gut endosymbionts and the effect that it has on the phenotype of the insect is explored in this case study.

- What is an "essential amino acid?" What is the implication for the weevil?
- Create a diagram to illustrate the relationship between the following terms: caeca, endosymbiont, bacteriocyte, bacteriome, and gut.
- Use your diagram to propose a model for how the bacterial endosymbiont might support digestion in the weevil.
- Explain the significance of the observation that endosymbionts are found within the reproductive tissues of the weevil.
- Attempts to culture the bacterial endosymbiont outside the weevil have not been successful. Discuss the implications of this finding.

METHODS

Insect rearing

Cereal weevils, *Sitophilus oryzae,* were grown on a diet of wheat grain at 27.5°C and 70% relative humidity. *Sitophilus* is naturally infected with the bacterial endosymbiont *Sodalis pierantonius*. A strain of aposymbiotic *S. oryzae,* was also used in this study. Weevil larvae and nymphs grow naturally inside wheat grains. After completing their metamorphosis with an ultimate insect molt

(UIM), the adults remain inside the grains for three more days before emerging. Adult lifespan is approximately 6 months.

For some experiments the weevils were fed starch pellets made using 5 g of rice starch in 250 μL of water. The rice starch mixture was placed into a stick-shaped mold and dried overnight at 37°C and then broken into pellets. Supplemented starch pellets were made using the same process but with the addition of 50 mg of either tryptophan, tyrosine, phenylalanine, DOPA, or a combination of the amino acids to the starch mixture prior to drying.

S. pierantonius DNA quantification with qPCR

DNA extraction from individual larvae, nymphs, adults, or ovaries was accomplished using the Genomic DNA from Tissue kit (Macherey–Nagel). The presence of *S. pierantonius* DNA was quantified by amplifying the sequence for the *S. pierantonius*-specific gene, *nuoCD*. Real-time PCRs were normalized to the host's β actin gene. The technique of real-time PCR uses fluorescent probes to quantify the amount of double-stranded DNA. While standard PCRs only look at the amplified product at the end of a reaction, real-time PCR measures the formation of that product over time.

Fluorescence microscopy

Several fluorescent probes were used in this study. **Fluorescence *in situ* hybridization (FISH)** was used to detect *S. pierantonius* DNA in whole samples of gut. Guts were dissected in buffer, permeabilized using 70% acetic acid at 60°C for 1 min, and rinsed with buffer. Samples were incubated with a fluorescently labeled probe that is specific to *S. peirantonius* and a 3 μg/mL DAPI. **YO-PRO** staining was used to detect apoptotic cells. Guts were dissected as before but incubated for 45 min in 1 mL of ice-cold buffer containing 1 μL of YO-PRO (100 μM in DMSO) and 1 μL of Hoechst 33342 (8.1 mM in water). Paraffin sections of the gut were prepared for incubation with an antibody specific to the activated form of the enzyme caspase 3. Caspase 3 plays an important role in triggering apoptosis in cells. Following incubation with the primary antibody, the sections were incubated with fluorescently labeled secondary antibody.

Electron microcopy

Guts were dissected as described earlier and fixed in a solution of 3% glutaraldehyde in buffer. Samples were washed in buffer and prepared either for embedding in epoxy resin for **transmission electron microscopy (TEM)** or mounted and fractured on holders and sputter-coated for use with **scanning electron microscopy (SEM)**. Measurements were made using image analysis software.

Cuticle coloration

Cuticle color of the thorax of adult weevils, was determined from images captured with a digital camera. All pictures were taken under the same conditions of lighting. The digital mean value of red was calculated using image analysis software. Digital red values range from 0 (black) to 255 (white). Insect cuticles darken during adult development causing the red values to decrease over time. For each day tested the thoracic colors of 10 adult weevils were measured. Values are reported as the inverse of the red values to illustrate the darkening process as a positive change.

- Find the definition of *aposymbiotic*.
- Convert the adult life span of a weevil into units of days.
- What is the chemical composition of starch? Discuss how the chemistry of a starch pellet compares with biological molecules you would find in a grain of wheat.
- Access the protocol instructions for the Genomic DNA from Tissue kit from Macherey–Nagel. Use the information on the Protocol-at-a-glance page to create a flow chart of how DNA is isolated.
- What concentration of YO-PRO stain was used in these experiments?
- Use the following data set to generate a graph of cuticle color change versus developmental stage.

Developmental Stage	Red Value
D0	0.02
D5	0.016
D10	0.014
D15	0.013

RESULTS

- Explain the logic of the experiment shown in Figure 15.2.1b.
- How does the pattern of endosymbiont population growth differ between the two samples shown in Figure 15.2.1b?
- At what point in development does the weevil begin to lose gut endosymbionts?
- Is the loss of gut endosymbionts the same for male and female weevils?
- Explain why FISH can be used as a specific stain for the presence of *S pierantonius* (Figure 15.2.1c).
- Use the information found in the images of Figure 15.2.1d to estimate the size of a caeca alone and the size of the apical bacteriome.
- Write a paragraph, using your own words, that synthesizes the data presented in Figure 15.2.1.
- Compare and contrast the cellular processes and outcomes of *apoptosis* and *autophagy*.
- Categorize the advantages associated with the elimination of endosymbionts through autophagy and apoptosis. Is one pathway "better" than another?
- Relate the data presented in Figure 15.2.2a–d to the information you summarized for Figure 15.2.1.

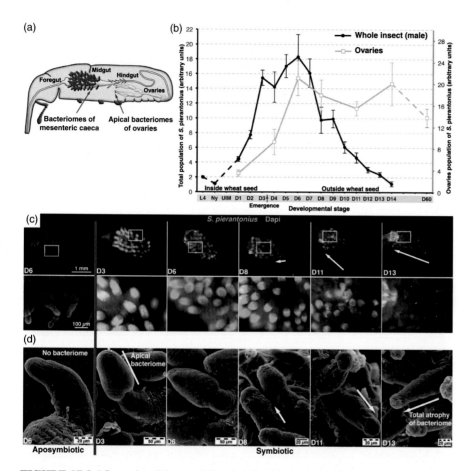

FIGURE 15.2.1 Dynamics of the population density of the endosymbiont *Sodalis pierantonius* in the weevil *S. oryzae*.

a. Diagram of the internal anatomy of an adult weevil showing the location of bacteriomes (red) in the caeca of the midgut and the ovaries. b. PCR quantification of endosymbiont DNA in a developing weevil from larval (L4), nymph (Ny), ultimate insect molt (UIM), and adult stages. Analysis was completed using isolated ovarian tissue (ovaries) or male weevils (whole insect). Data points represent the mean (±SEM) of five independent replicates. c. FISH staining of endosymbionts in the caecae from day 3 to day 13. *S. pierantonius* (green) progressively disappears from the posterior to the apical end of the midgut (orientation indicated by arrows). Caecae remain morphologically similar in aposymbiotic insects (far left panel). Boxes in the upper panels indicate area shown at high magnification in the lower panels. Nuclei were stained with DAPI and appear blue. d. SEM of adult caeca. Bacteriomes are present at the apical tip of the caeca in symbiotic insects at day 3 (D3), but are gradually lost from day 6 (D6) to day 13 (D13). Bacteriomes degenerate from the base of the caeca to the apical tip (see arrows). Bacteriomes are not present on the caecae of aposymbiotic insects.

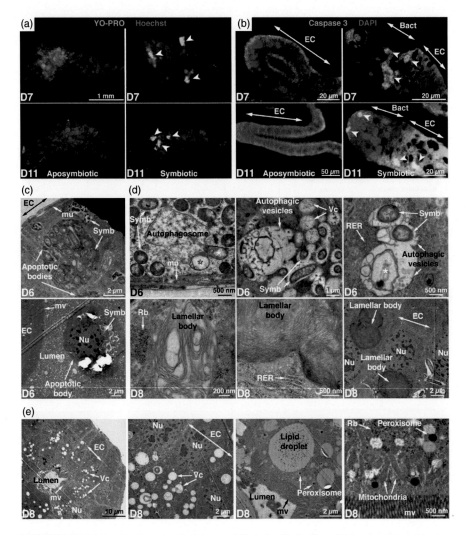

FIGURE 15.2.2 Apoptosis and autophagy contribute to elimination of the gut endosymbiont population.

a. YO-PRO staining of whole guts in symbiotic adults (right-hand panels) revealed apoptotic cells (green, arrowheads) in the posterior caeca at day 7 (top, arrowheads) and in the anterior caeca at day 11 (bottom, arrowheads). No apoptosis was detected in aposymbiotic insects (left-hand panels). Nuclei are stained blue (Hoechst). b. Activated caspase-3 immunostaining (green) of gut sections indicates apoptosis at the base of the bacteriome (Bact) in symbiotic adults at day 7 (top right panel, arrowheads) and in fewer cells, mainly located at the apical tip of the caeca, at day 11 (bottom right panel, arrowheads). No signal was detected in the epithelial cells (EC) of the aposymbiotic insects (left-hand panels). Nuclei are stained blue by DAPI. c. TEM images of symbiotic insects at day 6. Apoptotic bodies are seen in epithelial cells (top) and in the gut lumen (bottom). d. TEM images of symbiotic insects at day 8. Autophagy is observed in bacteriocytes and epithelial cells. Autophagosomes (*) are capable of encapsulating and fusing with the *S. pierantonius* endosymbionts in a bacteriocyte (top). Autophagy leads to the accumulation of lamellar bodies covering the majority of the cell's cytoplasm (bottom). e. TEM of gut epithelium from aposymbiotic adult insects at day 8. The cytoplasm of these cells is free of endosymbionts and contains a dense vesicular network including lipid droplets and peroxysomes, but lacking lamellar bodies. EC, epithelial cell; Vc, vesicle; Symb, symbiont; Nu, nucleus; mu, muscle; mv, microvilli; Lumen, gut lumen; RER, rough endoplasmic reticulum; Rb, ribosomes.

- Create a labeled drawing that illustrates the structure and organization of a generic gut epithelial cell. Study the TEM images of aposymbiotic gut cells in Figure 15.2.2e. Identify similarities and differences with your drawing.
- Which of the experiments (lines) in Figure 15.2.3 represent a control condition?
- How do the data for insect survival compare with the expected lifespan of a weevil?
- Can starch alone support the survival of a weevil? Justify your answer using the data in Figure 15.2.3.
- Develop an argument to support or refute the following statement: "The presence of endosymbionts creates a physiological burden that impacts host survival."

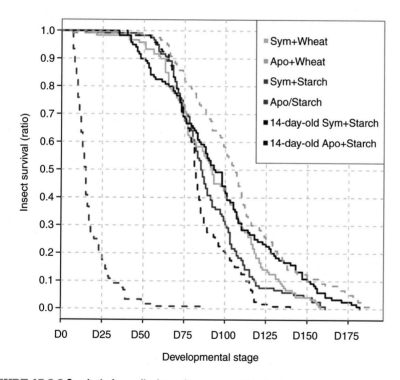

FIGURE 15.2.3 Survival of weevils depends upon a combination of food source and endosymbiont.
Percent survival of symbiotic (solid blue lines) and aposymbiotic (red dashed lines) on wheat grains (pale colors) or starch pellets (bright colors). Weevils were either left on wheat grains or transferred to starch pellets immediately after emerging from grain. Alternatively, symbiotic and aposymbiotic weevils were left for 14 days on wheat grains before being transferred to starch pellets.

■ Propose an explanation for the difference in survival in weevils switched immediately to starch compared with those switched to starch pellets after 14 days on wheat grains.

■ Analyze the structure of the weevil exoskeleton (Figure 15.2.4b). Describe the similarities and differences you observe between symbiotic and aposymbiotic animals.

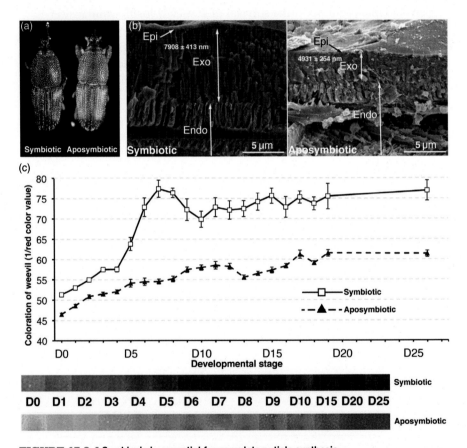

FIGURE 15.2.4 Symbiosis is essential for complete cuticle synthesis.

a. The cuticle of a 26-day-old symbiotic weevil (left) is darker than that of an aposymbiotic insect (right). b. SEM of cuticle sections from symbiotic (left) and aposymbiotic (right) 15-day-old adults. The cuticle consists of three layers; the epicuticle (epi), the exocuticle (exo, double arrow), and the endocuticle (endo). The thickness of the exocuticle layer represents the average (±SEM) of 11 measurements. c. Progression of cuticle colors in adult weevils. The inverse of a digital red color value was used to quantify changes in cuticle color for symbiotic (solid line) and aposymbiotic (dashed line) adults. Corresponding colors are indicated further the graph. Coloration of the cuticle is an indicator of sclerotization and melanization, processes that involve biosynthesis, and hardening of the cuticle.

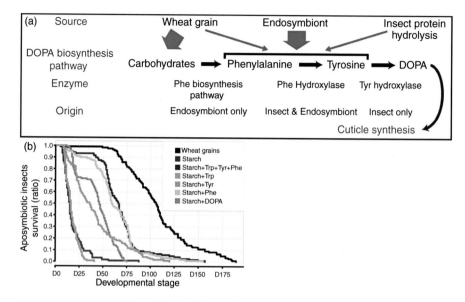

FIGURE 15.2.5 DOPA is a critical nutrient during early adulthood for a weevil.
a. Summary of the possible sources of the amino acids phenylalanine (Phe) and tyrosine (Tyr) used by the weevil and the DOPA synthesis pathway. Phenylalanine is an essential amino acid. However, insects are capable of converting Phe to Tyr and Tyr into DOPA. b. Survival of male aposymbiotic weevils on wheat grains (black), starch pellets (red), starch pellets supplemented with the amino acid tryptophan (Trp, green), Phe (orange), Tyr (blue), all three amino acids (Trp + Tyr + Phe, purple), or DOPA (pink).

- Calculate the approximate rate of cuticle biosynthesis from day 0 to day 6 for symbiotic and aposymbiotic insects using the data in Figure 15.2.4c.
- Is cuticle biosynthesis and sclerotization a predictor of insect survival? Justify your answer.
- Develop a series of predictions about insect survival based on the biochemical pathway illustrated in Figure 15.2.5a.
- Explain why male aposymbiotic weevils can survive by feeding on wheat grains, but not on starch-only particles.
- Summarize the conclusion supported by the data shown in Figure 15.2.5b.
- Predict whether the results of the experiment shown in Figure 15.2.5b would differ if an aposymbiotic female weevil were used.
- Generate a hypothesis relating the presence of endosymbionts, the pattern of cuticle biosynthesis, and insect survival.

Parvovirus: Hijacking Endocytosis [3]

INTRODUCTION

Viruses are obligate intracellular parasites that rely on the ability of a host cell to replicate genetic information and synthesize proteins in order for the virus to reproduce. Viruses infect all types of cells: prokaryotic, eukaryotic, animal and plant. All viruses contain a viral genome, although the nature of the nucleic acid is quite variable. All viruses enclose their genetic information inside a protein shell known as a **capsid**. Some, but not all, viruses surround the capsid with a phospholipid bilayer "borrowed" from the previous host cell. Viruses with an outer membrane are **enveloped viruses** while viruses that only have capsids are **nonenveloped viruses**.

The life cycle of a virus starts with attachment and entry of the virus into a host cell. Viruses take advantage of proteins or other molecules that are normally present on the surface of the cell as points of attachment. Once a virus has attached to the surface of a cell, the viral capsid can gain entry into the cell's cytoplasm either through membrane fusion, in the case of enveloped viruses, or through **endocytosis**, in the case of nonenveloped viruses. The capsid delivers viral genetic information into the cell where it can be replicated, transcribed, and translated. Production of viral proteins occurs in the same manner as translation of the host cell's own proteins. Newly synthesized capsid and other viral proteins assemble with newly replicated viral genetic information, forming new viral particles which then exit the cell, spread, and infect new host cells.

- Conduct a search for each of the viruses listed below. Identify the type of genetic information each contains, its host organism, and whether it is enveloped or nonenveloped.
    - Hepatitis B virus
    - Tobacco mosaic virus
    - Lambda (λ) phage
    - Avian flu virus
- Access the Center for Disease Control website to learn about current and past viral outbreaks (http://www.cdc.gov/outbreaks/).
- Discuss the evolutionary implications of the ability of a virus to reproduce using the biosynthetic machinery of a cell.

BACKGROUND

Canine parvovirus (CPV) is a single-stranded, negative-sense DNA, nonenveloped virus that infects dogs causing loss of appetite, fever, and dehydration leading, in some cases, to death. "Parvo" is highly contagious, but can be prevented by vaccination.

The CPV capsid consists of 60 protein subunits that assemble to form a 26 nm icosahedron. Approximately 90% of the capsid is made up of a 67 kDa protein known as viral protein 2 (**VP2**) with the remaining 10% being composed of the 83 kDa viral protein 1 (**VP1**). The amino acid sequence of VP1 is almost identical to that of VP2 with the exception of a 143 amino acid sequence at the amino-terminal domain. This unique amino-terminal sequence shares some homologies with a **nuclear localization signal (NLS)** and **phospholipase A2**, an enzyme that catalyzes the breakdown of phospholipids. Protein folding of VP1 sequesters the amino-terminal domain; however, when exposed to high temperatures or acidic pH, VP1 changes its conformation, exposing the amino-terminus.

CPV gains entry into cells through an attachment to **transferrin receptors (TfRs)**. TfRs function to bind and transport the iron-binding protein transferrin into the cell. TfRs are brought into the cell through the process of **receptor-mediated endocytosis**. Following internalization from the cell surface, TfRs are transported through the endocytic pathway, eventually recycling back to the cell surface. Viruses that enter a cell through endocytosis must be able to exit out of the membrane compartments of the endocytic pathway and enter the cytoplasm of the cell in order for them to gain access to the cellular components they need to reproduce.

- What is the relationship between a "negative-sense" DNA sequence and the sequence of the mRNA required to synthesize viral proteins?
- Outline the steps involved in viral protein synthesis starting with a viral mRNA and ending with a capsid protein.
- Explain how exposure to high temperatures or acidic pH would alter protein folding.
- Describe how a NLS functions.
- Research/review the steps involved in TfR-mediated endocytosis. Follow the receptor from the cell surface, into the cell, and back to the cell surface.

METHODS

Cells and viruses

Norden Laboratories feline kidney (NLFK) cells were grown in culture. Canine parvovirus type 2 (CPV-d) was grown in NLFK cells in 175 cm^2 cell culture flasks for 5–7 days then stored at $-20\,^{\circ}$C. To harvest viruses, 300 mL of thawed culture medium was centrifuged at 3000g for 30 min and the supernatant was concentrated with a 500 kDa filter. Viruses were pelleted by centrifugation at 173,000g for 2 h and resuspended in a saline buffer.

To generate heat-treated CPV, capsids were incubated for 2 min on a 65 $^{\circ}$C heat block. Samples were allowed to gradually cool to room temperature before being used for infection studies.

Lysosomal drugs

Various compounds known to influence endocytosis and lysosomal function were used in these studies. Bafilomycin (BFLA) is an antibiotic that is known to block the activity of H^+-ATPases. Similarly, amiloride (AMI) blocks sodium channels and the activity of sodium/hydrogen antiporters. Monensin (MON) is an ionophore and can disrupt protein transport in the cell. Finally, Brefeldin A (BFA) inhibits protein transport from the rough endoplasmic reticulum to the Golgi complex. Concentrations used of each compound are indicated in the experiments.

Microscopy

Immunofluorescence microscopy studies were performed on 80% confluent NLFK cells grown on coverslips. Cells were inoculated with CPV particles in the absence or presence of lysosome-inhibiting drugs (see further sections). Drugs were added to the cell 30 min prior to infection and maintained until fixation. Cells were fixed at different post infection times. Fixed cells were permeabilized with 1% Triton X-100 for 15 min and then incubated with primary antibody for 45 min at room temperature. Samples were rinsed well before addition of a fluorescently labeled secondary antibody. CPV was imaged using a polyclonal antibody to VP1 and VP2 capsid proteins. The amino-terminus of VP1 was visualized using a monoclonal antibody specific to that domain of the protein. Monoclonal antibodies to lysosomal membrane protein-2 (LAMP-2) and TfR were also used in these studies.

Immunoelectron microscopy was used to identify the subcellular localization of CPV capsid protein. Cells were incubated in the presence or absence of lysosome-inhibiting drugs 30 min prior to infection with CPV particles. Cells were incubated for 20 h then fixed and permeabilized using a combination of 0.01% saponin and 0.05% Triton X-100. Cells were incubated with primary antibody to CPV capsid protein for 1 h, washed, and then incubated with gold-labeled secondary antibody. Cells were washed again and postfixed in preparation for the dehydration, infiltration, and embedding steps required for electron microscopy. Embedded samples were cut into 50 nm sections and viewed using a transmission electron microscope.

Assay of CPV infectivity

Infected cells were incubated for 20 h, after which the number of fluorescent nuclei were determined using a monoclonal antibody to CPV. The level of infectivity was quantified as the ratio of cells with fluorescent nuclei to total number of fluorescent cells ($n = 300$). Infectivity was assayed for native and heat-treated CPV as well as heat-treated CPV in the presence of lysosome-inhibiting drugs.

Dextran particle release assay

NLFK cells were grown on coverslips as described earlier. Cells were exposed to 3 mg/mL rhodamine-conjugated dextran (M_r 3000 or 10,000) for 15 min in the presence or absence of lysosome-inhibiting drugs. The dextran-containing media was removed and replaced with fresh media without dextran or CPV, but containing the same concentration of drugs found in the original solution. Cells were maintained for 8 or 20 h post infection. Controls were made without the addition of CPV. Cells were fixed and imaged using a fluorescence microscope. On its own, dextran cannot easily pass through the membranes of the endosome or lysosome. The percentage of cells showing cytoplasmic and/or nuclear fluorescence was calculated based on approximately 1500 cells/sample.

- What conclusion(s) can you draw from the fact that CPV can infect feline cells?
- Connect the life cycle of a virus with the ability of the researchers to harvest parvovirus from tissue culture media.
- Relate how the mode of action of each of the drugs used in the study would ultimately influence lysosomal function.
- Why is it necessary to permeabilize cells for the immunofluorescence and immunoelectron microscopy protocols, but not for the dextran protocol?

RESULTS

- What is the earliest time point during endocytosis at which the N-terminus of VP1 becomes accessible to antibody binding (Figure 15.3.1a).
- Summarize the information provided in Figure 15.3.1a. What additional information would have improved this figure?
- What effect did the various drug treatments have on the appearance of the VP1 N-terminus?
- The drug BFLA functions by inhibiting the activity of H^+-ATPases. Why would this drug influence lysosomal function? How might inhibiting these pump proteins affect the exposure of the VP1 N-terminal region?

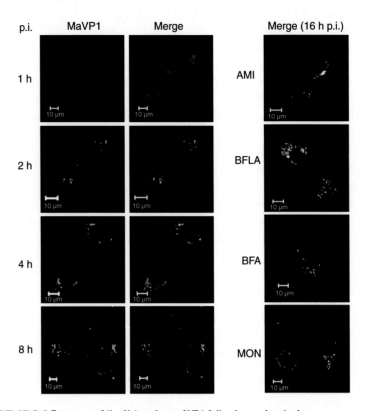

FIGURE 15.3.1 Exposure of the N-terminus of VP1 following endocytosis.
a. Double-labeled confocal immunofluorescence of CPV capsids (red) and VP1 N-terminus (MaVP1; green) in NLFK tissue culture cells. Time indicates hours post infection (p.i.). Appearance of yellow in the merged images (right-hand column) reflect colocalization of CVP capsid with VP1 proteins with exposed N-terminal domains. b. CPV were chased for 16 h post infection in the presence of drugs that impact lysosomal function (AMI, 0.4 mM; BFLA, 500 nM; BFA, 20 nM; MON, 500 nM) before double-labeling as described in (a). Only merged images are shown. Scale bar = 10 μm.

- Develop arguments for *and* against the design of the experiments shown in Figure 15.3.1b.
- Describe the distribution of CPV 20 h after infection in a control cell (Figure 15.3.2e).
- How does the distribution of CPV in Figure 15.3.2a–d compare with the control image in Figure 15.3.2e?
- What effect might exposure to the environment of a lysosome have on VP1?
- The researchers conclude that the various drug treatments investigated in the experiment shown in Figure 15.3.2 "inhibit the escape of CPV from lysosomes. However, they did not prevent binding of CPV to

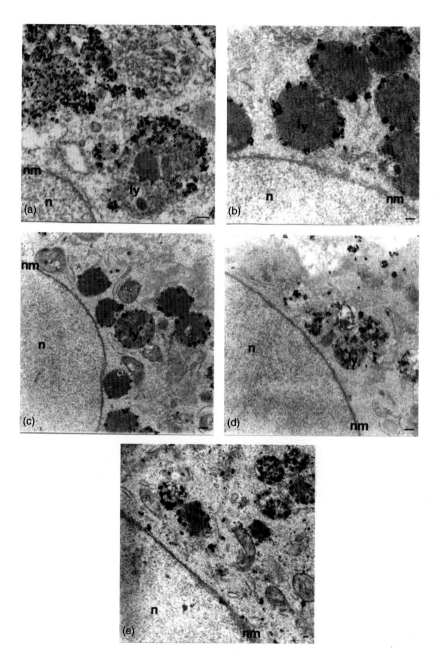

FIGURE 15.3.2 Immunoelectron microscopy of CPV endocytosed by NLFK cells in the presence of various inhibitory compounds.

Distribution of CPV in tissue culture cells in the presence (a–d) or absence (e) of drugs that can alter lysosomal function. Gold particles indicate the location of CPV capsid proteins 20 h post infection. a. AMI (0.4 mM) treated cell. b. BFLA (20 nM) treated cell. c. BFA (20 nM) treated cell. d. MON (500 nM) treated cell. e. Control infected cell. Nucleus (n), nuclear membrane (nm), and lysosome (ly) are indicated in the images. Scale bar = 100 nm.

the lysosomal membrane." Discuss whether the data presented in Figure 15.3.2 support these conclusions.

■ Refer back to your outline of the endocytic pathway of TrF. How did CPV end up in a lysosome if it is entering the cell attached to that receptor?

■ How would incubation at 65°C influence the VP1 protein?

■ Why was nuclear fluorescence used in the experiment shown in Figure 15.3.3 (see the properties of VP1 described in the Background)?

■ Did heating CPV prior to infection provide an advantage to the virus in the experiment shown in Figure 15.3.3?

■ Summarize how treatment with the various drugs tested in Figure 15.3.3 influenced the infectivity of CPV?

■ Explain how the results presented in Figure 15.3.3 relate to the results in Figure 15.3.2?

■ Describe the results presented in Figure 15.3.4a in context with what was already known about the mechanism of CPV infection.

■ What does Figure 15.3.4b add to your previous description?

■ Can you conclude that the fluorescent structures in Figure 15.3.4a are lysosomes? Justify your answer.

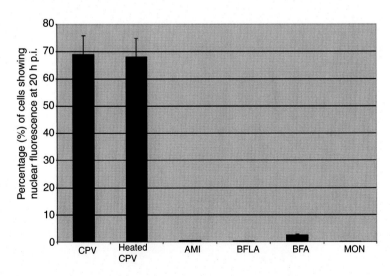

FIGURE 15.3.3 Infectivity of heat-treated CPV in the presence of various inhibitory compounds.
CPV capsids were pretreated by exposure to 65°C. Cells were infected with heat-treated CPV in the absence (heated CPV) or presence of various drugs that impact lysosomal function (0.4 mM AMI, 500 nM BFLA, 20 nM BFA, and 500 nM MON). Cells were fixed 20 h post infection and immunolabeled with VP1 N-terminal-specific antibody. Infectivity was determined as the proportion of cells with a fluorescent nucleus relative to the total number of fluorescent cells ($n = 300$).

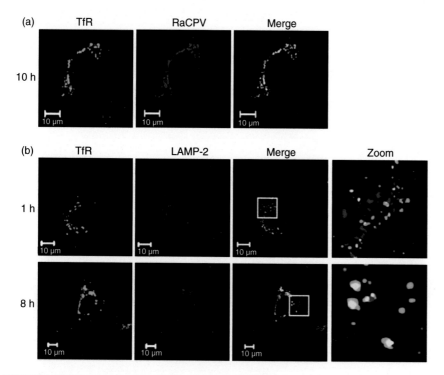

FIGURE 15.3.4 Colocalization of CPV with TrFs in NLFK cells.
a. Double-labeled confocal immunofluorescence microscopy of CPV (RaCPV, red) and transferrin receptor (TfR, green) 10 h post infection. Merged image reveals areas of colocalization. b. Distribution of transferrin receptors (TfR, green) and lysosomal membrane protein-2 (LAMP-2, red) in cells 1 and 8 h post CPV infection. Areas marked by white squares are shown at high magnification in the column labeled "ZOOM." Scale bars = 10 μm.

■ Is there a difference in the distribution of different sizes of dextran in the noninfected cells in Figure 15.3.5a?

■ How does the distribution of the 3000 and 10,000 dextrans compare with the distribution of CPV 20 h post infection?

■ Provide an explanation of how dextran molecules are entering the cell. What is meant by the term "coendocytosis?"

■ Use the data in Figure 15.3.5 to support the conclusion that CPV makes lysosomal membranes permeable to small molecules.

■ Did the presence of lysosome-inhibiting drugs alter the ability of the 3000 dextran to exit the lysosome? Justify your answer.

■ The amino-terminus of VP1 shares sequence homology with the enzyme phospholipase 2A. Propose a model for the results presented in Figure 15.3.5 that incorporates this fact.

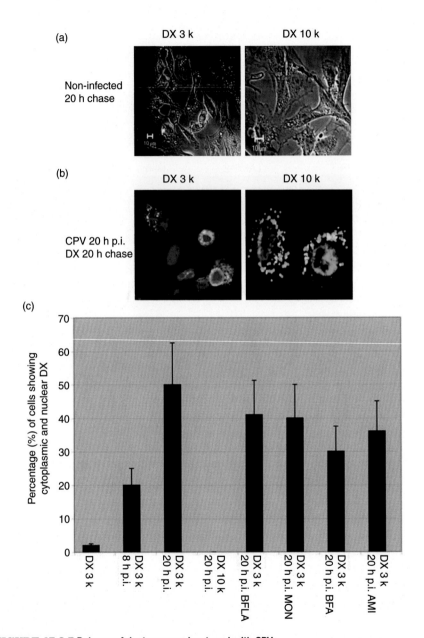

FIGURE 15.3.5 Release of dextran coendocytosed with CPV.
a. Dextrans sized 3000 kDa (DX 3k) and 10,000 kDa (DX 10k) were endocytosed by noninfected cells and chased for 20 h. b. Double-labeling confocal immunofluorescence staining of CPV (green) and dextran (red) 20 h post infection/endocytosis. Yellow indicates areas of colocalization. c. Quantitative analysis of cells showing cytoplasmic and nuclear dextran staining in the absence and presence of lysosome-inhibiting drugs.

- Does your model answer the question of how CPV moves from the lysosome into the cytoplasm to cause infection? Why or why not?
- Develop an argument in favor or against the following statement, "A cell is not infected if the virus cannot replicate."
- It is common medical practice NOT to administer antibiotics in response to viral infection. The drugs BFLA and MON are characterized as antibiotics. Discuss whether administration of these drugs should be considered for treatment of CPV.

References

[1] Sadd BM, Schmid-Hempel P. Insect immunity shows specificity in protection upon secondary pathogen exposure. Curr Biol. 2006;16:1206–10.

[2] Vigneron A, Masson F, Vallier A, Balmand S, Rey M, Vincent-Monégat C, Aksoy E, Aubailly-Giraud E, Zaidman-Rémy A, Heddi A. Insects recycle endosymbionts when the benefit is over. Curr Biol. 2014;24:2267–73.

[3] Suikkanen S, Antila M, Jaatinen A, Vihinene-Ranta M, Vuento M. Release of canine parvovirus from endocytic vesicles. Virology 2003;316:267–80.

Subject Index

A

Ab1/Ab2 antibodies, 84
Acceptor fraction, 204
Acidification, of phagosomes, 218
Acrylamide concentration, 24
Actin, 25
Actin-binding proteins, 142
Actin cytoskeleton, 142
Actin microfilaments (MF), 127, 142
Aequorin, 159
Agglutinins, 291
Agrobacterium-mediated transformation, 339
Alexa488-InlA beads, 219
  fluorescence intensity, 219
Algae, 11
Algal cultures, 271
Algal strains/growth conditions, 271
Alpha helix, 47
American Type Culture Collection (ATCC), 2
Amiloride (AMI) blocks sodium channels, 364
Amino acid sequences, Kyte-Dolittle analysis of, 49
Amphipathic, 105
Anaerobic organisms, 17
Anaphase, checkpoints, 333
Anaphase-promoting complex (APC), 333
  CENP-E, 333
  Mad1 and Mad2 bind, 333, 336
Ankryin, 25
Antagonistic force, 328
Antibiotic resistance
  background, 242
  bacterial strains, 242

minimal bactericidal concentration for biofilm-grown cells (MBC-B), 243
results, 243–245
Antibiotic, sensitivity of biofilm-grown and planktonic *P. aeruginosa* strains, 244
wild-type and 45E7 biofilms, 244, 245
Antibodies, 56
  12G10, 248
  Rtn4A inhibit network formation, 162
  vesicular stomatitis virus (VSV), 195
Anti-FLAG antibody, 128
Antigens, on foreign cells, 345
Anti-GFP antiserum, 33
Anti-Nup85, 75
Anti-Nup133 depleted nuclei, 76
Anti-PKG antibodies, 292
APC. *See* Anaphase-promoting complex (APC)
AP2 functions, 229
Aposymbiotic, 356
APs proteins, 229
A. punctulata, 300, 301
AQP1 monomers, 48
Aquaporin proteins, amino acid sequence alignment of, 50
Aquaporins, 48
ARFFT, 32
Arp2/3 binding, 143
Arp2/3 proteins, 142
Ascorbic acid, 209
ATCC. *See* American Type Culture Collection (ATCC)

ATPases, 54
  nonhydrolyzable, 141
ATP-regenerating system, 65, 205
AUG start codon, 31
  in-frame AUG sequence, 34
AUG1–UAG mutant cells, 37
Autocrine secretion, 246
Autoradiography, 300
Auxilin, 229
Axoneme, 150, 290
  of primary cilium, 150
Axoplasm
  of giant axon, 134
  preparation of, 135
  vesicles/supernatant preparation, 135

B

Bacillus subtilis, 275
Bacteriome, 354
Bafilomycin (BFLA), 364
Bands, 24
Beta mercaptoethanol (BME), 23
BFLA. *See* Bafilomycin (BFLA)
BGS. *See* Bovine growth serum (BGS)
Biofilm
  dental plaque, 241
  formation of, 241
Biotinylated proteins, 159
BME. *See* Beta mercaptoethanol (BME)
Bombus terrestris, 346
Botulinum toxin, 173
Bovine growth serum (BGS), 247
Bovine serum albumin (BSA), 3, 82, 314
BPAG1/NF-l double mutation, 131
BPAG1 null axons, 129

373

β-pleated sheet. *See* Pleated fold
Brefeldin A (BFA) inhibits protein transport, 364
Brush border, bundling
  background, 253–254
  bead aggregation assay, 255
  calcium-dependent heterophilic adhesion complex protocadherin, 260
  cell culture, 254
  IBB cells cluster, microvilli of, 257
  identical cadherins, 253
  intermicrovillar links, 258
  microscopy, 254
  microvillar clustering, analysis of, 255
  PCDH24 expression abolishes microvillar clustering, knock-down of, 259
  polarized, 253
  results, 256–258
BS4, 13
BS5, 13
BSA. *See* Bovine serum albumin (BSA)
BSA-coated gold particles nuclear uptake of, 8
Budding yeast. *See* Saccharomyces cerevisiae
Buds, 229
Bumblebees. *See* Bombus terrestris

C

CACO-2$_{BEE}$ cell lines, 254
CACO-2$_{BEE}$ cells, 254, 256
Caco-2 tissue culture cells, 219
Cadherin proteins, 258
Cadherins, 253
CA domain. *See* Central acidic (CA) domain
Caeca, 354
Cage-like scaffold, 25
Calcium transport, in flagella, 120
  background, 120–121
  CAV2 antibodies, 122
  CAV2 expression, RNA interference suppression of, 121
  cell cultures, 121
  Chlamydomonas, localization of, 124
  gated channels, 120
  membrane potential, 120
  results, 122–125
  reverse transcriptase–PCR, 121

Calvin cycle, 270
cAMP. *See* Cyclic AMP (cAMP)
CAMs. *See* Cell adhesion molecules (CAMs)
Canine parvovirus (CPV), 363
Capping proteins, 142
Carbon fixation, 270
CAT. *See* Chloramphenicol acetyltransferase (CAT)
CAV2 antibodies, 122
  localization of, 122
  polyclonal antibodies, 122
Caveolae, 114
CAV2 gene expression, 121
CAV2 knockdown cells, 123
CAV2 protein, localization of, 124
CAV2 voltage-gated calcium channel, 124
CDC9, amino acid sequence of, 34
CDC9 gene, 32
CDC9–GFP fusions, 32
  protein constructs, 35
CDC9–GFP protein, 35
CDC9–HA
  AUG2–GCG mutant cells, 37
  mitochondrial fraction, protected against protease digestion, 36
  nuclei and mitochondria, 37
CDC9 mutagenesis, 32
Cdk proteins, 299
Cdks. *See* Cyclin-dependent protein kinases (Cdks)
Cdk triggers, 305
Cell adhesion, 253
Cell adhesion junctions, 253
Cell adhesion molecules (CAMs), 253
Cell cultures, 3, 13
  calcium transport, in flagella, 121
Cell cycle
  background, 300, 306
  Cdk2 by cyclin E, 299
  Cdk triggers, 305
  cell ploidy, 305
  cleavage index, determination of, 301
  cyclin-dependent protein kinases (Cdks), 299
  cyclin level with cell division cycle, 303
  cyclin proteins bind, 305
  egg protein synthesis, 302
  eggs/embryos, preparation and incubation of, 301

fluorescent mitosis biosensor (MBS) tracking nuclear envelope breakdown, 307
  HeLa cells, synchronization of, 307
  immunoblot analysis, 307
  Lytechinus eggs, 304
  maternal mRNAs, 300
  ploidy determination by flow cytometry, 307
  results, 302–304, 308–310
  sample preparation/analysis, 301
  siRNA, cyclins knockdown, 307
Cell division, 8
Cell-free systems, 97, 203, 230
  for clathrin binding, 231
Cell lysates, 339
Cell migration, 246
Cell ploidy, 305
Cell's cytoskeleton, 134
Cell signaling
  adaptor, phosphorylation of, 290
  assays, 285
  background, 284–285, 291
  cell culture, 285, 292
  cells/flagella, fractionation of, 292
  Chlamydomonas cGMP-dependent protein kinase, 294, 295
  IFT
    Chlamydomonas PKG, 296
    flagellar adhesion–induced signaling, 296
  immunoblotting/ immunofluorescence, 292
  ligand binds, 290
  northern/immunoblot analysis, 285
  phosphorylation cascade, 290
  PKG, flagellar adhesion signals, 295
  receptor protein, 290
  results, 286–288, 293–297
  RpkA
    Dictyostelium, 286
    G$_α$1 regulated CMF signaling, 289
  rpkA$^-$ cells produce, 287, 288
  signal cascade, 283
  signal transduction pathway, 290
  in vitro protein kinase assay, 292
Cellular diversity, 11
Cellular response, 283
CENP-E depletion, 336
Center for Disease Control website, 362

Central acidic (CA) domain, 143
Centrifuge tube, 17
Centriole separation
  background, 313
  cyclin-dependent kinases, 318
  embryo extract modifications, 314
  fluorescent labeling, 314
  frog embryo injections, 314
  microtubules, 312
  pericentriolar material (PCM), 312
  replicate, 312
  results, 315
  in vitro centriole separation assay,
    314, 317
  in vivo injection experiments, 316
Centrosomes, 312, 327
  duplication, 313
Cereal weevils, 354
cGMP-dependent protein kinases,
  293
Chaperones, 62
Chick embryo fibroblast cells,
  primary culture of, 211
Chlamydomonas, 152, 273
  cell body, 125
  cells, 270
  genomic database, 121
  mutants, 270
  PKG protein, 293
  strains, 122
  swimming, 122
Chlamydomonas reinhardtii, 120, 270,
  291, 292
  strains, 151
Chloramphenicol acetyltransferase
  (CAT), 320
Chloramphenicol migrates, 322
Chlorophyll *a*, 11
CHO 15B cells, 203
CHO cell-free system, 206
Cholesterol, 114
  density gradient centrifugation,
    distribution of, 118
  lipid analysis, 114
Chromatic adaptation, 12
Chromatin, 335
Cilium. *See* Primary cilium
Ciprofloxacin (Cip)
  bacterial DNA-specific enzymes,
    242
Circularly permuted yellow
  fluorescent protein (cpYFP),
  275
Cisternal maturation, 209, 210

background, 210
chick embryo fibroblast cells,
  primary culture of, 211
endomembrane transport in PC
  release, 214
immunoelectron microscopy, 211
immunofluorescence, 211
PC aggregates within lumen, 213
PC release, assay of, 212
PC "wave" protocols, 211
results, 212–215
Clathrin, 229
  background, 230–231
  cell-free system for binding, 231
  electron microscopy, 231
  endocytosis, 229
  to liposomes and clathrin-coated
    buds *in vitro* formation, 232
  results, 234
  sedimentation assay, 231
  of uncoating proteins, 233
Clathrin binding
  cell-free system, 231
  liposomes and clathrin-coated
    buds *in vitro* formation, 232
Clathrin-coated bud formation, 232
Clathrin-coated pits, 224, 229
Cleavage furrow, 338
Clostridium botulinum bacteria, 173
CLUSTAL W5, 49
Coated pit, 229
Coated vesicle, 172
  acceptor membrane, 172
  adaptor proteins, 172
  background, 173
  cell-free vesicle-budding
    assay, 174
    energy required, 176
  cell-free vesicle fusion to Golgi
    membranes, 175
  electron microscopy, 175
  gel filtration chromatography, 175
  ER-derived vesicles, isolation,
    178
  gradient-purified ER-derived
    vesicles, protein composition
    of, 177
  immunocytochemistry, 175
  results, 176–177
  Sec protein localization, 179
Cofilin, 142
Coimmunoprecipitation (co-IP), 165
co-IP. *See* Coimmunoprecipitation
  (co-IP)

Collagen, 209, 246
  fibers, 246
Colocalization, 194
Column chromatography, 63
Column fractions, 175
Competition, 13
Complete cuticle synthesis
  symbiosis, 360
Confluent, 2
Confluent cells, size distribution of
  intermediate, 9
Confocal fluorescence microscopy,
  189
Contractile ring, 338
Coomassie Blue, 24
Coomassie Blue R250, 26
COPI–coated vesicles, 173
COPII vesicles, 193
COP I protein, 176
COPS
  ts-G-GFP$_{ct}$–positive structures
    colocalizing, quantification
    of, 199
Cortactin-expressing cells rescues
  motility defects
  autocrine-produced extracellular
    matrix, 251
Cortactin expression
  adhesion formation, 250
  suppression of, 247
CortactinKD. *See* Cortactin
  expression
Cortactin-KD cells, 250
Cortactin-knockdown cells, 251
  motility, 249
Cortactin promotes fibronectin
  recycling, 252
Cotranslational translocation, 39
  in vitro assays of, 40
CPV. *See* Canine parvovirus (CPV)
CPV endocytosed, immunoelectron
  microscopy of, 367
cpYFP. *See* Circularly permuted
  yellow fluorescent protein
  (cpYFP)
C. reinhardtii, 121
Crude cell lysates, 204
Crude cytoplasmic, 80
Crude membranes, 144
Cryosections, of frozen pellet, 175
CSF. *See* Cytostatic factor (CSF)
Cultured ET fibroblast cells, 211
Cutaneous, 25
Cuticle, 354

Cyanobacteria, 11
Cyclic AMP (cAMP), 284
Cyclic GMP-dependent protein
        kinase (PKG), 291
Cyclin A2 regulates G2/M phase, 310
Cyclin B1 behavior, centrosome
        separation, 311
Cyclin-dependent protein kinases
        (Cdks), 299
Cyclin proteins, 299
Cyclins, 299
Cycloheximide (CHX), 205, 314
Cytokinesis, in plant cell, 338
    background, 339
    cell plate, 338
    contractile ring, 338
    immunofluorescence, 340
    immunoprecipitation, 339
    kinase assay, 339
    kinesin-like protein NACK1
        binds, 341
    light microscopy, 340
    NACK1/NPK1 in BY-2 tobacco
        tissue culture cells, 342
    NACK1, truncated overexpression,
        343
    plant/cell material, 339
    results, 340–342
Cytoplasmic dynein, 291
Cytoplasmic filaments, 73
Cytoplasmic viscosity, variation in
        nuclear uptake, 6
Cytoskeletal-associated proteins
    axonal MT cytoskeleton,
        disruption, 130
    background, 128
    BPAG1, neuron-specific isoform
        of, 131
    interactions, 127
    mice, 128
    mouse neurons, primary culturing
        of, 128
    nBPAG1 stability, 133
    results, 129–132
    transfections/immunofluorescence,
        128
    in vitro microtubule-binding
        assays, 129
Cytoskeletal filaments, 134
Cytoskeleton, 127
Cytosolic proteins, 80
Cytosolic side, of membrane, 105
Cytostatic factor (CSF), 333

D

Data, experimental design, 1
Deflagellation, 292
Demembranated sperm chromatin,
        76
Denature, protein, 23
Deoxythymidine triphosphate
        (dTTP), 308
3,3′-Diaminobenzidine tetrachloride
        (DAB-HCl), 211
Dictyostelium discoideum, 284
Differential centrifugation, 17
Diffraction patterns, 48
Dihydroxyphenylalanine (DOPA),
        354
Dithiothreitol (DTT), 23, 91, 159
DNA ligase, 31
    functions, 31
DNA ligase I, 31
DNA replication, 76, 305
    assay, 76
DNA sequence, 32
Docking complex, 181
Donor fraction, 204
DOPA. See Dihydroxyphenylalanine
        (DOPA)
DOPA synthesis pathway, 361
Dorsal root ganglia, 128
Double-stranded siRNA (d-siRNA)
        molecules, 307
Drp1 protein, 187
DTT. See Dithiothreitol (DTT)
dTTP. See Deoxythymidine
        triphosphate (dTTP)
Dye, coomassie, 24
Dyneins, 328

E

45E7 bacteria, 244, 245
EDTA. See Ethylenediaminetetra-
        acetic acid (EDTA)
Effector proteins, 283
Eg5, end-directed motor, 328, 329
Egg extract protein concentrations,
        99
45E7, in antibiotic Tb, 242
Electron microscopy, 3
Elutants, 63
Embryo tendons (ET), 211
Emission wavelength, 275
Endocuticle, 354
Endocytic vesicles, 229

Endocytosis, 229, 362
Endomembrane system, 193, 202
Endomembrane transport
    background, 203–204
    CHO cells, 204
    Golgi transport, cell-free system
        for, 205
    membrane compartments, 202
    protein synthesis, in vivo inhibition
        of, 205
    results, 205–208
    subcellular fractions,
        homogenization/gradient
        purification of, 204
    VSV, 204
Endoplasmic reticulum (ER),
        106, 157
Endosymbionts, 353
Endosymbionts, bug out, 353
    background, 353–354
    cuticle coloration, 355
    electron microcopy, 355
    fluorescence microscopy, 355
    insect rearing, 354–355
    results, 356–361
    S. pierantonius DNA quantification
        with qPCR, 355
Endosymbiont Sodalis pierantonius
    population density of, 357
Endosymbiont, survival of
        weevils, 359
Endosymbiotic theory, 17
Energy-regenerating systems, 81
ENTH phospholipid-binding
        domain, 230
Epicuticle, 354
Epidermal growth factor (EGF), 224
EPS. See Extracellular polymeric
        substance (EPS)
Epsin, 229
ER. See Endoplasmic reticulum (ER)
Escherichia coli, 81, 143, 242
ET. See Embryo tendons (ET)
Ethylenediaminetetra-acetic acid
        (EDTA), 26
Eukaryotic cells, 106
Excitation wavelength, 275
Exocuticle, 354
Exocytic membrane markers
    COPI/COPII, ts-G-GFP
        colocalization, 197
    microtubules facilitate, 201
    ts-G-GFP$_{ct}$, colocalization of, 196

Exoplasmic, 105
Experimental design, 1
Extracellular polymeric substance (EPS), 241
Extruded, of giant axon, 134

F

F actin polymers, assembly, 142
Fast axonal transport, 134
FBA. *See* Fructose-bisphosphate aldolase (FBA)
Fibronectin, 246, 251
Fibronectin secretion
    background, 247
    cell migration, 246
    cells/cell culture, 247
    extracellular matrix (ECM), 246
    12G10 antibody, 248
    immunodectection procedures, 248
    internalized fibronectin, 251
    lamellipodia persistence assay, 247
    results, 248–252
    single-cell motility assay, 247
FISH. *See* Fluorescence *in situ* hybridization (FISH)
FITC. *See* Fluorescent dye fluorescein isothiocyanate (FITC)
FITC emission, 218
FITC –pH exposure, 221
fla10-1 cells, 296
Flagellar membranes, 292
Flagellar PKG, 295
Flagellar protein tyrosine kinase, 291
Flippase, 106
Flow cytometer, 13, 305
Fluorescein isothiocyanate (FITC)-labeled dextran, 314
Fluorescence *in situ* hybridization (FISH), 355
Fluorescence recovery after photobleaching (FRAP), 264
Fluorescent dye fluorescein isothiocyanate (FITC), 218
Fluorescent mitosis biosensor (MBS), 307
Fluorescent Rex (Frex)
    background, 275
    biosensor, 275
    cell culture, 276
    fluorescence microscopy, 276
    glucose, subcellular NADH levels, 279

live cell fluorescence measurements, 276
mitochondrial transport, inhibition of, 280
NADH fluorescent biosensor, properties, 277
nicotinamide adenine dinucleotide (NAD), 275
oxidized mitochondrial NADH, electron transport chain of, 278
results, 277–281
in vitro characterization, 276
Flu virus, 346
Fractions, centrifuge tube, 203
Frex. *See* Fluorescent Rex (Frex)
Fructose-bisphosphate aldolase (FBA), 124
293FT cells, 276
Fusion protein, 34

G

G actin monomers, assembly, 142
Gal4-binding site, 320
Gas exchange measurements, 271
GDP. *See* Guanosine diphosphate (GDP)
Gel filtration, 63
    chromatography, 175
Gelsolin, 142
Genetic dissection, 203
Genomic DNA from Tissue kit, 356
Gentamicin (Gm), bacterial protein synthesis, 242
GFP. *See* Green fluorescence protein (GFP)
Giardia, 18
    cell fractions
        Isc proteins, distribution of, 20
    infections, 19
    intestinalis, 18
    subcellular fractionation of, 19
    trophozoites, IscS/IscU localization, 21
G. intestinalis trophozoites, 19
GlcNAc transferase 1, 203
GlcNAc transferases, 202
Glucans, 242
Glucose depletion
    ATP dependence, of Myo2 distribution, 268
    ATP levels, measurement of, 265
    background, 264

carbon depletion, 264
fluorescence recovery, after photobleaching, 265
Myo2 forms rigor complex, 269
Myo2 localization, 267
Myo2 rapidly relocalizes to actin cables, 266
organelles, intracellular transport of, 263
results, 266–269
yeast cell permeabilization, 265
yeast growth, 264
Glycolysis, 263
Glycosphingolipids, 114
Glycosylated proalpha factor (³⁵SgpαF), 174
Glycosylation, 202
Glycosylphosphatidylinositol (GPI)-anchored proteins, 113
Gold particles
    nucleoplasmin protein/nonnuclear protein BSA, 7
    proliferating cells, 4
Golgi cisternae, fusion protein aggregates, 51
Golgi complex, 164, 193, 202, 209
    background, 164
    BMCC binding, 165
    cell cycle –dependent manner, 170
    cell-free disassembly/reassembly assays, 166
    cis to trans, 164
    coimmunoprecipitation (co-IP), 165
    electron microscopy, 167
    immunoblots/immunoprecipitation, 167
    immunoprecipitation (IP), 165
    membrane/cytosol isolation, 166
    mitotic Golgi fragments (MGFs), 164
    p65, GM130, and mannosidase II (Mann II), 169
    physical stacking, 164
    p65 influences stacking, 171
    protease/extraction treatments, 166
    protein, identification of, 168
    results, 167–170
Golgi membranes, 205
    pellets, 167
    proteins, 166, 168
Golgi vesicles, 338

G protein, 203
    Golgi compartment donates
        protein *in vitro*, 208
    Golgi vesicles as transmembrane
        protein, 207
    ^{3}HGlcNAc, CHO cell-free
        system, 206
G protein-coupled receptors
    (GPCRs), 283
G protein–independent signaling
    pathway, 284
G proteins, 283
G protein-specific antibody, 205
Green fluorescence protein (GFP),
    32, 188, 194, 264
Guanosine diphosphate (GDP), 283
Gut endosymbiont population,
    apoptosis and autophagy
    contribution, 358

H

Hallmarks
    of confluent cell culture, 8
    of scientific study, 3
HA-tagged CDC9, in yeast, 36
HA tagging, 32
^{3}H-chloramphenicol, 322
[^{3}H] dibutyrylphosphatidylcholine
    ([^{3}H]diC$_4$PC), 106
HeLa cells, 306, 307
Heterophilic binding, 253
High-speed supernatant (HSS), 81
His tag, 230
His$_6$-tagged RanQ69LGTP, 90
Histidine, 230
Histidine-repeat sequence, 90
Histone deacetylase (HDAC), 326
Histone H1 mix, 334
HLD, 115, 117
    lipids, analysis of, 119
[^{3}H]mannose-treated membranes,
    108
Homophilic binding, 253
Homotetramer forming, 48
Horseradish peroxidase (HRP), 166
    conjugated secondary
        antibody, 211
HRP. *See* Horseradish peroxidase
    (HRP)
Hsc70 catalyze, 229
Hsp60 staining, 35
HSS. *See* High-speed supernatant
    (HSS)
HT1080 cells, 252

Human bone cancer cell line
    U2OS, 189
Human exportin(tRNA), 89
Hypothesis, 1

I

IBB. *See* Intestinal brush border (IBB)
IBB-labeled quantum dots
    (IBB–Qdots), 98
IBB-Qdots. *See* IBB-labeled quantum
    dots (IBB–Qdots)
Ice-cold buffer, 109
IF. *See* Intermediate filaments (IF)
IgG–Sepharose beads, 90
Immunizations. *See* Vaccinations
Immunoblot analysis, of mutant
    cells, 37
Immunoblotting, 17, 33, 75, 91
    photosystems, 271
Immunoelectron microscopy, 211, 364
Immunofluorescence, 292
    cisternal maturation, 211
    microscopy, 364
Immunoprecipitate, 225
Importin α, 96
Importin β, 96
    background, 80
    immunoblotting, 82
    immunodepletion from cytosol, 81
    importin 60, 85–87
    importin 90, 85–87
    importin 60 binds, to second
        subunit, 83
    NLS binding assay, 82
    nuclear envelope, 80
    nuclear protein import assay, 81
    protein gel electrophoresis, 82
    Ran-GTP binds, 80
    results, 82
    Xenopus eggs
        high-speed supernatant,
            preparation, 81
In-frame AUG sequence, 34
InlA, 218
Insect immunity, 345, 347
    adaptive immune system
        functions, 345
    background, 346
    bacteria, 347
    bumblebees, 346
    exposure combinations
        and bacterial clearance, 348
        and survival, 347
    results, 349–352

Integral membrane proteins, 105
Integrins, 246
Intermediate filaments (IF), 127
Intestinal brush border (IBB), 253
Intraflagellar transport (IFT),
    150, 291
Intraflagellar transport particle, 291
Ionophore A23817, 300
Iron-chelating compound 2,
    2′-dipyridyl (DPD), 211
Iron–sulfur (Fe–S) cofactors, 18
Isc proteins, 18, 19
IscS antibodies, 19
IscU antibodies, 19
Isolated mitochondria, proteinase K
    treatment of, 33

K

KD-generated ECM, 250
Kilodaltons (kDa), 24
Kinesins, 328

L

Lamellipodia, 247
Lamellipodium, 246
    protrusion of, 246
LAMP-2. *See* Lysosomal membrane
    protein-2 (LAMP-2)
LDL receptor. *See* Low-density
    lipoprotein (LDL) receptor
Leaflets, of phospholipid bilayer, 105
Ligand, 283
    molecule, 224
Lipid microdomains, 114
    extraction of, 117
    mouse 3T3 cells, 116
    from mouse 3T3 cells, 116
Lipid rafts
    background, 113
    caveolae, 114
    electron microscopy, 115
    glycosylphosphatidylinositol
        (GPI)-anchored proteins, 113
    lipid analysis, 114
    membrane lipid domains,
        extraction of, 114
    microdomains, 113
    plasma membrane purification, 114
    results, 115–117
    sucrose density gradient
        centrifugation, 114
Liposomes, 106, 144, 231
Listeria monocytogenes, 142, 218
    rocketing of, 143

LLC-PK1 cells, 328
LLD, 115, 117
  lipids, analysis of, 119
Loaded, protein, 24
Low-density lipoprotein (LDL)
  receptor, 224
Loxoceles envenomation, 25
Loxoceles spider
  bite of, 28
  venom, 26, 28
L. pealeii, 135
L. pictus, 301
Luciferase, 64, 65
  misfolded proteins, 64, 65
Lymphocytes, 346
Lysosomal membrane protein-2
  (LAMP-2), 364
Lysosome activates, acidic
  environment of, 218
Lysosome-inhibiting drugs, 365
Lyso-Tracker accumulates, 220
Lytechinus pictus, 300

M

Maleimide biotin (MB), 159
Maleimide PEG (MP), 159
Maltose binding protein (MBP), 55
Mating-type (MT) genes, 291
  kinetochore, 333
  motor proteins, 328
  nonkinetochore, 327
  phragmoplast, 338
  polymerization, 136
  salt-washed squid optic lobe, 136
Mature peroxisome, 181
MB. *See* Maleimide biotin (MB)
Medium-speed supernatant
  (MSS), 174
Melanization, 354
Membrane cytoskeleton, 25
  protein composition of, 27
Membrane proteins, integral and
  peripheral, 157
Membrane-spanning domains, 48
Messenger RNAs (mRNAs), 31, 88
Metalloenzymes, 18
Metalloprotease, 28
Metaphase checkpoint, 333
MF. *See* Actin microfilaments (MF)
Microdomains, 113
Microinjection, 3
Microsomal membrane fraction,
  differential centrifugation, 106
Microsomes, 106

Microtubule (MT) cytoskeleton, 150
Microtubule motor protein, 134
  background, 134
  dissociated axoplasm, preparation
    of, 135
  electron microscopy, 136
  eukaryotic cell, cytoplasm of, 134
  fast axonal transport, 134
  intracellular motility, 134
  motor protein, 134
  MT-based motility *in vitro*, 141
  purified MTs, movement of, 140
  results, 137–141
  single MT
    isolated organelles/carboxylated
      beads movement on, 139
    squid axoplasm, organelle and
      high-speed supernatant
      from, 138
    squid optic lobe microtubules,
      preparation of, 136
    taxol-polymerized MTs, 137
    vesicles/supernatant preparation
      from axoplasm, 135
  in vitro motility assay, 136
Microtubule organizing center
  (MTOC), 312
Microtubules (MTs), 127, 312, 327
  motor proteins, 328
Microvilli, 253
Minimal bactericidal concentration
  (MBC), 242
  for biofilm-grown cells, 243
  for planktonic bacteria, 243
Misfolded proteins
  amino acid sequence, 62
  background, 63
  β-galactosidase, 70
  elute, 63
  fractions, 63
  gel filtration, 63, 65
  GroEL chaperonin, 62
  heat shock protein 70 (Hsp70), 62
  Hsp40, 68
  Hsp70, 68
  Hsp104, 68
  Hsp40, purification of, 64
  Hsp70, purification of, 64
  Hsp104, purification of, 64, 66, 67
  luciferase, 63
  molecular chaperones, 62
  refolding assays, 65
  results, 65
  S. cerevisiae, 63

turbidity assay, 64
  yeast Hsp40, 68
  yeast Hsp104, 68
  yeast lysates, preparation of, 64
Mitochondrial fission, 188
mitochondrial fraction
  CDC9-HA, protected against
    protease digestion, 36
Mitochondrial presequence peptidase
  (MPP), 32
Mitochondrial presequences, 32
Mitochondrial protein (*mito-dsRed*),
  189
Mitochondria, powerhouse of
  cell, 187
Mitochondria, squeeze
  background, 188
  blebbistatin treatment, increases
    mitochondria length, 190
  cell lines/labeling, 189
  microscopy, 189
  myosin II accumulates, at
    mitochondrial restriction sites,
    192
  myosin II depletion, increases
    mitochondrial length, 191
  myosin II, inhibition of, 189
  powerhouse of cell, 187
  results, 190–191
  transfection, 187
Mitosome, 20
Mitotic checkpoint, 333
  background, 333–334
  CENP-E antibodies, 337
Mock depletion, 317
Modeling membrane fission
  actin, 142
  background, 235–236
  dynamin assays, 236
  dynamin protein, 236
  endocytosis, 235
  membrane fission, biochemical
    analysis
    sedimentation assay, 240
  negative staining, 236
  protein dynamin, 235
  results, 237–240
  SUPER templates, 236
    membrane tubules, formation,
      238
  in vitro membrane fission, 239
Molecular chaperones, 62
Molecular weight standards, 24
Monastrol, 329

Monensin (MON), 364
Monoculture, 13
Mouse cell culture line B82L, 225
Mouse fibroblast 3T3 cell cultures, 3
Mouse intestine, 255
Mouse 3T3 cells, 2
    cell cultures, 3
        lipid microdomains, 116
        nuclear transport, 8
Mouse tissue culture 3T3 cells, 114
MPP. *See* Mitochondrial presequence
        peptidase (MPP)
MSS. *See* Medium-speed supernatant
        (MSS)
MT genes. *See* Mating-type (MT)
        genes
Myo2-GFP fusion protein, 266
Myosin, 188
Myosin 2 (Myo2), 264
Myo2 staining, 268

N

N-acetylglucosamine, 202
NACK1 motor function, 342
NAD. *See* Nicotinamide adenine
        dinucleotide (NAD)
Nascent chain, 39
ndvB gene, 242
NEM. *See* N-ethylmaleimide (NEM)
N-ethylmaleimide (NEM), 26, 159
N-ethylmaleimide sensitive factor
        (NSF) binds, 54
    protein fragments, 55
Nicotiana tabacum, 339
Nicotinamide adenine dinucleotide
        (NAD), 275
    NAD⁺/NADH, 275
Ni-NTA purification, 143
Nitrocellulose membrane, 17
NLS. *See* Nuclear localization signals
        (NLS)
Nocodazole, 333
Nonpermissive temperature, 194
Norden Laboratories feline kidney
        (NLFK) cells, 363
NPK1:NACK1 complex, 339
Ntf2-recombinant proteins, 99
Nuclear assembly, 98
Nuclear envelope (NE), 73, 96, 306
Nuclear envelope breakdown
        (NEB), 306
Nuclear lamina, 96
Nuclear localization signals (NLS),
        73, 75, 363

Nuclear pore complex (NPC), 2, 73,
        74, 96
    background, 74
    DNA replication assay, 76
    immunoblotting, 75
    immunodepletion, 75
    NLS-dependent targeting, 80
    nuclear import, 75
    nuclear reconstitution assay, 75
    Nup107-160 complex inhibits
        DNA replication, 78
    Nup107-160 complex lack
        NPCs, 79
    Nup85, depletion of, 77
    results, 76–79
    scanning electron microscopy, 76
Nuclear proteins, 2
    import assay, 81
Nuclear reconstitution assay, 74, 75
Nuclear transport, nucleoplasmin-
        coated gold particles, 7
Nuclear uptake, calculation of, 3
Nucleating proteins, 142
Nucleoplasmin, 2, 73
    nuclear uptake of, 8
Nucleoplasmin-coated gold particles
    in confluent cells, 7
    of intermediate size range, 8
    intracellular distribution,
        differences, 5
    nuclear uptake, 6
    in proliferating and confluent cells,
        7, 9
    size distribution of intermediate, 9
Nucleoporins, 73
    Nup62, 58, 54, and 45, 74
Nucleus–size matters, 96
    background, 97
    cell-free systems, 97
    fluorescence microscopy, 98
    image analysis, 98
    immunoblots, 99
    immunodepletions, 98
    importin α/Ntf2 regulate, 102
    nuclear assembly, 98
    recombinant proteins, 98
    results, 99–103
    in vitro replication, 98
    Xenopus egg extracts, 98
    X . laevis and *X . tropicalis*, 100
        proteins associated with nuclear
        import, 101
Null hypothesis, 2
Nup85, 77

Nup133, 77
Nup160, 77

O

OptiPrep gradient, 151
Oxidative phosphorylation, 263

P

pAb–anti-LexA, 322
Paenibacillus alvei, 346
Paenibacillus larvae, 346
P. aeruginosa, 242–244
P. alvei (DSM 29), 347
Paramethylsulfonyl fluoride (PMSF),
        26
Parvovirus
    capsid, 362
    cells/viruses, 363–364
    CPV, colocalization of, 369
    CPV, dextran coendocytosed, 370
    CPV endocytosed by NLFK cells,
        367
    CPV infectivity, assay of, 365
    dextran particle release assay, 365
    endocytosis, 362
    enveloped viruses, 362
    heat-treated CPV, infectivity of, 368
    lysosomal drugs, 364
    microscopy, 364
    nonenveloped viruses, 362
    results, 365–371
    VP1, N-terminus exposure, 366
Pathogenic bacteria, 241
PCDH24, extracellular adhesion
        domains of, 255
Pellet, 17
Peripheral membrane proteins, 105
Permissive temperature, 194
Peroxisomal assembly, 181
Peroxisomal enzyme luciferase, 180
Peroxisomal membrane proteins
        (PMPs), 180
Peroxisomal targeting sequence
        (PTS), 180
Peroxisomes, 180
    background, 181
    fluorescence pulse chase, 180
        galactose induction, 182
    Gal1 inducible promoter, 180
    mating assays, 182
    membrane-bound organelles, 180
    microscopy, 182
    peroxisomal membrane proteins
        (PMPs), 180

peroxisomal targeting sequence (PTS), 180
preperoxisomal vesicle fusion assayed, by fluorescence pulse chase, 185
preperoxisomal vesicle into mature new peroxisomes, 186
protein subunits, 181
results, 183
split GFP assay, 180
yeast strains/DNA, 182
Pex1p protein, 181
Pex2p protein, 181
Pex10p protein, 181
Pex14p protein, 181
P. fluorescens (DSM 50090), 347
Phagocytosis, 217, 218
Phagolysosome, 217
Phagosome
    acidification of, 217, 218
    Alexa488 antibody
        quantitative changes in fluorescence intensity, 221
    Alexa488-InlA bead
        internalization, determination, 219
    background, 218
    CaCo-2 cells
        internalization, acidification, and phagosome/lysosome fusion, 222
        internalization/acidification/ lysosome fusion, rates of, 223
    FITC-InlA beads, acidification, 220
    InlA-coated beads, 219
    internalization of, 217
    lysosome fusion, determination, 220
    phagocytosis, kinetics of, 220
    results, 220–223
Phenylalanine, 361
Phosphoglucose isomerase (PGI), 268
Phospholipase A2, 363
Phospholipid transport, analysis of, 112
Phosphorylation cascade, 291
Phosphorylation, to S phase, 319
    background, 320–321
    CAT assay, 322
    E2F, inactive, 319
    immunoprecipitation, 322
    plasmid constructs/transfection, 321

protein E2F, 319
Rb protein, 319
    C-terminal domain regulates repression of gene expression, 323
    E2F depends on cyclin E–Cdk4/6, 324
    E2F regulation, 325
    results, 322–326
    Western blot, 322
Phosphotidylcholine, 105
Photoautotrophs, 11. *See also* Phytoplanktons
Photobleaching, 264
Photosynthetic pigments, characterization of, 13
Photosystem I (PSI), 270
Photosystem II (PSII) functions, 270
Photosystems, 270
    Algal strains/growth conditions, 271
    background, 270–271
    Chlamydomonas, CO_2 consumption, 273
    exogenous carbon sources
        PSI mutant cell lines, growth of, 274
    gas exchange measurements, 271
    immunoblotting, 271
    light intensity on O_2 production and consumption, 272
    P700 pigment, 270
    results, 272–273
Phragmoplast, 338
Phycocyanin, 11
Phycoerythrin, 11
Phytoplanktons, 11. *See also* Photoautotrophs
Picocyanobacteria, 11, 12
Picocyanobacteria BS4/BS5
    filamentous cyanobacterium *Tolypothrix*, 16
    monoculture and competition experiments, 15
    optical characteristics of, 14
Picoeukaryotes, 11
Pig brain, phospholipids isolated, 231
Pigments, 11
PKD2, vesicle membrane association, 156
^{32}P-labeled Golgi membranes, 166
Planktonic, 242
P. larvae (DSM3615), 347

Plasma membrane (PM), 105
    background, 106–107
    [^{3}H]diC$_4$PC, into vesicles transport, 108
    leaflets, 105
    microsomes/submicrosomal membrane fractions, 107
    phosphotidylcholine, 105
    proteoliposomes
        protease treatment of, 108
        reconstitution of, 108
    results, 109–110
    salt-washed rough microsomal membranes, preparation of, 107
Plasma membrane purification, 114
Plasmids carrying gene deletions, 182
Pleated fold, 47
PMSF. *See* Paramethylsulfonyl fluoride (PMSF)
Polyclonal antibody, 271
Polycystic kidney disease (PKD), 150, 290
Poly-D-lysine, 136
Polypeptide. *See* Nascent chain
Population density, 13
ppr2 gene, 121
ppr2 mutant, characterization of, 123
Predictions, 1
Preperoxisomal vesicles, 181
Preprocollagen, 209
Preprolactin (pPL), 40
    nascent chains, 41
    radioactively labeled fragments, 41
Primary cilium, 150
    α/β tubulin, 151
    axoneme of, 150
    background, 151
    blindness and mental retardation, 290
    cell culturing, 151
    in cell signaling, 290
    ciliary membrane and axonemal proteins, 154
    ciliary membrane/axonemal proteins, *in situ* immunogold labeling, 155
    cytoplasm extraction, 151
    cytoplasmic compartment, 151
    cytoplasmic vesicles, isolation of, 151, 153
    FMG-1/PKD2 proteins, 151
    immunoelectron microscopy, 152
    immunofluorescence microscopy, 152
    RSP1-3 proteins, 151

Primary literature, 1
Primary structure, 47
Procollagen (PC), 209
Professional phagocytic cells, 217
Profilin, 142
Proliferating, 2
    N/C ratios of, 10
Prolyl hydroxylase, 209
Propidium iodide, 305
Protease K digestion, 33
Proteases, 25
Protease treatment blocks
        phospholipid transport, 111
Proteasome, 333
Protein A-coated beads, 98
Protein A–sepharose beads, 225
Protein 4.1 bind, 25
Protein conformation
    ATPases, 54
    background, 54–55
    electron microscopy, 55
    NSF, 57
        changes conformation, 59
        domain structure, 58
    protein expression/purification, 55
    results, 56
    SNARE complex labeling, 56
    visualizing, 54
    v-SNAREs and t-SNAREs, 60
Protein-containing fractions, 81
Protein density gradient
        centrifugation, distribution
        of, 118
Protein expression/purification, 55
Protein function, 47
    AQP1/GlpF channels, effective
        pore diameter/hydrophobicity
        of, 52
    aquaporin proteins, amino acid
        sequence alignment of, 50
    background, 48
    bovine AQP1, isolation/structural
        analysis of, 48
    chain of amino acids, 47
    conformation of, 54
    data processing, 49
    Golgi cisternae, fusion protein
        aggregates, 51
    results, 49
Protein gamma tubulin (γ tubulin),
    312
Protein gels, 24
    electrophoresis, 26
    samples, 23

Protein Ntf2, 96
Protein–protein binding, 319
Proteins α/β spectrin, 25
Proteins dihydrofolate reductase
        (DHFR), 90
Protein synthesis, 263
Protein tyrosine kinase (PTK), 291
    pathway, 290
Proteoliposomes, 106, 108
    protease treatment of, 108
    reconstitution of, 108
Protozoa, 18
    Dictyostelium discoideum, 284
psaB mutants, 271
Pseudomonas aeruginosa, 242
Pseudomonas fluorescens, 346
PTS. See Peroxisomal targeting
        sequence (PTS)
P value, 2

Q

Quaternary structure, 47

R

RanGDP, 88
RanGDP to RanGTP, conversion, 88
RanGTPase proteins, 88
RanGTP triggering, 96
Ran protein, 92, 96
RanT24NGDP, 90
Rat livers, 166
    homogenized, 107
RBC ghost band 3 protein. See Red
        blood cells (RBC) ghost band
        3 protein
Rb gene sequence, 321
Receptormediated endocytosis,
        224, 363
Receptor proteins, 283, 290
Receptors, 224
    AP-2, 225
    AP2 to EGF-R, 228
    background, 224
    cell culture, 225
    EGF-R protein binding using SPR
        detection, 227
    epidermal growth factor
        (EGF), 224
    immunoprecipitation, 225, 227
    results, 226
    surface plasmon resonance
        (SPR), 226
Reconstitution, of liposome, 106
Red blood cell ghosts, 25

Red blood cells (RBC) ghost band 3
        protein
    background, 24
    Loxosceles gaucho spider, 29
    methods, 25–26
    results, 26–30
Replicated, 1
Replication, 1
RER membrane
    amino acid chain influences
        interactions, 42
    background, 40
    cotranslational translocation, 42
    crosslinking assay, 41
    flotation assay, 41
    nascent pPL, binding of, 45
    pPL fragments, transport of, 43, 44
    protease resistance assay, 41
    results, 41
    in vitro translation, 40
Resonance units (RU), 226
Retic, in endoplasmic reticulum, 157
    aequorin Ca²⁺ assay, 159
    endoplasmic reticulum (ER), 157
    integral/peripheral membrane
        proteins, 157
    IP3 receptor (IP3R), 158
    membrane proteins, integral and
        peripheral, 157
    protein isolation, 159
    results, 160–163
    reticulons, 158
    rough ER (RER), 158
    smooth ER (SER), 158
    sulfhydryl modification, 159
    vesicle isolation, 159
    in vitro network formation, 159
Reticulons, 158
    localization of, 163
Rex protein, 275
R groups, 47
Rhizobia bacteria, 353
Ribonucleases, 306
Ribophorin 1, 109
Ribosome, reading frame, 31
RING finger complex, 181
RNA interference (RNAi), 187
    knock down, 306
RNA molecules, double-stranded,
        187
Rotor, 17
Rough endoplasmic reticulum (RER),
        32, 193, 202
Rough microsomes, 107

Rtn4A inhibit network formation, 162
Rtn4a/NogoA protein, 162
Rubisco enzyme, 270

S

Saccharomyces cerevisiae, 31, 96, 182, 264, 338
Salt-washed membranes, 159
Salt-washed microsomal membranes, 107
Salt-washed rough microsomal membranes
  phospholipid transport, analysis of, 112
  Triton X-100 extract of, 108
Salt-washed small vesicle membranes, 159
Sample size, 1
Scanning electron microscopy (SEM), 355
S. cerevisiae, 32, 63, 264, 339
Scientific method, 1
Sclerotization, 354
SDS-PAGE. *See* SDS polyacrylamide gel electrophoresis (SDS-PAGE)
SDS-PAGE gel, 146
SDS-polyacrylamide gel, 26, 307
SDS polyacrylamide gel electrophoresis (SDS-PAGE), 17, 75, 300
Secondary structure, 47
Second messenger, 284
Securin, 333
Securin-encoding plasmid, 334
Sedimentation assay, 231
Selective permeability, 47
Sensor protein, 275
Sensors, spectrophotometer measure, 13
Serum starvation, 8
Serum-starved cells, 10
  N/C ratios of, 10
Short-hairpin RNAs (shRNAs), 247
SH reagents, *in vitro* network formation, 161
Signal pathway, 283
Signal peptidases, 39
Signal peptide, 39
Signal recognition particle (SRP), 39
Signal transduction pathway, 290
siRNA inhibits nuclear envelope breakdown
  cyclin expression knockdown, 309

Sitophilus oryzae, 354
Size distribution, calculation of, 3
Small interfering RNAs (siRNAs), 306
Small transport vesicles, 202
SNARE complexes, 56
  labeling of, 56
Sodalis pierantonius, 354, 357
Sodium dodecyl sulfate (SDS), 23
S. oryzae, 354, 357
Sperm, *Xenopus* male, 75
S phase–specific proteins, 319
Spheroplasts, 33
S. pierantonius DNA, 355
Spindle assembly, push-pull model of, 332
Spindle formation, 327
Spindle, motor proteins, 327
  aster, 327
  background, 328
  bipolar spindle, 327
    mathematical model, 332
  cell culture, 328
  Eg5, end-directed motor, 328, 329
  inhibitors, 329
  microtubules (MTs), 327
  monopolar spindle formation, predictor of, 331
  monopolar spindles, 327
  nocodazole treatment, 328
  nocodazole washout, bipolar spindle assembly, 330
  results, 329–332
  spindle motor proteins Eg5 antagonistic forces, 329
  time lapse microscopy, 329
Spindle poles, 327
Squid axoplasm, 135
Squid optic lobe microtubules, preparation of, 136
Staphylococcus aureus z domain sequence, 89
Starch pellets, supplemented, 355
Start codon, 31
Statistical tests, use of, 2
Subcellular fractionation, 33
Submicrosomal membrane fraction, characterization of, 109
Sucrose gradient density centrifugation, 203
Sulfhydryl (SH) groups, 159
Supernatant, 17
SUPER templates, 236
Surface plasmon resonance (SPR), 226

SV40 promoter/enhancer, 320
Synechococcus genus, 12
Synthetic complete (SC) media, 264

T

3T3 cell growth
  cytoplasm to nucleus transport, 4–7
Temperature-sensitive (ts)
  Ts-045-G, 194
Tertiary structure, 47
TfRs. *See* Transferrin receptors (TfRs)
Tissue culture, 187
  cell lines, 2
Tobramycin (Tb), bacterial protein synthesis, 242
Tolypothrix, 12
  cell cultures, 13
  genus, 12
Total internal reflection fluorescence (TIRF) microscopy, 248
Transfection, 187
Transferrin receptors (TfRs), 363
Transfer RNAs (tRNAs) transport, 88
  background, 89
  exportin (tRNA) shuttles
    binds to RanGTP, 93
    binds to tRNA, 94
    nucleus and cytoplasm, 92
    stimulates transport of tRNAs, 95
  gel electrophoresis, 91
  immunoblotting, 91
  ^{35}S-labeling, 89
  small nuclear RNAs (snRNAs), 88
  in vitro
    binding assays, 90
    translation, 89
  Xenopus frog eggs, 90
Trans Golgi network (TGN), 193
Translation, protein, 31
Translocon, 39
Transmembrane proteins, 224
Transmission electron immunomicroscopy, 19
Transmission electron microscopy (TEM), 355
Transported G protein, 208
Transport proteins, 47
Trichloroacetic acid (TCA), 301
Triskelion proteins, 173
Triton X-100, 159, 225, 254, 340, 364
Triton X-114 (TX-114), 166

Trophozoites, 18, 19
ts-G-GFP$_{ct}$ colocalization of, 196

U

Ubiquitin ligase, 333
Ubiquitin–proteasome pathway, 305
Ultimate insect molt (UIM), 354
Unconventional myosin, 264
Unicellular green algae
    Chlamydomonas reinhardtii, 120
UV irradiation, 41

V

Vaccinations, 346
V-ATPases, 217
Venom digestion assays, 26
Venom exposure
  protease inhibitors, 29
  on RBC membrane
    cytoskeleton, 27
  of envenomated patients, 28
Vesicular stomatitis virus (VSV), 194
  antibodies, 195
  background, 194
  cell culture, 194
  gene expression, 194
  immunofluorescence, 195
  results, 195–198
  temperature-sensitive (ts)
    version, 194
  ts-G-GFP$_{ct}$ *in vivo*, vesicular
    structures movement, 200
Viral protein 1 (VP1), 363
Viral protein 2 (VP2), 363
Viscerocutanous, 25

Voltage-gated calcium channel, 123
Voltage-gated channels, 120
VP1, N-terminus of, 366
VSV-G protein, 203

W

WASP
  actin filament capture assay, 144
  actin MFs, 142
  attachment zone, 143
  attachment zone localization
    assay, 144
  background, 143
  central acidic (CA), 143
  comet tail, 143
    attachment, 149
  fluorescent labeling, 143
  membrane movement
    reconstitution using purified
    components, 148
  minus end, 142
  N-WASP, WASP homology 2
    (WH2) domain of, 145
  plus end, 142
  protein expression, 143
  purification, 143
  results, 144–147
  in vitro membrane motility,
    reconstitution of, 144
  WH2 domain, 143
    uncouple actin nucleation, 146
Weevil exoskeleton, 360
Weevils, 353
  survival of, 354
Western blotting, 17, 225

WT knockdown cells, photophobic
    response of, 123

X

Xenopus, 3
Xenopus cytosol, 76, 78
Xenopus egg extracts, 75, 144
  CENP-E depletion, 335
  vesicle motility reactions, 144
Xenopus eggs, 75, 76, 97
  high-speed supernatant,
    preparation, 81
  membranes, 160
Xenopus fibroblast tissue culture
    (XTC) cells, 314
Xenopus frog egg, 333
Xenopus laevis, 74, 96, 98
  frog eggs, 158
X. tropicalis, 98

Y

Yeast. *See S. cerevisiae*
Yeast cells, 181, 265
  cultures, 182
  mid-log phase, 265
Yeast cultures, conditions for, 32
Yeast Hsp40 (Ydj1), 64
Yeast nuclei, 33
Yeast *Saccharomyces cerevisiae*, 96
Yellow fluorescent protein (YFP), 307
YO-PRO staining, 355

Z

Ziplocs®, 253
Zymolase 5000, 33

OTHER VOLUMES IN THE SERIES
PROBLEM SETS IN BIOLOGICAL AND BIOMEDICAL SCIENCES

Case Studies in Pharmacology
Edited by Terry Kenakin

Case Studies in Physiology
Edited by Mary Cotter

Case Studies in Biochemistry
Edited by Stanley Lo